Advances in Comparative and Environmental Physiology 12

Volumes already published

Vol. 1: NaCl Transport in Epithelia. Edited by *R. Greger* (1988)

Vol. 2: From the Contents: Hibernation – Vapor Absorption. Nutrient Transport – Stable Pollutants (1988)

Vol. 3: Molecular and Cellular Basis of Social Behavior in Vertebrates. Edited by *J. Balthazart* (1989)

Vol. 4: Animal Adaptation to Cold. Edited by *L. C. H. Wang* (1989)

Vol. 5: From the Contents: Osmoregulation – Chemoreception – Temperature – Spectroscopy – Metalloproteins – Prostaglandins (Eicosanoids) – Maternal-Fetal Relationships (1989)

Vol. 6: Vertebrate Gas Exchange: From Environment to Cell. Edited by *R. G. Boutilier* (1990)

Vol. 7: From the Contents: Protein Turnover – Xenobiotic Metabolism Lipoproteins (1991)

Vol. 8: From the Contents: Dormancy in Aquatic Invertebrates-Root Effect – Red Blood Cell Functions in Fish Diving Mammals (1991)

Vol. 9: Volume and Osmolality Control in Animal Cells. Edited by *R. Gilles, E. K. Hoffmann,* and *L. Bolis* (1991)

Vol. 10: Comparative Aspects of Mechanoreceptor Systems. Edited by *F. Ito* (1992)

Vol. 11: Mechanics of Animal Locomotion. Edited by *R. McN. Alexander* (1992)

Vol. 12: Muscle Contraction and Cell Motility: Molecular and Cellular Aspects. Edited by *H. Sugi* (1992)

Volumes in preparation

Vol. 13: Blood and Tissue Oxygen Carriers. Edited by *Ch. P. Mangum*

Vol. 14: Interaction of Cell Volume and Cell Function. Edited by *F. Lang* and *D. Häussinger*

Biomechanics of Feeding in Vertebrates.
Edited by *V. L. Bels, M. Chardon,* and *P. Vandewalle*

Ion Transport in Vertebrate Colon. Edited by *W. Clauss*

Electrogenic Chloride Transporters in Biological Membranes.
Edited by *G. A. Gerencser*

Effects of High Pressure on Biological Systems. Edited by *A. G. Macdonald*

Mechanisms of Systematic Regulation. Vol. 1: Respiration and Circulation.
Vol. 2: Acid-Base Regulation, Ion Transfer and Metabolism. Edited by *N. Heisler*

Advances in

Comparative and Environmental Physiology 12

Muscle Contraction and Cell Motility

Molecular and Cellular Aspects

Guest Editor: Haruo Sugi

With Contributions by

A. M. Gordon · T. J. Itoh · R. Kamiya · I. Mabuchi
G. Pfitzer · J. C. Rüegg · T. Shimmen · J. M. Squire
H. Sugi · S. Winegrad · L. D. Yates

With 77 Figures

Springer-Verlag
Berlin Heidelberg New York
London Paris Tokyo Hong Kong
Barcelona Budapest

Guest Editor:

Prof. Dr. *Haruo Sugi*
Department of Physiology
School of Medicine
Teikyo University
2-11-1 Kaga
Itabashi-ku
Tokyo 173, Japan

ISBN 3-540-54542-5 Springer-Verlag Berlin Heidelberg New York
ISBN 0-387-54542-5 Springer-Verlag New York Berlin Heidelberg

Library of Congress Cataloging-in-Publication Data. Muscle contraction and cell motility: molecular and cellular aspects / guest editor, Haruo Sugi; with contributions by A. M. Gordon ... [et al.]. p. cm. – (Advances in comparative and environmental physiology; 12) Includes bibliographical references and index. ISBN 3-540-54542-5 (alk. paper). – ISBN 0-387-54542-5 (alk. paper) 1. Muscle contraction. 2. Muscle cells. 3. Cells – Motility. I. Sugi, Haruo, 1933– . II. Series. [DNLM: 1. Cell Movement – physiology. 2. Muscle Contraction – physiology. W1 AD545 v. 12 / WE 500 M9845] QP33.A38 vol. 12 [QP321] 591.1 s–dc20 [591.1'852] DNLM/DLC for Library of Congress 92-2173

Printed in the United States of America

Typesetting: Best-set, Hong Kong, and K+V Fotosatz GmbH, Beerfelden
31/3145-5 4 3 2 1 0 – Printed on acid-free paper

Foreword to the Series

The aim of the series is to provide comprehensive, integrated reviews giving sound, critical, and provocative summaries of our present knowledge in environmental and comparative physiology, from the molecular to the organismic level.

Living organisms have evolved a widespread range of basic solutions to cope with the different problems, both organismal and environmental, with which they are faced. A clear understanding of these solutions is of course of fundamental interest for all biologists, zoologically or medically oriented. They can be best comprehended in the framework of the environmental and/or comparative approaches. These approaches demand either wide surveys of animal forms or a knowledge of the specific adaptive features of the species considered. This diversity of requirements, both at the conceptual and technological level, together with the fact that physiology and biochemistry have long been mainly devoted to the service of medicine, can account for the fact that these approaches emerged only slowly amongst the other new, more rapidly growing disciplines of the biological sciences.

The field has now gained the international status it deserves and the organization of a series devoted to it appeared timely to me in view of its actual rapid development and of the interest it arouses for a growing number of biologists, physiologists, and biochemists, independently of their basic, major orientation.

Liège, Belgium, Spring 1988 *Raymond Gilles*

List of Editors

Those interested in contributing to this series should contact Professor R. Gilles or one of the Series Editors. Review proposals should include a short outline – and a Table of Contents – briefly defining the aims and scope of the review.

Preface

This volume intends to provide a comprehensive overview on the mechanisms of muscle contraction and non-muscle cell motility at the molecular and cellular level, not only for investigators in these fields but also for general readers interested in these topics. A most attractive feature of various living organisms in the animal and plant kingdoms is their ability to move. In spite of a great diversity in the structure and function of various motile systems, it has frequently been assumed since the nineteenth century that all kinds of "motility" are essentially the same. Based on this assumption, some investigators in the nineteenth century thought that the mechanisms of motility could better be studied on primitive non-muscle motile systems such as amoeboid movement, rather than on highly specialized muscle cells. Contrary to their expectation, however, the basic mechanisms of motility have been revealed solely by investigations on vertebrate skeletal muscles, since a monumental discovery of Szent-Györgyi and his coworkers in the early 1940s that muscle contraction results from the interaction between two different contractile proteins, actin and myosin, coupled with ATP hydrolysis.

Muscles are classified into three types, skeletal, cardiac and smooth muscles, according to their structure and function. In vertebrate animals, skeletal muscle produces body movement, cardiac muscle is responsible for the function of the heart as a pump, and smooth muscle is distributed among various visceral organs and blood vessels. The mechanical activity of muscle is regulated primarily by changes in the transmembrane potential of muscle cells. From the standpoint of muscle physiology, therefore, an important question is how muscle contraction (ATP-dependent actin-myosin interaction) is controlled by the electrical phenomena taking place at the surface membrane. It is now clear that the above "regulatory" mechanism of muscle contraction is generally mediated by Ca ions and various "regulatory" proteins, though a number of problems still remain to be solved. Recent progress in research work on the regulatory mechanisms in skeletal, cardiac and smooth muscles is described in Chapters 1, 2 and 3, respectively.

Another monumental discovery in the field of muscle research, made by H. E. Huxley, A. F. Huxley and their coworkers in the early 1950s, is that muscle contraction results from relative sliding between the thick (myosin) and thin (actin) filaments. The establishment of the "sliding filament" mechanism in muscle contraction led to the question, what makes the filaments slide past each other? In spite of intense physiological, biochemical and structural studies, however, the above question is not yet answered clearly. In Chapters 4 and 5, up-to-date information about the mechanism of actin-myosin sliding coupled with ATP hydrolysis is presented to give readers an idea about muscle as a machine converting chemical energy into mechanical work.

Recent remarkable progress in the research field of non-muscle motile systems is obviously based on the accumulated knowledge on the mechanism of muscle contraction. As described in Chapters 6 and 7, non-muscle cell motility can generally be divided into the actin-myosin based and the tubulin-dynein based motilities. The actin-myosin based motility is responsible for cytoplasmic streaming and amoeboid movement, while ciliary and flagellar movement results from the tubulin-dynein based motility. In Chapter 8, various aspects of cell division (mitosis and cytokinesis) is described, in which the structures of microtubules (tubulin) and contraction ring (actin) are extremely elusive and exhibit dynamic changes.

Finally, I would like to emphasize that future progress of muscle and cell motility research will lead not only to the understanding of the mechanism of chemo-mechanical energy conversion but also to the understanding of the mechanism of energy transduction in biological systems in general.

Tokyo, January 1992 *Haruo Sugi*

Contents

Chapter 7

Chapter 8

Chapter 1

Regulatory Mechanism of Contraction in Skeletal Muscle

A.M. Gordon and *L.D. Yates*

Contents

1 Introduction

Contraction of skeletal muscle is an explosive event whereby flaccid muscle is transformed within a few milliseconds into stiff muscle capable of generating large forces or shortening quickly, lifting loads and doing work. The process by which the action of this tissue is so quickly mobilized involves four component steps: (1) membrane excitation; (2) excitation-contraction coupling, the process by which membrane excitation leads to elevated intracellular calcium; (3) calcium regulation of contractile protein interaction; and (4) contractile protein

Department of Physiology and Biophysics, SJ-40, University of Washington, Seattle, Washington 98195, USA
and Department of Animal Science, 104 Animal Science Building, Oklahoma State University, Stillwater, Oklahoma 74078, USA

Advances in Comparative and Environmental Physiology
Vol. 12. Edited by H. Sugi

interaction and contraction. In this chapter we will discuss the middle two steps, how intracellular calcium is regulated and how calcium regulates contractile protein interaction. Membrane excitation is discussed elsewhere (see Hille 1992, for a good review of this subject) and contractile protein interaction is covered in Chapter 5 of this volume.

From the early studies of Ringer (1883) and later Heilbrunn and Wiercinski (1947) it was realized that calcium played a central role in controlling muscle contraction. This was suggested also by data from other laboratories (Weber and Winicur 1961), but the data were not conclusive. The definitive studies establishing a major role for calcium were done by Ebashi and his colleagues (see Ebashi and Endo 1968, for a review of these studies leading to this conclusion). They demonstrated that certain agents had an ability to relax contracted myofibrils through the removal of calcium (Ebashi 1960). Along with Hasselbach and Makinose (1961, 1962), Ebashi and Lippmann (1962) found that the sarcoplasmic reticulum from skeletal muscle could actively sequester calcium to achieve this relaxation physiologically. Furthermore, the purified contractile proteins, actin and myosin, did not require calcium for actin to stimulate the ATPase activity of myosin, the energy-liberating event in contraction, but that control proteins, troponin and tropomyosin, components of the thin filament, inhibited the actin stimulation of the myosin ATPase and that this inhibition was relieved by calcium. Thus, not only were changes in intracellular calcium regulated by the sarcoplasmic reticulum but in turn this calcium regulated contractile protein interaction through an additional set of proteins.

For the most part, decisions on the strength and contration duration are made in the nervous system and relayed to skeletal muscle. The muscle then responds faithfully and rapidly to these commands. In most cases, contractile strength is achieved by the nervous system modulating both the number of active muscle motor units and the rate of stimulation of individual motor axons and thus the individual muscle fibers in that unit. In each muscle fiber calcium is the intracellular second messenger whereby membrane excitation controls contraction. In skeletal muscle this control is achieved rapidly and directly through calcium binding. The direct regulation by calcium makes this tight, rapid motor unit control possible.

Many details have been worked out on the mechanisms which regulate intracellular calcium in skeletal muscle and by which this regulated calcium in turn regulates contractile protein interaction. This chapter will discuss these two processes as they have unfolded in skeletal muscle, focusing primarily on vertebrate, fast twitch muscle. The similar processes in cardiac muscle are discussed in Chapter 2 and show some changes consistent with the different properties of this muscle.

2 Excitation-Contraction Coupling

2.1 Excitation

Changes in the surface membrane potential lead to contraction. These changes can either be membrane action potentials that propagate over the muscle surface membrane or slower depolarizations such as would occur with synaptic potentials at neuromuscular junctions (which occur over much of the muscle surface in those muscle fibers that do not show action potentials) or they can be elicited experimentally using a voltage clamp or elevated extracellular potassium. In most fast vertebrate skeletal muscle fibers, membrane depolarization is conducted into the interior of the muscle fiber by the transverse tubule system, invaginations of the surface membrane which occur at regular intervals with respect to the repeating sarcomere structure. This was shown some time ago by Huxley and Taylor (1958). The inward conduction along these transverse tubules occurs rapidly (Gonzáles-Serratos 1971) and involves changes in the conductance of sodium channels as in the surface membrane action potential (Nakajima and Gilai 1980). Furthermore, disruption of these T-tubules by glycerol treatment interrupts excitation-contraction coupling without blocking excitation (Gage and Eisenberg 1967). Thus, the electrical signal for calcium release spreads rapidly over the through a skeletal muscle fiber to the internal sites where calcium is stored and released. In some smaller, slower invertebrate muscle, surface membrane depolarization may allow calcium entry to initiate calcium release and/or contraction.

2.2 Calcium Regulation

It has been known for sometime that fast twitch skeletal muscle is rather insensitive to changes in extracellular calcium as long as care is taken to minimize the effects of changes in extracellular calcium on the surface membrane either by restricting the decrease in calcium or by substituting another divalent cation such as magnesium (Curtis 1963). From these studies it was surmised that the calcium responsible for activating contraction in these relatively large cells with a comparatively small surface/volume ratio was of intracellular not extracellular origin. Furthermore, measurements of radioactive calcium fluxes across the surface membrane during activation and contraction (Winegrad and Shanes 1962) showed that the transmembrane flux was only a very small fraction of what was needed for direct contractile activation. The major fraction of calcium needed for direct binding to the control contractile proteins comes from the major intracellular storage site, the sarcoplasmic reticulum. The sarcoplasmic reticulum is the smooth endoplasmic reticulum which Hasselbach and Makinose (1961) and Ebashi and Lippmann (1962) had shown earlier to contain a calcium pump which could actively transport (sequester) calcium and relax contracted myofibrils. The structure of the sarcoplasmic reticulum (SR), illustrated diagrammatically in Fig. 1, is made up of three regions functioning primarily but not exclusively in the roles of: (1) a junctional region with close approximation

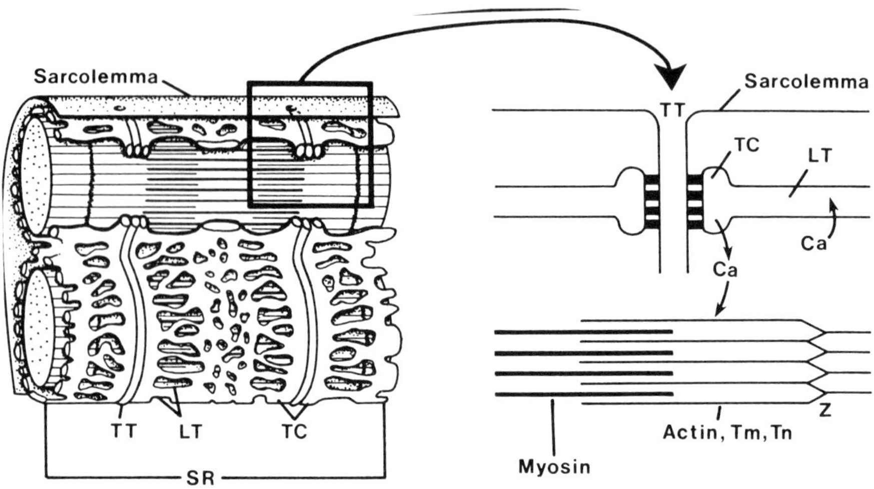

Fig. 1. Internal membrane system in striated muscle and structures involved in excitation-contraction coupling. *Left* drawing of the internal membrane system surrounding the myofibrils in a mammalian skeletal muscle cell. *TT* are the transverse tubules, invaginations of the sarcolemma into the muscle cell. The sarcoplasmic reticulum (*SR*) is composed of the terminal cisternae (*TC*) and the longitudinal tubules (*LT*) which run along the myofibrils. The two terminal cisternae butt up against either side of the transverse tubules forming a structure which is called the triad. The triad is shown schematically on the *right* with the foot structures coupling the transverse tubules (*TT*) and the terminal cisternae (*TC*) of the SR. Depolarization of the surface membrane (or sarcolemma) brings about calcium release from the TC through calcium-releasing channels. This calcium interacts with the thin filament [*actin*, tropomyosin (*Tm*), and troponin (*Tn*)] bringing about interaction of the actin and myosin. Ca is transported back into the longitudinal tubules of the SR by a calcium pump. (After Rüegg 1986)

to the transverse (T) tubules and specialized "foot" processes for signaling tubule membrane depolarization to the SR and releasing calcium; (2) a terminal cistern or sack region for storing and releasing calcium; and (3) a longitudinal tubule region that sends tubules along the myofibrils for transporting calcium into the SR lumen via a calcium ATPase. These various functions have been discovered using preparactions in which different portions of SR membranes are isolated and through morphological or physiological studies on skinned or intact fibers. Thus, the question of how membrane excitation leads to contraction in skeletal muscle divides into the three questions of (1) how is calcium stored and released from the SR; (2) how is membrane excitation signaled to the SR; and (3) how is calcium pumped back into the SR?

2.2.1 Calcium Storage

Calcium is present in the sarcoplasmic reticulum to a total of about 0.7 mM/kg muscle (1.0 mM in the fiber water if it is 70%). Since the sarcoplasmic reticulum occupies about 4% of the muscle volume, this would amount to a concentration of about 25 mM if it were all free. Present in the SR is a highly acidic protein calsequestrin, molecular weight 44 kD, which binds with a relatively low affinity (K_D about 1 mM) 40–50 mol of calcium per mole of calsequestrin (MacLennan and Wong 1971). This protein forms a network visible in electron micrographs in the terminal sack of the SR (Franzini-Armstrong et al. 1987). The calsequestrin provides a site for the storage of a sizable pool of calcium near the releasing sites. It also keeps the free SR luminal concentration low and thus prevents precipitation of insoluble calcium salts but still allows rapid access to a large pool through rapid dissociation from the many low affinity binding sites.

2.2.2 Preparations Used to Study Calcium Release

Calcium release from the sarcoplasmic reticulum has been studied in a number of preparations from whole muscle preparations down to such simple but elegant preparations as artificial membrane bilayers having incorporated calcium channels.

In the intact muscle calcium transient measurements have been measured using intracellular optical indicators such as the calcium-sensitive luminescence of the photo-protein aequorin (Ashley and Ridgway 1970; Blinks et al. 1978), the calcium-sensitive fluorescence of fluorophors such as Fura-2 or Indo-1 (Tsien 1980), and the calcium-sensitive absorption of dyes such as arsenazo or antipyrylazo III (Baylor et al. 1982). Each technique has its strengths and weaknesses in ease of use, ability to quantify, and relative freedom from artifacts. Importantly, all show that membrane depolarization rapidly elevates the free intracellular calcium concentration preceding muscle contraction. To quantify calcium release, one must make assumptions and estimations of calcium binding and uptake out of this free pool of calcium to calculate the calcium released (see Melzer et al. 1987; Baylor et al. 1982) as well as assumptions about the response of the given calcium indicator. From these measurements of the time course of free calcium, one can estimate the time course and maximum rate of calcium release in vertebrate muscle.

For a steady depolarization, the estimated calcium may rise to a peak of 2 μm and the derived rate of calcium release may rise to a peak of 6 μM/ms in some 20 ms before falling to a lower steady level of 1 μM/ms in about 100 ms in frog twitch muscles at about 10 °C (Melzer et al. 1987). Because of the highly derived nature of these calculations, the conclusions must be considered as tentative. The importance of these studies lies in the fact that the intact fiber must be the ultimate testing ground for all hypotheses on the calcium-releasing mechanism.

Another preparation that has been utilized extensively is the skinned muscle fiber. The sarcolemma can be removed selectively using either mechanical treatment, as first done by Natori (1954), or chemical treatment with specific detergents such as saponin (Endo and Iino 1980) which act first on the sarcolemma before solubilizing the SR. In the case of mechanical skinning (or pealing, see Donaldson 1985), the transverse tubule seals over reforming a polarizable T-tubule with the intact T-tubule-SR junction thereby allowing calcium release from the SR to be studied with and without T-tubule connections. Because the surface membrane is no longer a barrier to diffusion, the effect of ions and agents acting on the SR or T-tubule directly can be determined. Calcium release is usually assayed using contraction of the calcium-sensitive myofibrils, calcium indicators such as aequorin with the intact fiber discussed above, or radioactive ^{45}Ca. The review by Endo and Iino (1988) gives a good summary of the techniques used and some results.

The third type of preparation that has been used for some time is the isolated, fragmented sarcoplasmic reticulum made as separated membrane fractions. These membrane fragments can be separated into relatively pure fractions containing either the longitudinal SR, the terminal sacks, or the junctional SR with portions of the terminal sacks and T-tubules. Calcium release from these fractions are typically assayed using calcium indicators discussed above or ^{45}Ca.

The fourth type of preparation more commonly used is a reconstituted system. It is possible to insert Ca channels or other isolated membrane fractions involved in calcium transport into artificial membrane bilayers. Channel opening is measured as the unitary change in resistance to ion flow through the artificial membrane.

2.2.3 Calcium Is Released Through Calcium Channels

Using these preparations, it has been found that calcium release from the SR takes place through calcium channels. Fragmented portions of the SR rich in terminal sacks, when reconstituted back into lipid bilayers, demonstrate the presence of calcium channels having properties pharmacologically similar to that of calcium release from isolated SR vesicles or skinned fibers: a release stimulated by micromolar free calcium, millimolar adenine nucletides (ATP, ADP, and AMP), and millimolar caffeine, inhibited by millimolar magnesium and micromolar ruthenium red and sensitive to micromolar calmodulin and ryanodine (Smith et al. 1985). The high affinity binding of ryanodine, a plant alkaloid, to calcium-releasing channels of isolated SR has been used to isolate this protein. Thus, this calcium-releasing protein is often referred to as the ryanodine receptor. Using this property, Meissner and colleagues (Lai et al. 1988) and Fleischer and colleagues (Inui et al. 1987) have isolated a large polypeptide complex with a relative molecular weight of about 400 kD isolated from the terminal sack-rich SR fraction. When reconstituted into a lipid bilayer, this protein forms a channel with similar pharmacological properties as the calcium-releasing channel with single channel conductance of about 100 pS in 50 mM Ca. The primary structure of this calcium channel (ryanodine receptor), determined from its cDNA sequence, has C-terminal sequence homology with the IP_3 (inositol 1,4,5-trisphosphate) sensitive calcium channels of brain neurons and the channel portion of the nicotinic acetylcholine receptor and thus may contain the calcium channel (Takeshima et al. 1989). However, there is little homology with the N-terminal regions which are thought to be cytoplasmic and contain the modulator binding sites for calcium, adenine nucleotides, and calmodulin (Furuichi et al. 1989).

These calcium-releasing channels appear to be located at the foot process coupling the terminal sacks of the SR to the T-tubule at the junctional SR (Block et al. 1988; Fleischer and Inui 1989). The actual ryanodine receptor may be an oligomer of the basic channel protein. Both the ryanodine receptor and the foot process have a similar fourfold symmetrical structure (Wagenknecht et al. 1989). If they are the same, this would place the hypothetical calcium-releasing channels in close proximity to the T-tubule, part of the foot process (see Fig. 2), thus in close proximity to the membrane to whose depolarization it

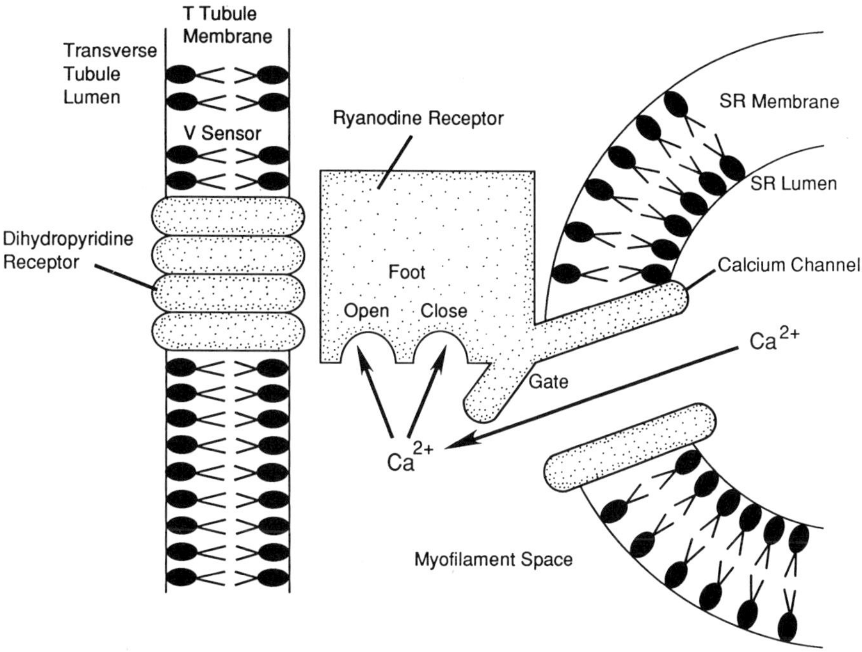

Fig. 2. Schematic representation of the coupling between the transverse tubule and the sarcoplasmic reticulum at the foot process and the mechanism of control of calcium release from the SR. Shown is the ryanodine receptor which forms the foot process physically coupling the two membranes. It is part of the calcium-releasing channel in the SR membrane. This channel can be gated by calcium binding to two controlling sites on the foot, one opening the channel and one closing it. The dihydropyridine receptor, which appears to function not as a calcium channel but a voltage sensor, is shown contacting the foot process in a position to transmit mechanically to the foot process the signal to open the SR calcium channel. The foot process really has a fourfold symmetry not shown for simplicity. Evidence favoring the mechanical model of coupling is discussed in the text. (Drawing made from an initial diagram of Martin Schneider)

responds. There is little evidence to imply or to exclude the presence of calcium-releasing channels at other sites in the SR.

2.2.4 How Membrane Depolarization Opens the Calcium-Releasing Channels

The central question concerns the physiological mechanism by which these sarcoplasmic reticulum calcium channels are opened by transverse tubule membrane depolarization. There is good evidence that these channels can be opened in at least two ways chemically and, because of their location at the foot process, possibly one way mechanically. The chemical stimuli found to open SR calcium channels are calcium and IP_3, but they may not be the physiological stimulus of skeletal muscle (although calcium may be the stimulus in cardiac muscle, see Chap. 2; and calcium and IP_3 in smooth muscle, see Chap. 3). Figure 2 summarizes the structure and control of these SR calcium channels as discussed in the following section.

Calcium channels can be opened by an elevation in sarcoplasmic calcium. In 1968 Endo and Ford and Podolsky (see Endo 1977, for a review) showed independently that in skinned fibers an elevation in the calcium concentration of the solution outside the SR caused the SR of the skinned fiber to release calcium. This release could be observed as a substantial contraction of the skinned muscle fiber to a small elevation in free calcium (called trigger calcium) which in the absence of the SR or for unloaded SR would not have caused muscle contraction and thus was subthreshold for direct myofilament activation. This process was termed calcium-induced calcium release and was suggested long before it was observed experimentally (see Bianchi and Frank 1982).

The existence of this calcium-induced calcium release led to the hypothesis that calcium entering the cell physiologically from the extracellular space or the T-tubule upon membrane depolarization triggered the release of calcium from the SR. Interestingly, the SR calcium-releasing channel observed by Smith et al. (1985) is also opened by calcium giving support to this hypothesis. Since this calcium-induced calcium release could be a regenerative process, it was questioned how this mechanism could produce the controlled, graded release seen experimentally. Fabiato (1985) found that calcium has a biphasic effect on the calcium release with opening induced rapidly at low levels of calcium, but closing more slowly and possibly at somewhat higher levels of calcium. Thus, calcium entry into the cell could first trigger calcium release, and then inhibit release, but not cause complete regenerative unloading of all the SR calcium on each membrane action potential. The transverse tubules of skeletal muscle have calcium channels (Sánchez and Stefani 1978; Almers et al. 1981), but their activation characteristics lead one to question whether a significant calcium entry occurs during an action potential. However, blockers of these calcium channels such as nifedipine (a dihydropyridine) and D600 (a phenylalkylamine) under special circumstances block contraction (Eisenberg et al. 1983; Rios and Brum 1987). Finally, since contraction can proceed in the total absence of extracellular calcium (Armstrong et al. 1972), calcium entry from the extracellular space per se does not appear to be the event triggering calcium release from the sarcoplasmic reticulum of skeletal muscle. It does not rule out a role for calcium-induced calcium release as an amplifier to enhance calcium release or as an inhibitor to inactivate calcium release, but suggests that it is not the initiating event.

Calcium channels may also be opened by IP_3. This was first discovered in non-muscle tissue by Berridge and coworkers (see Berridge and Irvine 1984, for a review of IP_3, its actions and metabolism). IP_3 is formed by the action of phospholipase C (a phosphodiesterase) on phosphatidyl inositol bisphosphate to produce diacylglycerol (an activator of protein kinase C) and inositol 1,4,5-trisphosphate (IP_3). In skeletal muscle, Volpe et al. (1985) and Vergara et al. (1985) found that injection of IP_3 into skinned muscle fibers caused calcium release. Furthermore, the enzymes for the metabolism of IP_3 are present in skeletal muscle although at reduced levels in comparison to other tissues where IP_3 is most likely the stimulator of calcium release. Under some conditions, there is a change in IP_3 content during stimulation (Vergara et al. 1985). Furthermore, presumed depolarization of the T-tubule membrane enhances the calcium-releasing effect of IP_3 (Donaldson et al. 1988). However, the physiological role for IP_3 in calcium release has been questioned on several grounds including the apparent slowness of action (Walker et al. 1987), the requirement of depolarized T-tubules for IP_3 action (Hannon 1989), and the insensitivity of EC coupling to inhibitors of IP_3 hydroysis such as heparin (Pape et al. 1988). The evidence on the slowness of action comes from experiments using "caged" IP_3 which releases IP_3 quickly by flash photolysis. Walker et al. (1987) showed that when calcium release from the SR was measured using the rise in tension, IP_3-stimulated SR calcium release in skinned skeletal muscle fibers was too slow to account for the rapid activation of skeletal muscle. Since Walker et al. (1987) measured force rather than calcium directly, IP_3 may have had a somewhat faster effect on calcium release in skeletal muscle. They also measured the rise of tension in smooth muscle upon calcium release using caged IP_3 and found that the rise was rapid enough to account for the slower activation of smooth muscle. Thus, IP_3 could account for calcium release from the SR in smooth muscle, but not in skeletal muscle. Taken together, this evidence does not support IP_3 as the sole coupling agent for skeletal muscle SR calcium release although it may play a modulatory role in both skeletal and cardiac muscle, and a primary role in smooth muscle.

The third possible coupling mechanism is mechanical. As discussed above, the ryanodine receptor or calcium channel of the SR is clearly identified as being part of the foot process coupling the T-tubule to the terminal sacks of the SR (Block et al. 1988; Fleischer and Inui 1989). Franzini-Armstrong and coworkers (Block et al. 1988) have identified particles with fourfold symmetry organized in a regular array in the T-tubule membrane. This regular array matches up with the arrangement of the ryanodine receptor foot processes in the SR junctional membrane except that there are only half as many T-tubule particles. These T-tubule structures are thought to be the T-tubule membrane calcium channels which are the classical L-type calcium channels blocked by dihydropyridine compounds. This 12-nm-long foot process spans most of the 16-nm gap between the T-tubule membrane and the SR and presumably links the dihydropyridine receptor, calcium channel voltage sensor in the T-tubule membrane with the ryanodine receptor, calcium-releasing channel of the SR terminal cistern.

Sometime ago, this linkage between the T-tubule and SR was proposed by Schneider and Chandler (1973) on the basis of their measurement of membrane charge movements associated with EC coupling. They proposed that on depolarization a charged structure in the T-tubule membrane moved (the charge movement) and that this movement was coupled mechanically to open calcium channels in the SR membrane. It is now clear that the charge movement is associated with the dihydropyridine calcium channel in the T-tubule membrane and that this channel is a voltage sensor which in some manner gates the opening of the SR ryanodine receptor/calcium channels. Separate estimates of the density of charge generators, foot processes, and dihydropyridine receptors give similar values of 500, 700, and 230 molecules/μm^2 of tubule area (see Rios and Brum 1987). Proof that this dihydropyridine calcium channel serves this role comes from the elegant studies of Tanabe et al. (1988). They used dysgenic mice, in which the skeletal muscle fibers are paralyzed, and both the L-type calcium channel currents and EC coupling are absent. They found that the introduction of plasmids carrying the gene for the alpha subunit of the L-type calcium channel into nuclei of primary culture myotubes from these dysgenic mice restores both the calcium currents and EC coupling. That the dihydropyridine calcium channel acts primarily as a voltage sensor and not as a calcium channel during EC coupling is also shown in this study of Tanabe et al. (1988) in that EC coupling is maintained even in the absence of external calcium and thus calcium currents. Further support for this conclusion comes from the studies of Rios and Brum (1987) who showed that only charge movement associated with these channels is required, and not calcium flux through the channels, for EC coupling to occur. To show this, they blocked charge movement, calcium current, and EC coupling with either dihydropyridine compounds or by replacing all external calcium with organic cations. Then in the case of block by use of organic cations, they showed that charge movement and EC coupling but not calcium currents could be restored by replacing these organic cations with non-calcium divalent or monovalent cations such as Mg^{2+}, Na^+, K^+, Rb^+, Cs^+.

Thus the morphological and physiological studies support the hypothesis that the primary initiating event in calcium release is a mechanical coupling between voltage sensor dihydrophyridine receptors in the T-tubule membrane and the ryanodine calcium-releasing channels in the SR membrane (see Fig. 2). However, there may be some additional calcium release from the SR from the free ryanodine receptors (not opposite dihydropridine receptor clusters) in the SR membrane (about half of them) which are not in a position to couple to the T-tubule dihydropyridine receptors. These releasing channels could be induced to open by the calcium released through the mechanical coupling. Furthermore, calcium release whether stimulated mechanically or by calcium may be inactivated by the action of calcium to inhibit calcium release, giving rise to a phasic calcium release even if T-tubule membrane depolarization were to continue (Schneider and Simon 1988). Thus, probably both mechanical and calcium-induced calcium release play roles in regulatory calcium release in skeletal muscle. What, if any, role IP_3 plays remains to be determined.

2.2.5 Calcium Transport by the Sarcoplasmic Reticulum

The sarcoplasmic reticulum membrane transports calcium from the sarcoplasm into the lumen of the sarcoplasmic reticulum to keep the free sarcoplasmic calcium low (about 0.1 μM at rest and to sequester the calcium released on activation to relax the muscle). This is accomplished by a Ca-ATPase; a calcium pump. The ability of the SR to take up calcium in the presence of ATP was first demonstrated in isolated vesicular fragments of the SR by Hasselbach and Makinose (1962) and by Ebashi and Lippmann (1962). This pump protein constitutes about 50% of the membrane dry weight and is the major membrane protein in the longitudinal SR. It has a molecular weight of about 115 kD but associates into larger oligomers between which there may be cooperativity. There are two high affinity calcium binding sites and one catalytic site per 115 kD, so it is not surprising that the stochiometry of transport is 2 calciums/ATP. Maximum pump rates (measured in a "stopped-flow" apparatus) are about 300 nmol per mg SR protein/s in isolated SR membrane fragments from fast muscle fibers (about 100–200 nmol/0.1 s in intact muscle). In skinned muscle fibers the SR is preserved by mechanical skinning. Studies using this preparation show that the calcium transport has a half time of 25 ms at 20 °C, fast enough to

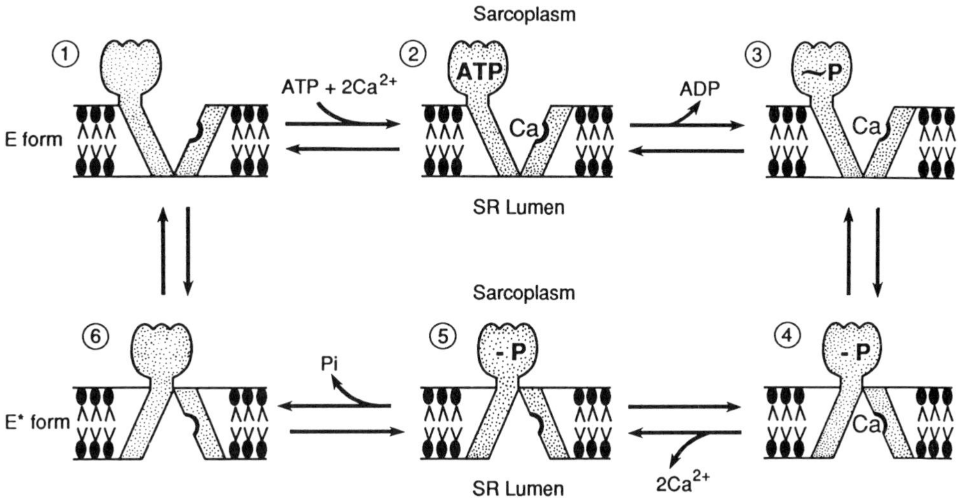

Fig. 3. Diagram of the calcium transport-ATPase cycle of the sarcoplasmic reticulum calcium pump. *E* and *E** are the two configurations of the transport protein oriented toward the sarcoplasm and SR lumen, respectively. The protein transports two calciums per ATP hydrolyzed, although only one calcium is shown for simplicity. Calcium binds with high affinity when the pump is in the E orientation in *step 2*. After formation of the phosphorylated intermediate (*step 3*), the protein changes to the E* form (*step 4*), reorienting to face toward the SR lumen. The Ca dissociates from these sites (*step 5*) which now have lower affinity and with the dissociation of phosphate (*step 6*) the protein can now change back to the E form reorientating toward the sarcoplasm. (After Rüegg 1986)

account for relaxation in an intact frog muscle (Podolsky and Costantin 1964). The concentration of high affinity calcium binding sites on the SR is about 100 nmol, or approximately enough to relax the muscle fiber with all of the troponin calcium-binding sites saturated. Thus, initial high affinity calcium binding by the pump may be enough to account for much of the decline in myoplasmic calcium in relaxation, but transport must take place to relax after repetitive stimulation and to recycle calcium for further release (Ebashi and Endo 1968).

The mechanism of calcium transport has been well studied. Basically, the process involves the Ca-ATPase enzyme being in two different states with different affinities for calcium and opposite orientations with respect to the SR membrane brought on by control of phosphorylation of an enzyme intermediate (see Fig. 3). The scheme first proposed by de Meis and Vianna (1979) is shown in Fig. 3. First calcium binds to the enzyme in the E form at the cytoplasmic surface in a cooperative two-step process such that the second calcium binds more tightly than the first (see Inesi 1985). The affinity is high, greater than $10^6\,M^{-1}$. Once the two calciums are bound and in the presence of magnesium and ATP, the enzyme forms a phosphorylated intermediate by transferring the terminal phosphate from ATP onto an aspartyl residue of the catalytic site. This then brings about a reorientation of the enzyme to the E* form so that the calcium-binding sites are occluded from exchange with the cytoplasm and exposed to the luminal surface of the SR membrane. Once exposed to the SR lumen, the affinity of calcium binding decreases to less than $10^3 M^{-1}$ allowing dissociation of the bound calcium into the SR lumen. Following the dissociation, there is a magnesium-dependent hydrolysis of the E* phosphorylated intermediate, a release of the phosphate formed, and a return of the enzyme to the E state in which it can bind calcium from the cytoplasm with high affinity and begin the cycle again. This process is reversible so that in the presence of an oppositely directed calcium gradient (from lumen to cytoplasm) and in the presence of ADP and P_i, the enzyme can synthesize ATP utilizing the energy in the calcium gradient as calcium moves down its gradient in a nonphysiological direction for this pump from lumen to cytoplasm (Barlogie et al. 1971).

Knowledge of other aspects of the molecular mechanism of calcium transport comes from the amino acid sequence obtained from the cDNA sequence (MacLennan et al. 1985). They have found two genes coding for the Ca-ATPase of slow and fast mammalian muscles with some 93% sequence homology. They have proposed the structure shown in Fig. 4 based upon the similarity with

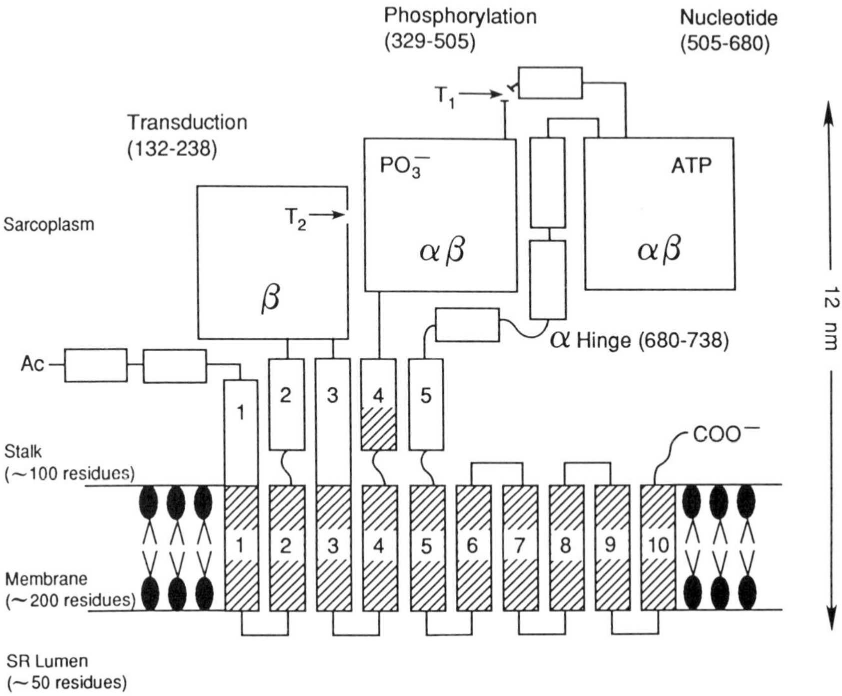

Fig. 4. Schematic drawing of the calcium ATPase derived from the amino acid residue sequence data by MacLennan et al. (1985). Shown are the ten hypothesized membrane-spanning regions, the five stalk regions that connect to the three hypothesized sarcoplasmic regions involved in nucleotide binding, phosphorylation, and transducing the signal from phosphorylation to the membrane domains and calcium binding (possibly in the membrane domain) to the nucleotide binding and phosphorylation domains. α and β denote that the amino acids are in an α-helical or β-sheet configuration. T_1 and T_2 denote the two sites for proteolytic cleavage. See the complete amino acid sequence and the rationale for this hypothesized structure in the paper by MacLennan et al. (1985)

known structures of kinases. This structure has three globular cytoplasmic domains, one nucleotide-binding domain, one phosphorylation domain and a so-called transduction domain. These correspond to the large cytoplasmic regions seen in electron micrographs of the Ca-ATPase. There are also ten possible hydrophobic transmembrane domains which could potentially form the channel through the membrane. Five helical regions connect the large cytoplasmic domains to the membrane domains in a stalk-like fashion. This stalk region was initially proposed to contain the calcium-binding site, but more recently it has been shown that mutations in this stalk region did not affect high affinity calcium binding to the pump, but that mutations in the presumed transmembrane regions did (Clarke et al. 1989). For this reason, the calcium-binding site has been assigned to one or more of the transmembrane regions. There is little homology between these presumed transmembrane domains and the well-known hydrophobic regions of classical Na and Ca channels. MacLennan et al. (1985) propose that there is coupling between the various domains so that calcium binding to the membrane site is signaled through the helical stalk region to the nucleotide binding-phosphorylation domains to bring about phosphorylation which in turn is transmitted back to the membrane calcium-binding region allowing it to change, exposing the calcium-binding sites to the SR lumen and decreasing the calcium affinity. Much of this is speculative but fascinating as the molecular mechanism of calcium transport unfolds.

In summary, the picture which emerges (Fig. 2) is that calcium is stored in the lumen of the SR partially bound to the low affinity chelator, calsequestrin. T-tubule membrane depolarization is sensed by the voltage sensor of a T-tubule calcium channel (dihydropyridine binding site, receptor), the T-tubule end of the foot process. This in turn is coupled to the SR component of the foot process, an SR calcium-releasing channel (and ryanodine-binding site, receptor). Although this SR calcium-releasing channel can be opened in several ways, by elevated calcium and by IP_3, it appears in skeletal muscle to be opened primarily by coupling through the foot process. In this manner, calcium is released. The release turns off either through repolarization of the T-tubule or inactivation of the releasing channel. Calcium is then taken up again back into the SR through the Ca-ATPase transporting protein in the SR membrane and stored for subsequent release.

3 Calcium Activation of Actomyosin Interaction

It was suspected for some time that calcium ions were important in regulating muscle contraction. Using isolated actomyosin preparations, Weber and Winicur (1961) observed that the ATPase activity, a measure of contractile activity in isolated proteins, was usually lower when calcium was chelated than when it was added. However, some preparations demonstrated equivalent ATPase activities both in the presence and absence of calcium. Ebashi and colleagues (see Ebashi and Endo 1968) found the difference between these preparations to be the presence and absence of the proteins tropomyosin and troponin and that these components regulated the thin filament. Actin and myosin from rabbit skeletal muscle show ATPase activity independent of calcium, while the addition of troponin and tropomyosin imparts a calcium dependence to the ATPase. Thus, the role of the regulatory proteins is to effectively inhibit the interaction between myosin and actin, the relaxed state, and to relieve inhibition in the presence of calcium. In rabbit striated muscle, the major regulation site is the thin filament with the regulatory proteins being components of the thin filament. Later, Andrew Szent-Györgyi and colleagues (Szent-Györgyi and Szentkiralyi 1973) showed that in a striated muscle from the sea scallop, regulation took place mainly through the light chains of the thick filament protein myosin, with the light chains serving an inhibitory role in the absence of calcium. Both the systems show regulation through direct calcium binding. In contrast, smooth muscles show regulation by calcium through phosphorylation of one of the myosin light chains by a calcium-calmodulin activated protein kinase (see Chap. 3, this Vol.). In skeletal muscle light-chain phosphorylation occurs, but serves a more modulatory function. Thus, regulation of actin-myosin interaction involves changes in either (or both) of the constituents, the thin (actin) and/or thick (myosin) filaments. It is not surprising that systems have evolved which achieve control in these different ways.

We will discuss first the thin filament-regulated skeletal muscle control; next the thick filament, myosin-linked, direct calcium-binding regulation; and finally the regulation achieved through calcium-calmodulin-myosin, light-chain, kinase-mediated phosphorylation.

3.1 Thin Filament Regulation

Calcium regulates the ATPase activity of myofibrils and force in skinned muscle fibers over a very narrow range of calcium concentrations (see Fig. 5). This regulation in vertebrate skeletal and cardiac muscle occurs predominantly through thin filament regulation.

3.1.1 Proteins Involved in Regulation

The typical muscle cell is approximately 20% protein of which 60% is in the contractile elements, the myofibrils. The two major contractile proteins myosin and actin compose 65% of the myofibril while the regulatory proteins, troponin and tropomyosin, constitute only 10% (see Yates and Greaser 1983). On a molar basis, the ratio of total concentrations of actin, myosin, tropomyosin, and troponin is 7:1:1:1 but in the region of overlap of thick and thin filaments the ratio is approximately 7:2:1:1 because of the filament structure and the lattice arrangement. This indicates that the action of the regulatory proteins is amplified to regulate the possible interaction between one myosin with one of 3–4 actins. How do the regulatory proteins control this interaction?

3.1.2 The Actin Filament

Actin, composing approximately 23% by weight of the myofibril, is the major structural unit of the thin filament so that its interaction with the regulatory proteins forms the basis of regulation (Fig. 6). As discussed in Chapter 4, the

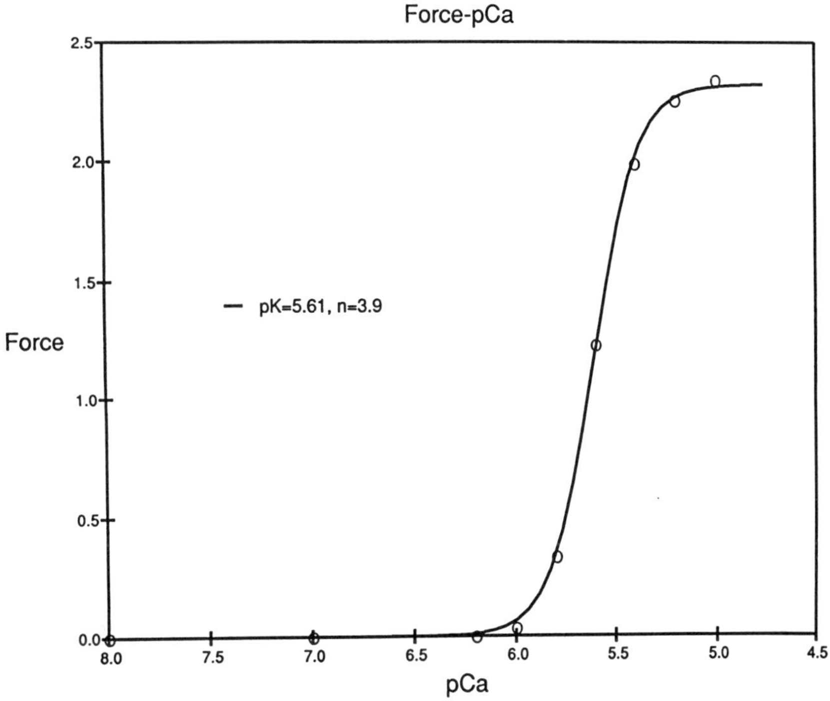

Fig. 5. Steady state force-calcium relationship in a skinned rabbit psoas muscle fiber in the presence of 130 mM K, 3 mM free Mg, 5 mM MgATP, 15 mM creatine phosphate, 10 units/ml of CPK, pH 7 with at least 20 mM pH buffer keeping ionic strength at 0.2 M, 15 mM EGTA with calcium added to adjust the pCa ($-\log_{10}[Ca]$) to the indicated value. The data points for the active force (normalized to unit cross-sectional area using the measured fiber diameter) vs pCa are fit using a nonlinear least square fit to the Hill equation $\{force = force_{max}[1/(1 + 10^{n(pCa-pK)})]\}$. The Hill parameters fitting these data are pK = 5.61 and n = 3.9. Note how steep this relationship is. A simple calcium-binding curve would go from 10 to 90% over a factor of 81 in calcium concentration. This curve does so in a factor of about 3. The apparent cooperativity in this curve is discussed in the text

structural backbone of the thin filament consists of polymerized actin with the two domain actin monomers being oriented at approximately 50° to the filament axis (Holmes et al. 1990). These actin monomers interact with one another to form a genetic helix and also interact along the filament to form a long pitch helix giving rise to the two-stranded helical appearance (Holmes et al. 1990).

Troponin. Troponin consists of three subunits each having quite distinct properties: troponin-C binds calcium, troponin-I inhibits the actomyosin ATPase in a calcium-insensitive manner on a one-to-one basis with actin, and troponin-T links the troponin complex to tropomyosin and actin (Greaser and Gergely 1973). TnC binds both TnI and TnT while TnI binds to actin, TnC, and tropomyosin. On the other hand, TnT binds to tropomyosin, TnC, and TnI. It is by means of these attachments and the calcium-induced conformational changes influencing these attachments that contraction can be initiated and terminated.

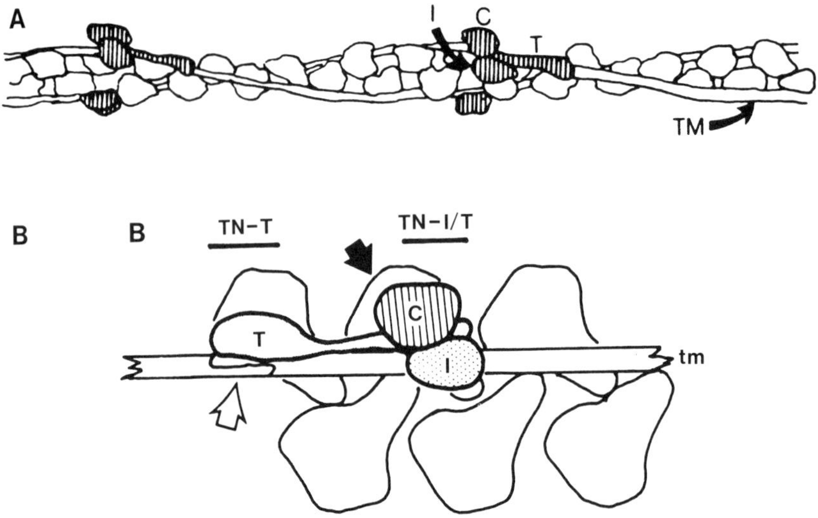

Fig. 6.A Diagram of the thin filament from vertebrate skeletal muscle showing the backbone of helically arranged actin monomers (not labeled) with tropomyosin (*TM*) lying along either side of the actin helix. Troponin with its three subunits TnT (*T*), TnI (*I*), and TnC (*C*) is shown interacting with actin and tropomyosin in **A** and near the region of overlap of neighboring tropomyosin molecules in **B**. The *solid arrow* in **B** shows the actin monomer to which the TnI binds; the *open arrow* indicates the overlap between neighboring tropomyosins. (Sheterline 1983)

TnC. TnC is composed of 159 amino acid residues with a molecular weight of 17846. It is the most acidic of the troponin subunits deriving its low isoelectric point from a large proportion of aspartic and glutamic amino acid residues. These residues play a key coordinating role in metal binding by TnC. TnC binds both TnI and TnT (see Fig. 8).

The calcium-binding subunit of troponin, TnC, is a symmetrically shaped protein having a dumbbell appearance in the crystal structure (Herzberg and James 1985; Sundaralingam et al. 1985; Fig. 7). Located at each end of the molecule are two divalent metal-binding sites. At the N-terminus end are two sites which bind only calcium with a relatively low affinity (K about $10^5\,M^{-1}$), termed calcium-specific sites, while at the other end of the molecule are two sites which can bind either calcium with higher affinity (K about $10^7\,M^{-1}$) or magnesium (K about $10^3\,M^{-1}$), termed the calcium-magnesium sites (Potter and Gergely 1974). These binding sites are formed in the classical helix-loop-helix E-F hand configuration (Kretsinger 1980) with calcium binding coordinated by amino acids in the loop flanked by alpha helical regions. Coordination of calcium occurs with the alpha helical flanking regions oriented perpendicular to one another (in the so-called E-F hand configuration of Kretsinger (1980). This can be seen in the C-terminal sites in the crystallized TnC (lower sites in Fig. 7) where calcium is bound. In contrast, under the acid conditions necessary for

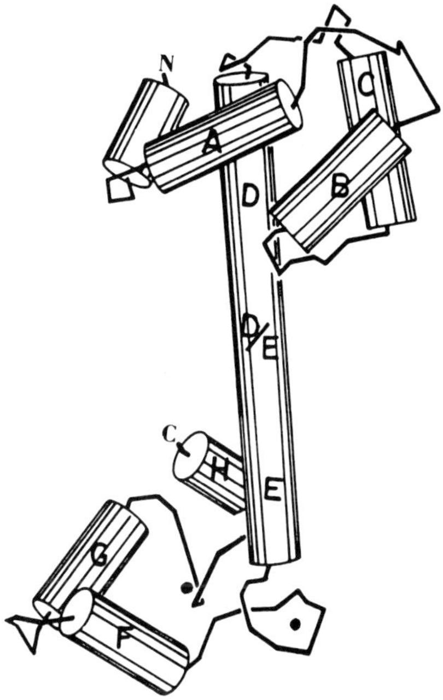

Fig. 7. Crystal structure of chicken fast skeletal TnC with the helical regions drawn as cylinders and labeled *A to H*. The two high affinity calcium-magnesium binding sites at the C terminal are indicated by *dots*. The two lower affinity calcium-specific binding sites in the N terminal are not shown since calcium dos not bind under the condition of crystallization. Note in the calcium-magnesium sites the divalent ion binding is coordinated by a loop with the flanking helices (such as the *E–F* or the *G–H*) oriented at right angles to one another. In the case of no metal binding (as in the C terminal), the flanking helices are oriented approximately parallel (*A–B* or *C–D*). Sundaralingam et al. (1985)

crystallization, calcium is not bound to the Ca-specific N-terminal sites. In the absence of calcium the flanking helices for these two sites are more parallel rather than perpendicular to one another.

The calcium-specific sites are referred to as the regulatory sites in that calcium binding to these sites allows the troponin complex in the presence of tropomyosin to regulate actin and myosin interaction (Potter and Gergely 1974). On the other hand, the calcium-magnesium sites appear to have more of a structural function in that when calcium or magnesium is removed from these sites, TnC can be easily removed from preparations of myofibrils or even skinned muscle fibers (Cox et al. 1981; Zot and Potter 1982). Since with TnC removal calcium activation is lost, this structural function is apparently vital for regulation. In the presence of physiological levels of magnesium (about 1 mM) the calcium-magnesium sites are occupied with magnesium. During muscle twitches, the exchange of calcium for magnesium is sufficiently slow so that these sites would primarily be occupied by magnesium. During a tetanus, however, exchange of calcium for magnesium could occur (Robertson et al.

1981) which would influence the TnC structure in the region near the calcium-specific regulatory sites (Allen et al. 1987), and thus affect regulation (see Grabarek et al. 1986).

TnI. TnI plays an inhibitory role. This protein, having a molecular weight of approximately 20864, consists of 179 amino acid residues. TnI binds to actin, TnT, TnC, and tropomyosin. The protein is the most basic of the troponin subunits with two regions, 1–21 and 96–116, containing many basic amino acid side chains, implicated in interaction with TnC and actin.

TnT. The largest of the troponin subunits, TnT has a molecular weight of 30503 and consists of 259 amino acid residues. The molecule is quite elongated, approximately 18.5 nm long by 2 nm wide. The N-terminal is oriented toward the Z line, the carboxy end toward the H zone. TnT binds to TnC and TnI at its carboxy terminal end and the carboxy terminal end of tropomyosin and the amino terminal portion of the next tropomyosin beyond the head-to-tail juncture of adjacent tropomyosin molecules. This latter interaction involves an unusual triple coiled-coil (Nagano et al. 1980).

Tropomyosin. Tropomyosin is a long (41 nm) rod-shaped molecule with a molecular weight of approximately 66000 from 284 amino acid residues. Tropomyosin is a dimer composed of two alpha helical strands wrapped about each other in a coiled-coil. Two types of subunits exist, α and β, which lead to the formation of heterodimers. The ratio of these two subunits is fiber type dependent with fast or white muscle fibers having predominantly the α type, while slow or red muscle fibers have approximately equal molar ratios of each (Cummins and Perry 1974).

This protein is found in the long groove of the actin helix of the thin filament in a head-to-tail fashion. Consecutive tropomyosin molecules overlap each other by eight to nine amino acid residues (McLachlan and Stewart 1975). Troponin binds to tropomyosin in a two-prong attachment approximately 125 Å from the carboxy terminal end of tropomyosin. As mentioned above, TnT covers a region adjacent to the tropomyosin overlap plus the region of head-to-tail overlap leading to a stabilization of the interaction between neighboring tropomyosins (Fig. 6). This overlap accounts for some of the observed cooperativity in S-1 binding to actin (Pan et al. 1989). This latter strong linkage appears to be calcium-insensitive thereby enabling the complex of troponin and tropomyosin a means of regulating the interaction of the cross-bridges to the actin filament both in the presence and absence of calcium.

The binding of tropomyosin to the actin filament involves a series of very interesting interactions. One tropomyosin binds seven actins, indicating that some sort of structural repeat may occur along the entire tropomyosin molecule. In fact, linear sequence analysis suggests the presence of two sets of seven regions termed a- and b-binding sites with the a-sites being primarily responsible for the interaction of tropomyosin with actin (McLachlan and Stewart 1976).

The b-sites may possibly be involved with the interaction of myosin with actin (see below).

3.1.3 Interaction of Regulatory Proteins and Tropomyosin Movement

Interaction between the proteins discussed above are modulated to bring about activation of muscle. As discussed below, following the sudden increase in intracellular calcium, X-ray diffraction patterns of whole muscle (Haselgrove 1972; Huxley 1972; Parry and Squire 1973) are consistent with the movement of tropomyosin from a position more on the periphery of the helical actin groove to a position more central. This movement would occur as a result of changes in the interactions amongst troponin, actin, and tropomyosin when calcium is bound to troponin.

Inhibition or relaxation is also characterized by weak associations between the troponin subunits. In the relaxed state, the carboxy terminal half of TnT is weakly bound to TnC and possibly actin but more strongly bound to tropomyosin with the remainder of TnT more firmly attached to tropomyosin and its overlap portion (Fig. 6). An important interaction for inhibition occurs through the attachment of region 96–116 of TnI to actin thereby affixing tropomyosin to the outer periphery of the long-pitched actin groove (Fig. 8). Under these circumstances, the flexibility of the actin filament and those of tropomyosin are reduced, reflecting not only the strength of the interactions between actin, tropomyosin, and troponin but possibly a mechanism of regulation by preventing actin from participating in the active stroke in this relaxed condition (Yanagida and Oosawa 1980).

Activation of the interaction of myosin with the thin filament begins through the binding of calcium to the calcium-specific sites of TnC. This appears to result in a rearrangement of the helices bounding the regulatory sites as the protein wraps itself about calcium following the initial collision complex of calcium for these sites. The rearrangement is hypothesized by Herzberg et al. (1986) to be the movement of the A-B and C-D helices of TnC to adopt the E-F hand configuration with the helices flanking the calcium-binding loop, being oriented at right angles to one another. This rearrangement is then hypothesized to expose a region on the central helix of TnC or the C helix that will interact with TnI. Support for this hypothesis comes from experiments in which TnC has been genetically modified. If a disulfide bridge is introduced into TnC, preventing this rearrangement, calcium binding will no longer lead to activation (Grabarek et al. 1990). If a salt bridge is established between the B-C linking loop and the D helix, making this rearrangement less favorable (Fujimori et al. 1990) and this mutant TnC substituted for the native TnC in a skinned muscle fiber, the calcium sensitivity is reduced as a higher calcium is required for activation. This change in the amino terminal half of TnC with calcium binding strengthens the interactions of TnC with TnI and TnC. Region 1–21 of TnI binds to helix C of TnC (Grabarek et al. 1981; Leszyk et al. 1990) and region 96–116 of TnI, which inhibits activation when bound to actin, now binds to helix E of TnC (Chong and Hodges 1981; Grabarek et al. 1981; Dalgarno et al. 1982). This latter interaction arises either through a change in TnC or a change in TnI or both.

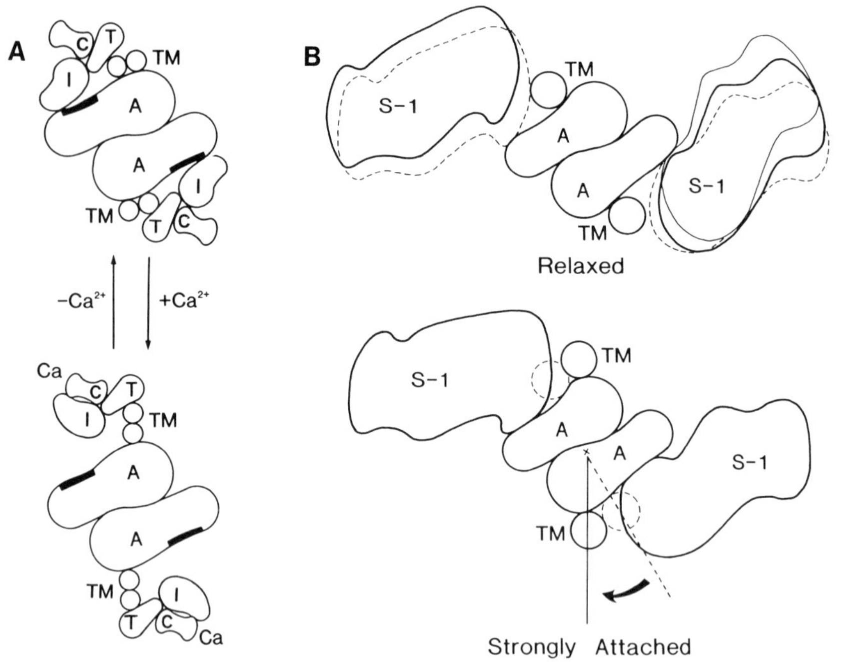

Fig. 8. Diagrams of calcium regulation of the thin filament of vertebrate striated muscle illustrating: **A** possible changes in the interactions of the regulatory proteins tropomyosin (*TM*) and troponin with its subunits TnI (*I*), TnT (*T*), and TnC (*C*) and their approximate relationship with respect to the cross-section of the thin actin (*A*) filament. This is meant to be diagrammatic and does not attempt to represent the precise molecular configuration. In the upper resting state, both tropomyosin and TnI bind to actin [TM to seven actins and TnI to one actin near the (*dark*) region where myosin can bind] stabilizing the regulatory complex on the thin filament. Calcium binding to TnC (calcium-specific sites) increases the affinity of TnC for the region of TnI that binds to actin, thus releasing the TnI from the actin allowing the regulatory complex to move to the position shown in the lower part of **A**. This allows changes in tropomyosin position (and possibly actin structures as well) which facilitate the changes in myosin binding from rapidly associating-dissociating but predominantly dissociated as in the relaxed state to more strongly attached as in the contracted state shown in **B**. In **B** the cross-section of the thin filament with interacting myosin subfragment 1 (*S-1*) is illustrated with the actin (*A*), myosin S-1, and tropomyosin (*TM*) outlines as indicted in the reconstruction of Milligan and Flicker (1987). The angular change in the position of tropomyosin is that calculated from the X-ray diffraction changes in the second actin layer line discussed in the text. The tropomyosin position is drawn to allow weak binding to S-1 in the relaxed state, but retard stronger attachment unless the TM is moved by calcium binding or the S-1 binding itself. This latter point is discussed in this chapter

Also strengthened is the binding of TnC to TnT through helix A of TnC binding for the carboxy terminal end of TnT. This transfer of activation information to TnT possibly influences the interaction of TnT with actin and the interaction between the carboxy terminal half of TnT for tropomyosin. In addition, there is a further weakening of the interaction between TnI and tropomyosin.

All these changes are thought to allow tropomyosin to move on the actin filament as demonstrated by X-ray diffraction studies (Phillips et al. 1986). The binding of tropomyosin to actin is still weak enough to give an increase in flexibility of tropomyosin on the thin filament (Yanagida et al. 1984). This flexibility may describe a state in which the thin filament is capable of forming strongly attached cross-bridges by either interfering with cross-bridge binding to actin or with some kinetic step in the actomyosin ATPase scheme (but biochemical evidence indicates that this may not be in a fully activated state, requiring cross-bridge attachment as well: see later discussion).

The distance between fluorescent probes attached covalently to known amino acids on the various troponin and actin components is consistent with the changes in interactions of the regulatory proteins discussed above that allow tropomyosin to undergo the movement inferred from the X-ray diffraction studies in living muscle. Gergely et al. (1988) measured the distances between probes on TnC (on Cys98), on TnI (on Cys133), and on actin (on Cys374) using the fluorescent energy transfer technique for the reconstituted regulated thin filament (actin, troponin, tropomyosin) either with or without myosin (S-1), and with or without calcium. They found that calcium induced a profound change in the distances, with TnI moving toward TnC (4 to 3.4 nm) with added calcium and away from actin (presumably away from the site of myosin binding, possibly the site where TnI inhibits actin-myosin interaction with a one-to-one stochiometry; 4 to 5.5 nm) with calcium binding. These distance changes are consistent with a loosening of the troponin attachment to actin, allowing the movement of tropomyosin and regulation of the actin-myosin interaction (see Fig. 8).

In addition to the effects of calcium, the attachment of S-1 to actin in rigor linkage also changes the distances between the regulatory proteins in a similar fashion: TnI moves closer to TnC and away from the N-terminus of actin. These changes demonstrate at the molecular level the similarity between calcium-induced changes in the thin filament and cross-bridge-induced changes.

3.1.4 Models of Regulation

Since the discovery of the regulatory proteins, several models of regulation have been proposed. The predominant models are based on the sliding filament model where the interaction of myosin with actin results in the mutual sliding of the thick and thin filaments (see Chap. 5). Thus, activation may be considered as the ability of the thick filament to interact with the thin filament in such a manner as to result in either force development, sarcomere shortening, or both. To date, the most popular models involve either a mechanism whereby the interaction between myosin and actin is physically prevented by tropomyosin or a mechanism whereby calcium shifts kinetic equilibria, thus enhancing the probability of interaction in the presence of calcium or decreasing the ability in its absence. Because of their strong support, both models will be reviewed before we discuss which step appears to be regulated and propose a combined model of regulation. In considering models of regulation, it must be again pointed out that the interaction between actin and myosin is being regulated.

This is dependent on both the structure of the thin filament, with the changes produced by calcium and cross-bridge binding, and the structure of the myosin head, with the changes produced by the bound nucleotide altering the interaction of myosin with actin.

Steric Blocking. Structural studies using electron microscopy and X-ray diffraction reveal that tropomyosin is located within the helical groove along the thin filament and that the position of tropomyosin changes upon activation. A steric blocking model was proposed as the major mechanism for the calcium control of skeletal muscle contraction (Haselgrove 1972; Huxley 1972; Parry and Squire 1973). At rest, tropomyosin is in a position to physically block the interaction of myosin with actin. Calcium binding to troponin allows tropomyosin to move deeper into the groove of the actin double helix thereby uncovering the myosin-binding sites. Actin is then free to interact with myosin. Thus, in the absence of calcium, troponin keeps tropomyosin in a blocking position, while, following calcium binding to troponin, troponin allows tropomyosin to unblock the myosin-binding sites on actin. This simple model of the control of muscle contraction provided a means of regulating contraction through a simple switch mechanism: calcium on, contraction occurs; calcium off, relaxation occurs (see Fig. 8).

This model elegantly accounted for many of the then known observations on the structure of the thin filament, dimensions of the regulatory proteins, and the observations from X-ray diffraction studies.

Allosteric Mechanism. During the late 1970s, several kinetic studies suggested that a more complex regulatory mechanism occurred. Myosin S-1 was found to hydrolyze ATP while bound to actin (Mornet et al. 1981; Stein et al. 1981). Furthermore, myosin·ATP existed in a rapid equilibrium with actin·myosin·ATP, as well as the other intermediate forms (Stein et al. 1979). The rapid equilibrium binding of myosin·ATP to regulated actin occurred in the absence of calcium for both thin filaments in solution (Chalovich and Eisenberg 1982) and for skinned fibers (Brenner et al. 1982). Furthermore, the strength of binding in the absence of calcium decreased with increasing ionic strength, becoming very weak at physiological ionic strengths. In addition, in both the presence and absence of calcium, myosin·ATP and myosin·ADP·P_n (where n represents intermediates of this form) were found to bind weakly to actin while myosin and myosin·ADP bound with a much greater affinity. Thus, all of the myosin states reversibly bind actin in the presence or absence of calcium. These results further imply that the presence of the terminal phosphate of ATP on myosin, either hydrolyzed or not, determines the binding state with its release transforming myosin into a state which binds more tightly. These studies suggest that a weak binding state precedes the strong binding state. Since the presence and absence of the ATP terminal phosphate on myosin appears to determine the binding state of the cross-bridge, it was suggested that calcium regulation controls this conversion through the regulation of a kinetic step, the release of this phosphate (Greene and Eisenberg 1980).

Several lines of evidence suggest that tropomyosin may not physically block the interaction of myosin with the thin filament. Rigor cross-bridges bind to actin in the absence of calcium and also increase the affinity of troponin for calcium (Bremel and Weber 1972). In the presence of ATP, the binding of S-1 or HMM to regulated actin is insensitive to calcium. In the presence of ADP and absence of calcium, the binding of myosin·ADP to regulated actin is cooperative (Greene and Eisenberg 1980). Furthermore, in the intact fiber, stiffness studies show that cross-bridges attach in a relaxed fiber, but are in rapid equilibrium with the detached states (Brenner et al. 1982). Other studies also support the idea that cross-bridges affect the regulatory proteins. The binding of tropomyosin to actin increases in the presence of rigor complexes (Eaton 1976; Rosenfeld and Taylor 1985). When troponin is absent, tropomyosin partially inhibits the ATPase as if tropomyosin were neither in a blocking or activating position (Eaton et al. 1975). Furthermore, the degree of inhibition was found to be dependent on the myosin concentration in the presence of troponin and activating levels of calcium. At high levels of S-1, full activation was observed while at low levels of S-1, the system became partially inhibited (Lehrer and Morris 1982). Furthermore, changes in the interaction between tropomyosin and actin are seen with the addition of rigor cross-bridges to isolated thin filaments, but not with calcium binding (Ishii and Lehrer 1987). Likewise, the increases in V_{max} of the actin-myosin ATPase (Bremel and Weber 1972) or the increase in binding of myosin S-1 or myosin·ADP (S-1 or HMM) to actin when tropomyosin is present suggest that changes may occur within the thin filament upon, or that a composite site of actin-tropomyosin is necessary for, contraction.

This sort of evidence led Hill et al. (1980) to propose that the strong binding of cross-bridges to the thin filament switches 7 actin-1 tropomyosin units into the strong binding state and that this effect spreads cooperatively along the thin filament through interactions between neighboring tropomyosins. Calcium binding does not activate directly by allowing myosin binding, but shifts the equilibrium between the weak and strong binding states of the 7 actin-1 tropomyosin-1 troponin units. Activation of the thin filament in the scheme of Hill et al. involves increasing the kinetic step of release of phosphate from actin·myosin·ADP·P_i to form the strong binding state of actin·myosin·ADP. This model can quantitatively fit the cooperativity in the myosin binding data.

3.1.5 What Step Is Regulated and How Is It Regulated?

The two hypotheses for regulation proposed above differ in two major ways: (1) Which step in the interaction of actin and myosin is regulated, cross-bridge attachment or a kinetic step in the actomyosin ATPase? (2) What regulates this step, calcium binding or cross-bridge attachment? The X-ray diffraction data provide strong evidence that the intensity changes associated with the movement of tropomyosin during activation of an intact muscle fiber precedes cross-bridge attachment as measured either by X-ray diffraction, or by force or stiffness measurements (Fig. 9). Furthermore, tropomyosin movement occurs with calcium activation at sarcomere lengths beyond filament overlap and thus cannot require strongly attached cross-bridges (Kress et al. 1986). This clearly implies

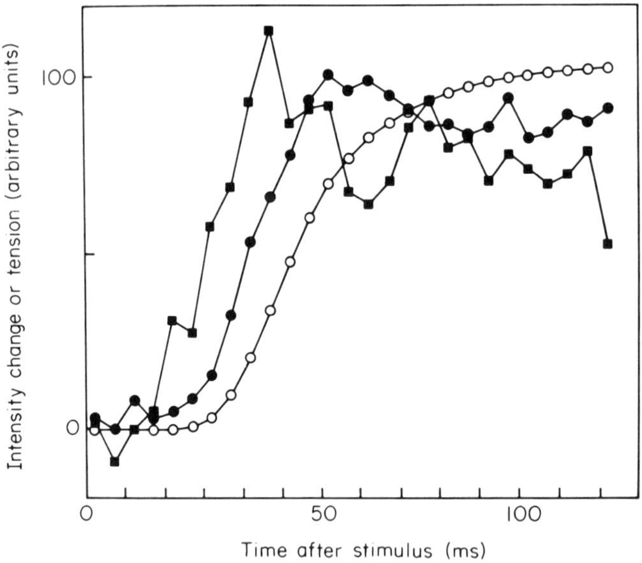

Fig. 9. Time course of the sequential changes in the X-ray diffraction patterns of frog sartorius muscle compared with the rise in tension after the first stimulus. Plotted as a function of time after this first stimulus are the second actin layer line intensity (*filled squares*; indicative from modeling studies of tropomyosin position), equatorial [1,1] reflection increase (*filled circles*; indicative of myosin S-1 movement toward actin), and tension (*open circles*) all in the same muscle at 5 °C (Kress et al. 1986). Note that the second actin layer line leads the equatorial changes which in turn lead the tension rise consistent with calcium-induced tropomyosin movement preceding strong cross-bridge attachment, in turn preceding tension generation

that the strong attachment of cross-bridges prior to force generation is regulated and brought about as a result of the movement of tropomyosin, presumably due to calcium binding and regulatory protein interactions and not by cross-bridge attachment itself. Of course, these conclusions are critically dependent on the interpretation of the X-ray diffraction changes as tropomyosin movement and that tropomyosin movement signals the change in the thin filament associated with activation.

As discussed above, there is strong evidence that myosin attaches to actin in a weakly bound state (low affinity, rapidly dissociating) in skinned fibers, particularly at low ionic strengths (Brenner et al. 1982). This weak binding also occurs in isolated contractile proteins even at physiological ionic strengths and under these conditions calcium may have little effect on this binding (El-Saleh et al. 1986). Furthermore, Brenner (1988), simplifying the multistep process into two steps with effective attachment and detachment rates based on the A.F. Huxley (1957) model of cross-bridge interaction, concluded that in skinned muscle fibers the number of cross-bridges capable of attaching was not affected by calcium, but that the effective rate for cross-bridge attachment was. This was based on measurements under similar conditions of force, stiffness, rate of force redevelopment after a shortening step, and ATPase rate in skinned rabbit skeletal muscle fibers. In particular, the rate of force redevelopment after a shortening step was strongly dependent on calcium. Since one can argue that this rate depends on the rate of cross-bridge attachment, this implies directly that this kinetic step depends on calcium. Previously, others found that the maximum shortening velocity was also a function of calcium, although this was somewhat controversial (Podolin and Ford 1983). Moss and colleagues (Hofmann et al. 1990), however, from their data implying a second calcium regulatory site, specifically the regulatory light chain of myosin, suggest that this is the site responsible for regulating the rate of force redevelopment. This point will be discussed later. In any case the data support calcium affecting some kinetic step associated with attachment. Parenthetically, the study of Brenner's has another interesting corollary that the steep relationship between force and calcium concentration does not necessarily imply cooperative binding of calcium but could arise if the effective rate of cross-bridge attachment is a noncooperative function of calcium.

A more complete model of regulation of contraction requires knowledge of the cross-bridge cycle. Although this is covered in Chapter 5, some discussion is necessary here. There have been many multistep kinetic schemes proposed for ATP hydrolysis by actomyosin. One of the more recent that sheds light on this dilemma is that of Geeves et al. (1984) who propose that each of the actin·myosin (AM) bound states, with and without nucleotide, occurs in two forms, a weak and a strong binding form, with the equilibrium constants favoring weak binding forms for AM·ATP, AM·ADP·P_i and favoring the strong forms for AM and AM·ADP. The occurrence of a weak and a strong binding form for each nucleotide state suggests that regulation of the transition from weak to strong binding does not necessarily imply regulation of a kinetic step involving phosphate release, but could regulate the transition from AM·ADP·P_i weak to AM·ADP·P_i strong binding forms (see Homsher and Millar 1990). Binding of myosin to actin without or with ATP and each of the hydrolysis products (ADP·P_i, ADP) occurs in two steps, a weakly bound, rapidly dissociating step and a strongly bound, slowly dissociating state formed by an isomerization. Further, data on the effects of the rapid release of P_i (using caged P_i) on the force transients of skinned rabbit skeletal muscle fibers by Homsher and colleagues (Millar et al. 1990) are more consistent with calcium regulating a weak to strong transition of the AM·ADP·P_i state (see Fig. 10). In other words, calcium regulates the strong binding of myosin. Since the next step in the cycle is phosphate release, it would be difficult to decide which of these two steps is regulated by calcium without the use of caged phosphate and rapid kinetic analysis.

These results of Homsher's combined with the model of Geeves et al. (1984) provide a resolution to the dilemma of the regulated step. The isomerization from weak to strong attachment is one of the kinetic steps in the cross-bridge cycle and immediately precedes P_i release. These results are also consistent with the data Brenner et al. (1982), and since they involve the regulation of the strong binding of myosin to actin, they are also consistent with the X-ray data. Control of this transition could lie in a structural change in the thin filament, movement of tropomyosin through either calcium binding to the troponin-tropomyosin-actin complex, or with strong cross-bridge attachment

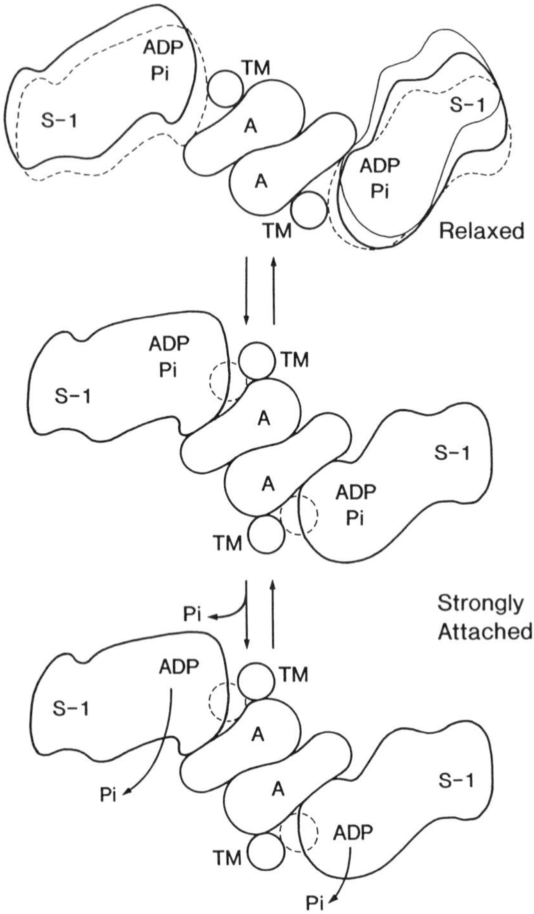

Fig. 10. Diagrams showing possible configurations of myosin S-1 heads with various attached nucleotides and hydrolysis products interacting with the regulated thin filament [shown as in Fig. 8 with actin (*A*) and tropomyosin (*TM*), only with no troponin]. At the *top* in the relaxed state S-1, with either ADP · P_i (shown) or ATP, binds only weakly to actin, prevented from strong attachment by the position of tropomyosin. With activation, tropomyosin moves so that the transition to strong attachment occurs (*middle*). This attachment leads to phosphate release (*below*) and force production. As in Fig. 8, the positions of actin, myosin S-1, and tropomyosin are as indicated in the reconstruction of Milligan and Flicker (1987). In this model calcium regulates the transition from weak to strong binding of myosin S-1 · ADP · P_i to actin, not the release of phosphate

per se. This could involve a change from myosin binding at one (weak) to two (strong) sites on the thin filament on one or two actins, although this is not necessary. Also, as pointed out later in the discussion on myosin-linked regulation, it could also occur through changes in the structure of the myosin head itself brought on by the specific nucleotide present or the binding of calcium in conjunction with the regulatory light chain.

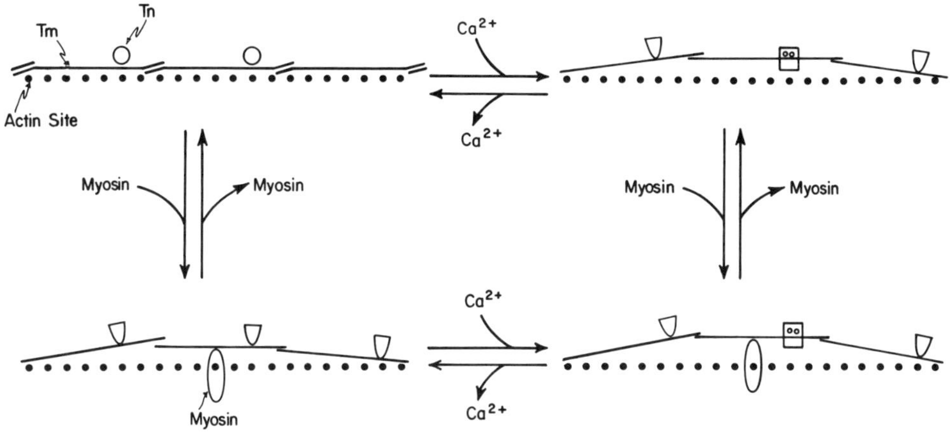

Fig. 11. Model of activation of the thin filament by both calcium and cross-bridges. In the thin filament, tropomyosin (*Tm*) binds to seven actins and overlaps the neighboring tropomyosins. Calcium binding to troponin (*Tn*; *top row*) causes movement of tropomyosin which is coupled to the neighboring tropomyosins, also partially moving them. This allows for the stronger attachment of myosin to a number of actins. Attachment of myosin in a strongly bound configuration (*left*) also moves tropomyosin and affects neighboring tropomyosins and affects troponin-increasing calcium binding. In this diagram, tropomyosin movement is meant to represent thin filament activation and increase with both calcium and cross-bridge binding, although this has not been proven for the intact fiber (see text). Diagram from Yates (1985), adapted from Hill et al. (1983)

3.1.6 Combined Model for Regulation

The results just presented lead to the combined model of regulation shown in Fig. 11. Both calcium binding to troponin and cross-bridge attachment can move tropomyosin and bring about the appropriate thin filament state in which the transition from weakly bound, rapidly dissociating to strongly bound AM·ADP·P_i is favored. Cooperativity along the thin filament can occur through three mechanisms: (1) calcium or strong cross-bridge attachment at one site can favor weak to strong transitions at all actin-myosin·ADP·P_i sites in that tropomyosin-7 actins unit; (2) calcium binding at one site can be transmitted along the thin filament to the neighboring tropomyosin-7 actins unit through the tropomyosin or TnT overlap; and (3) strong cross-bridge attachment at one actin site not only effects the weak to strong transition at all actin-myosin·ADP·P_i sites in that tropomyosin-7 actins unit, but in neighboring units through the overlap between tropomyosins (Pan et al. 1989).

This model, through the principle of detailed balancing, accounts for the observations of the effect of cross-bridge attachment to increase calcium binding to troponin seen in myofibrils first by Bremel and Weber (1972) and in intact muscle fibers by Ridgway and Gordon (1984) and others. If calcium binding increases strong actin-myosin binding, cross-bridge formation, cross-bridge formation in turn should increase calcium binding. These reciprocal changes have been seen using molecular probes by a number of workers (Trybus and Taylor 1980 and Greene 1986, in isolated proteins; Guth and Potter 1987 and

Schulte et al. 1987, in skinned muscle fibers) and are part of the allosteric model of regulation discussed above. They have also been used in a model to explain cooperativity of thin filament activation by Tobacman and Sawyer (1990). Thus, during sustained contraction of muscle with many cross-bridges attached, the calcium affinity of the troponin calcium-binding sites should be increased over the resting level.

This combined model of regulation produces highly cooperative activation of muscle through the three mechanisms suggested above. In the physiological condition of a large calcium release at the beginning of stimulation, the calcium concentration could be sufficient to activate directly without the participation of strongly attached actin-myosin states. During sustained contraction, however, strongly attached actin-myosin states (cross-bridges) could then sustain activation both through direct effects and by increasing the calcium affinity of troponin and by increasing the effectiveness of the myoplasmic calcium. During this period, cross-bridge attachment could be cooperative, sustained in part by cross-bridge attachment. Another source of cooperativity in production of force could be that the transition of a cross-bridge from strong binding to force producing might depend on the number of attached cross-bridges (as proposed by Bagni et al. 1988). This would give enhanced force at high levels of attachment, sustaining contraction.

Relaxation could occur both with a decrease in calcium due to uptake by the sarcoplasmic reticulum but the increased calcium affinity would mitigate this to some extent. Finally, cross-bridge attachment itself could sustain activation prolonging relaxation. X-ray evidence of Kress et al. (1986) supports this point showing a slowing of the return of tropomyosin at the end of stimulation if there is filament overlap in muscle and thus cross-bridge attachment. If this positive feedback should occur during relaxation, it could be terminated by detaching cross-bridges. This detachment appears to happen during relaxation in skeletal muscle brought on by nonuniformities in activation. In single fibers the end sarcomeres appear to relax, first allowing for rapid shortening of the central sarcomeres (Huxley and Simmons 1970), breaking cross-bridges, and eliminating this positive feedback.

This dual regulatory scheme, in addition to providing for highly cooperative, rapid activation and efficient force maintenance, also allows for additional control points in the sequential steps of activation. In other words, contraction can be modulated now by modulating cross-bridge interaction as well as modulating calcium concentration. This may be the scheme used in the thick filament regulation of scallop striated muscle discussed below and in the modulation of contraction by calcium binding to or phosphorlyation of the myosin regulatory light chain.

3.2 Thick Filament Regulation

3.2.1 Invertebrate Muscle – Direct Regulation by Calcium

After the discovery of thin filament regulation, it was thought that all striated muscles might be regulated through this mechanism. In this system, the interaction of actin with myosin (measured by ATPase activity) is not calcium-sensitive, but requires the thin filament regulatory proteins to confer calcium sensitivity and calcium binding to the thin filament. Andrew Szent-Györgyi and his colleagues (see reviews by Szent-Györgyi 1987; Kendrick-Jones and Scholey 1981) found in muscles from the sea scallop and clam that, in contrast, actin and myosin from these animals required calcium for high ATPase activity; furthermore, calcium bound to the myosin, not the thin filament. Calcium regulated by binding to the thick filament myosin, not to the thin filament. It is the light chains of myosin that confer this regulation on the myosin. There are two light chains (molecular weight about 20 000) per myosin head, one of the "regulatory" type and one of the "essential" type. Removal of the regulatory light chain from myosin or from a skinned scallop muscle fiber produces ATPase or force in the absence of calcium (Simmons and Szent-Györgyi 1978). The role of the essential light chains is less clear since conditions necessary for their removal denature the myosin. Thus, the regulatory light chain inhibits the actin-myosin interaction with the inhibition removed physiologically by calcium. As with the thin filament-regulated system, the muscle is kept relaxed by the regulatory proteins which inhibit interaction in the absence of calcium and this inhibition is removed by calcium.

The light chains lie close to one another and on the more distal part of the head away from the actin binding site, close to the head-rod junction (Vibert and Craig 1982). Their position has been established by antibody labeling studies and by cross-linking studies (Ashiba and Szent-Györgyi 1985). Their interaction with this head-rod junction must be important in regulation since myosin, single-headed myosin, and HMM, all with this head-rod junction, are regulated, while S-1 even with both light chains intact is not regulated (Stafford et al. 1979). Cross-linking studies show that the light chains move, presumably accompanying regulation. The amino terminal third of the regulatory light chain can be cross-linked with the essential light chain in rigor and with activation (calcium and ATP), but not in the relaxed state (Hardwicke et al. 1983). When it is cross-linked, the system is "locked" in the "on" state. Cross-linking the carboxy terminal half of the regulatory light chain to the essential light chain does not interfere with regulation. The positioning of the light chains far from both the ATPase site on the myosin S-1 head and the site of actin binding (>10 nm, Craig et al. 1980) implies that information on the state of the light chains must be transmitted some distance through the myosin head.

Calcium activation of the myosin does not require the presence of actin so that this movement of the light chains does not require actin (Ashiba et al. 1980). This specific step of the myosin ATPase accelerated by calcium in isolated myosin is the phosphate release step. The role of actin and the mechanochemical cycle in these muscles are not as well understood as in vertebrate striated muscle.

There is much similarity between the light chains from various species. Since the regulatory light chain can be removed from the sea scallop myosin, studies can be done to test the functional differences. It is found that removal of the regulatory light chain removes the calcium sensitivity and gives ATPase activity in the absence of calcium. Substitution of a similar light chain from vertebrate thin filament-regulated striated muscle shows that these light chains bind and keep the system relaxed, inhibiting the myosin ATPase activity in the absence of calcium but not allowing calcium to activate (Scholey et al. 1980). Substitution of a light chain from any other thick filament-regulated system, even those in smooth muscle, where the primary regulation is through phosphorylation of this light chain, restores full calcium regulation, relaxed in its absence, contracted in its presence (Sellers et al. 1980). This ability to exchange light chains will aid greatly in understanding how they regulate contraction in these thick filament-regulated muscles.

3.2.2 Vertebrate Muscle – Possible Modulation by Light Chains

Due to the similarities of the light chains from vertebrate skeletal muscle with the light chains from these thick filament-regulated muscles, the question arises as to whether these light chains play any role in the regulation of vertebrate skeletal muscle. Evidence is accumulating (see discussion by Moss and colleagues, in Hofmann et al. 1990) that calcium binding to the regulatory light chain in rabbit skeletal muscle modulates the rate of cross-bridge attachment. The data are most consistent with the regulatory light chains decreasing the rate of cross-bridge attachment, with this inhibition being removed by calcium binding. These studies were done by utilizing the ability to extract the regulatory light chains selectively (actually along with TnC, but TnC can then be added back). Earlier studies showed that the calcium-binding sites on myosin from these muscles showed slow exchange of calcium and thus they were not a viable candidate for the main regulatory site, however, slow exchange would be consistent with the modulatory role (Bagshaw and Reed 1977). Further evidence is needed to establish this role of the myosin light chains in calcium regulation of vertebrate skeletal muscle, although it appears that they do regulate cross-bridge kinetics.

3.2.3 Modulation by Light Chain Phosphorylation

Because the light chain 2 of vertebrate skeletal muscle (similar to the regulatory light chain of scallop or smooth muscle) can be phosphorylated by a calcium-calmodulin-dependent protein kinase, the role this played in activation or modulation of contraction in vertebrate skeletal muscle has been questioned. Earlier studies reported little effect on calcium regulation. More recently, careful studies have shown that although light chain phosphorylation did not change the maximum calcium-activated tension or the maximum unloaded shortening velocity, it does enhance force and velocity at lower levels of calcium activation. Sweeney and Stull (1990) and Moss and colleagues (see Metzger et al. 1989) showed that this was due to an enhancement of the effective rate of cross-bridge attachment. Thus, light chain phosphorylation enhances skeletal muscle performance at low levels of calcium activation, but does not influence maximum performance.

Acknowledgments. We would like to thank Professor Haruo Sugi for his patience during the writing of this chapter. The ideas on the activation of contraction presented have evolved through discussions with many of our colleagues here at the University of Washington such as Taylor Allen, Donald Martyn, Bo-Sheng Pan, C.-K. Wang, and Frank Brozovich and particularly with Ellis B. Ridgway at the Medical College of Virginia in Richmond, VA. They should be held blameless for the manner in which the data and ideas are presented. For the discussion of excitation-contraction coupling, we benefited greatly from the many recent reviews and symposia on this subject. We would like to acknowledge the assistance of Martin Schneider in conceiving the structure in Fig. 2. We particularly appreciate the assistance of Martha Mathiason, programmer and secretary, for help with the figures and manuscript. This work has been supported by grants from the National Institutes of Health, NS-08384, AR-39728, and HL-31962, and grants from the Oklahoma Center for the Advancement of Science and Technology, HN8-002 and RA9-002.

References

Allen T, Yates LD, Gordon AM (1987) Coupling between the two domains of muscle fiber TnC. Biophys J 51:331a

Almers W, Fink R, Palade PT (1981) Calcium depletion in frog muscle tubules: the decline of calcium current under maintained depolarization. J Physiol (Lond) 312:177–207

Armstrong CM, Bezanilla FM, Horowicz P (1972) Twitches in the presence of ethylene-glycol-bis-(β-amino-ethyl ether)-N, N′-tetraacetic acid. Biochim Biophys Acta 267:605–608

Ashiba G, Asada T, Watanabe S (1980) Calcium regulation in clam foot muscle. Calcium sensitivity of clam foot myosin. J Biochem (Tokyo) 88:837–846

Ashiba G, Szent-Györgyi AG (1985) Essential light chain exchange in scallop myosin. Biochemistry 24:6618–6623

Ashley CC, Ridgway EB (1970) On the relationship between membrane potential, calcium transient and tension in single barnacle muscle fibres. J Physiol (Lond) 209:105–130

Bagni MA, Cecchi G, Schoenberg M (1988) A model of force production that explains the lag between crossbridge attachment and force after electrical stimulation of striated muscle fibers. Biophys J 54:1105–1114

Bagshaw CR, Reed GH (1977) The significance of the slow dissociation of divalent metal ions from myosin "regulatory" light chains. FEBS Lett 81:386–390

Barlogie W, Hasselbach W, Makinose M (1971) Activation of calcium efflux by ADP and inorganic phosphate. FEBS Lett 12:267

Baylor SM, Chandler WK, Marshall MW (1982) Use of metallochromic dyes to measure changes in myoplasmic calcium during activity in frog skeletal muscle fibres. J Physiol 331:139–177

Berridge MJ, Irvine RF (1984) Inositol trisphosphate, a novel second messenger in cellular signal transduction. Nature (Lond) 312:315–321

Bianchi CP, Frank GB (1982) Excitation-contraction coupling in skeletal, cardiac and smooth muscle. Can J Physiol Pharmacol 60:415–588

Blinks JR, Rüdel R, Taylor SR (1978) Calcium transients in isolated amphibian skeletal muscle fibres: detection with aequorin. J Physiol (Lond) 227:291–323

Block BA, Imagawa T, Campbell KP, Franzini-Armstrong C (1988) Structural evidence for direct interaction between the molecular components of the transverse tubule/sarcoplasmic reticulum junction in skeletal muscle. J Cell Biol 107:2587–2600

Bremel RD, Weber A (1972) Cooperation within actin filament in vertebrate skeletal muscle. Nature 238:97–101

Brenner B (1988) Effect of Ca^{2+} on cross-bridge turnover kinetics in skinned single rabbit psoas fibers: implications for regulation of muscle contraction. Proc Natl Acad Sci USA 85:3265–3269

Brenner B, Schoenberg M, Chalovich JM, Greene LE, Eisenberg E (1982) Evidence for cross-bridge attachment in relaxed muscle at low ionic strength. Proc Natl Acad Sci USA 79:7288–7291

Chalovich JM, Eisenberg E (1982) Inhibition of actomyosin ATPase activity by troponin-tropomyosin without blocking the binding of myosin to actin. J Biol Chem 257:2432–2437

Chong PCS, Hodges RS (1981) A new heterobifunctional cross-linking reagent for the study of biological interactions between proteins. II. Application to the troponin C-troponin I interaction. J Biol Chem 256:5071–5076

Clarke DM, Loo TW, Inesi G, MacLennan DH (1989) Location of high affinity Ca^{2+}-binding sites within the predicted transmembrane domain of the sarcoplasmic reticulum Ca^{2+}-ATPase. Nature 339:476–478

Cox JA, Compte M, Stein EA (1981) Calmodulin-free skeletal-muscle troponin C prepared in the absence of urea. Biochem J 195:205–211

Craig R, Szent-Györgyi AG, Beese L, Flicker P, Vibert P, Cohen C (1980) Electron microscopy of thin filaments decorated with a Ca^{2+}-regulated myosin. J Mol Biol 140:35–55

Cummins P, Perry SV (1974) Chemical and immunochemical characteristics of tropomyosins from striated and smooth muscle. Biochem J 141:43–49

Curtis BA (1963) Some effects of Ca-free choline-ringer solution on frog skeletal muscle. J Physiol 166:75–86

Dalgarno DC, Grand RJA, Levine BA, Moir AJG, Scott GMM, Perry SV (1982) Interaction between troponin-I and troponin-C. Definition of the topography by proton magnetic-resonance spectroscopy. FEBS Lett 150:54–58

de Meis L, Vianna A (1979) Energy interconversion by the Ca^{++}-dependent ATPase of the sarcoplasmic reticulum. Annu Rev Biochem 48:275–292

Donaldson SKB (1985) Peeled mammalian skeletal muscle fibers. Possible stimulation of Ca^{2+} release via a transverse tubule-sarcoplasmic reticulum mechanism. J Gen Physiol 86: 501–525

Donaldson SK, Goldberg ND, Walseth TF, Huetteman DA (1988) Voltage dependence of inositol 1,4,5-trisphosphate-induced Ca^{2+} release in peeled skeletal muscle fibers. Proc Natl Acad Sci USA 85:5749–5753

Eaton BL (1976) Tropomyosin binding to F-actin induced by myosin heads. Science 192: 1337–1339

Eaton BL, Kominz DR, Eisenberg E (1975) Correlation between the inhibition of the acto-heavy meromyosin ATPase and the binding of tropomyosin to F-actin: effects of Mg^{2+}, KCl, troponin I, and troponin C. Biochemistry 14:2718–2725

Ebashi S (1960) Calcium binding and relaxation in the actomyosin system. J Biochem 48: 150–151

Ebashi S, Endo M (1968) Calcium ion and muscle contraction. Prog Biophys Mol Biol 18:125–183

Ebashi S, Lipmann F (1962) Adenosine triphosphate-linked concentration of calcium ions in a particulate fraction of rabbit muscle. J Cell Biol 14:389–400

Eisenberg RS, McCarthy RT, Milton RL (1983) Paralysis of frog skeletal muscle fibres by the calcium antagonist D-600. J Physiol (Lond) 341:495–505

El-Saleh SC, Warber KD, Potter JD (1986) The role of tropomyosin-troponin in the regulation of skeletal muscle contraction. J Muscle Res Cell Motil 7:387–404

Endo M (1977) Calcium release from the sarcoplasmic reticulum. Physiol Rev 57:71–108

Endo M, Iino M (1980) Specific perforation of muscle cell membranes with preserved SR functions by saponin treatment. J Muscle Res Cell Motil 1:89–100

Endo M, Iino M (1988) Measurements of Ca^{2+} release in skinned fibers from skeletal muscle. Methods Enzymol 157:12–26

Fabiato A (1985) Time and calcium dependence of activation and inactivation of calcium-induced release of calcium from the sarcoplasmic reticulum of a skinned canine cardiac Purkinje cell. J Gen Physiol 85:247–289

Fleischer S, Inui M (1989) Biochemistry and biophysics of excitation-contraction coupling. Annu Rev Biophys Biophys Chem 18:333–364

Franzini-Armstrong C, Kenney LJ, Varriano-Marston E (1987) The structure of calsequestrin in triads of vertebrate skeletal muscle: a deep-etch study. J Cell Biol 105:49–56

Fujimori K, Sorenson M, Herzberg O, Moult J, Reinach FC (1990) Probing the calcium-induced conformational transition of troponin C with site-directed mutants. Nature 345:182–184

Furuichi T, Yoshikawa S, Miyawaki A, Wada K, Maeda N, Mikoshiba K (1989) Primary structure and functional expression of the inositol 1,4,5-trisphosphate-binding protein P_{400}. Nature 342:32–38

Gage PW, Eisenberg RS (1967) Action potentials without contraction in frog skeletal muscle fibers with disrupted transverse tubules. Science 158:1702–1704

Geeves MA, Goody RS, Gutfreund H (1984) Kinetics of acto-S1 interaction as a guide to a model for the crossbridge cycle. J Muscle Reo Cell Motil 5:351–361

Gergely J, Grabarek Z, Leavis PC, Strasburg G, Tao T, Wang CLA (1988) Transmission of the Ca^{2+}-regulatory signal in skeletal muscle thin filaments. Adv Exp Med Biol 226:155–163

González-Serratos H (1971) Inward spread of activation in vertebrate muscle fibres. J Physiol (Lond) 212:777–799

Grabarek Z, Drabikowski W, Leavis PC, Rosenfeld SS, Gergely J (1981) Proteolytic fragments of troponin-C. Interactions with the other troponin subunits and biological activity. J Biol Chem 256:13121–13127

Grabarek Z, Grabarek J, Leavis PC, Gergely J (1986) Calcium binding to the low affinity sites in troponin C induces conformational changes in the high affinity domain. J Biol Chem 261:608–613

Grabarek Z, Tan R-Y, Wang J, Tao T, Gergely J (1990) Inhibition of mutant troponin C activity by an intra-domain disulphide bond. Nature 345:132–135

Greaser ML, Gergely J (1973) Purification and properties of the components from troponin. J Biol Chem 246:4226–4233

Greene LE (1986) Cooperative binding of myosin subfragment one to regulated actin as measured by fluorescence changes of troponin I modified with different fluorophores. J Biol Chem 261:1279–1285

Greene LE, Eisenberg E (1980) Cooperative binding of myosin subfragment-1 to the actin-troponin tropomyosin complex. Proc Natl Acad Sci USA 77:2616–2620

Guth K, Potter JD (1987) Effect of rigor and cycling cross-bridges on the structure of troponin C and on the Ca^{2+} affinity of the Ca^{2+}-specific regulatory sites in skinned rabbit psoas fibers. J Biol Chem 262:13627–13635

Hannon JD (1989) Inositol trisphosphate releases Ca^{++} in intact skeletal muscle fibers when the membrane potential is near the contracture threshold. FASEB J 3:A551

Hardwicke PMD, Wallimann T, Szent-Györgyi AG (1983) Light-chain movement and regulation in scallop myosin. Nature (Lond) 301:478–482

Haselgrove JC (1972) X-ray evidence for a conformational change in the actin-containing filaments of vertebrate striated muscle. Cold Spring Harbor Symp Quant Biol 37:341–352

Hasselbach W, Makinose M (1961) Die Calciumpumpe der "Erschlaffungsgrana" des Muskels und ihre Abhängigkeit von der ATP-Spaltung. Biochem Z 333:518–528

Hasselbach W, Makinose M (1962) ATP and active transport. Biochem Biophys Res Commun 7:132–136

Heilbrunn LV, Wiercinski FJ (1947) The action of various cations on muscle protoplasm. J Cell Comp Physiol 29:15–32

Herzberg O, James MNG (1985) Structure of the calcium regulatory muscle protein troponin-C at 2.8 Å resolution. Nature 313:653–659

Herzberg O, Moult J, James MNG (1986) A model for the Ca^{2+}-induced conformational transition of troponin C. A trigger for muscle contraction. J Biol Chem 261:2638–2644

Hill TL, Eisenberg E, Greene L (1980) Theoretical model for the cooperative equilibrium binding of myosin subfragment 1 to the actin-troponin-tropomyosin complex. Proc Natl Acad Sci USA 77:3186–3190

Hill TL, Eisenberg E, Green L (1983) Alternate model for the cooperative equilibrium binding of myosin subfragment-1-nucleotide complex to actin-troponin-tropomyosin. Proc Natl Acad Sci USA 80:60–64

Hille B (1992) Ionic channels of excitable membranes. 2nd edition Sinauer Associates, Sunderland, MA

Hofmann PA, Metzger JM, Greaser ML, Moss RL (1990) Effects of partial extraction of light chain 2 on the Ca^{2+} sensitivities of isometric tension, stiffness, and velocity of shortening in skinned skeletal muscle fibers. J Gen Physiol 95:477–498

Holmes KC, Popp D, Gebhard W, Kabsch W (1990) Atomic model of the actin filament. Nature 347:44–49
Homsher E, Millar NC (1990) Caged compounds and striated muscle contraction. Annu Rev Physiol 52:875–896
Huxley AF (1957) Muscle structure and theories of contraction. Prog Biophys 7:255–318
Huxley AF, Simmons RM (1970) Rapid "give" and the tension "shoulder" in the relaxation of frog muscle fibres. J Physiol 210:32P
Huxley AF, Taylor RE (1958) Local activation of striated muscle fibres. J Physiol (Lond) 144:426–441
Huxley HE (1972) Structural changes in the actin- and myosin-containing filaments during contraction. Cold Spring Harbor Symp Quant Biol 37:361–376
Inesi G (1985) Mechanism of calcium transport. Annu Rev Physiol 47:573–601
Inui M, Saito A, Fleischer S (1987) Purification of the ryanodine receptor and identity with feet structures of junctional terminal cisternae of sarcoplasmic reticulum from fast skeletal muscle. J Biol Chem 262:1740–1747
Ishii Y, Lehrer SS (1987) Fluorescence probe studies of the state of tropomyosin in reconstituted muscle thin filaments. Biochemistry 26:4922–4925
Kendrick-Jones J, Scholey JM (1981) Myosin-linked regulatory systems. J Muscle Res Cell Motil 2:347–372
Kress M, Huxley HE, Faruqi AR, Hendrix JJ (1986) Structural changes during activation of frog muscle studied by time-resolved X-ray diffraction. J Mol Biol 188:325–342
Kretsinger RH (1980) Structure and evolution of calcium-modulated proteins. CRC Crit Rev Biochem 8:119–174
Lai FA, Erickson HP, Rousseau E, Liu Q-Y, Meissner G (1988) Purification and reconstitution of the calcium release channel from skeletal muscle. Nature 331:315–319
Lehrer SS, Morris ED (1982) Dual effects of tropomyosin and troponin-tropomyosin on actomyosin subfragment 1 ATPase. J Biol Chem 257:8073–8080
Leszyk J, Grabarek Z, Gergely J, Collins JH (1990) Characterization of zero-length cross-links between rabbit skeletal muscle troponin C and troponin I: evidence for direct interaction between the inhibitory region of troponin I and the NH2-terminal, regulatory domain of troponin C. Biochem 29:299–304
MacLennan DH, Wong PTS (1971) Isolation of a calcium-sequestering protein from sarcoplasmic reticulum. Proc Natl Acad Sci USA 68:1231–1235
MacLennan DH, Brandl CJ, Korczak B, Green NM (1985) Amino-acid sequence of a Ca^{2+} + Mg^{2+}-dependent ATPase from rabbit muscle sarcoplasmic reticulum, deduced from its complementary DNA sequence. Nature 316:696–700
McLachlan AD, Stewart M (1975) Tropomyosin coiled-coil interactions: evidence for an unstaggered structure. J Mol Biol 98:293–304
McLachlan AD, Stewart M (1976) The 14-fold periodicity in alpha-tropomyosin and the interaction with actin. J Mol Biol 103:271–298
Melzer W, Rios E, Schneider MF (1987) A general procedure for determining the rate of calcium release from the sarcoplasmic reticulum in skeletal muscle fibers. Biophys J 51:849–863
Metzger JM, Greaser ML, Moss RL (1989) Variations in cross-bridge attachment rate and tension with phosphorylation of myosin in mammalian skinned skeletal muscle fibers. Implications for twitch potentiation in intact muscle. J Gen Physiol 93:855–883
Millar N, Lacktis J, Homsher E (1990) The effect of calcium on the phosphate release step of the actomyosin ATPase in skinned psoas muscle fibers. Biophys J 57:397a
Milligan RA, Flicker PF (1987) Structural relationships of actin, myosin, and tropomyosin revealed by cryo-electron microscopy. J Cell Biol 105:29–39
Mornet D, Bertrand R, Pantel P, Audemard E, Kassab R (1981) Structure of the actomyosin interface. Nature (Lond) 292:301–306
Nagano K, Miyamoto S, Matsumura M, Ohtsuki T (1980) Possible formation of a triple-stranded coiled-coil region in tropomyosin-troponin T binding complex. J Mol Biol 141:217–222

Nakajima S, Gilai A (1980) Radial propagation of muscle action potential along the tubular system examined by potential-sensitive dyes. J Gen Physiol 76:751–762

Natori R (1954) The property and contraction process of isolated myofibrils. Jikeikai Med J 1:119–126

Pan B, Gordon AM, Luo Z (1989) Removal of tropomyosin overlap modifies the cooperativity of myosin S-1 binding to reconstituted thin filaments in the presence of ADP. Biophys J 55:278a

Pape PC, Konishi M, Baylor SM, Somlyo AP (1988) Excitation-contraction coupling in skeletal muscle fibers injected with the $InsP_3$ blocker, heparin. FEBS Lett 235:57–62

Parry DA, Squire JM (1973) The structural role of tropomyosin in muscle regulation: analysis of the X-ray diffraction patterns from relaxed and contracting muscles. J Mol Biol 75:33–55

Phillips GN Jr, Fillers JP, Cohen C (1986) Tropomyosin crystal structure and muscle regulation. J Mol Biol 192:111–131

Podolin RA, Ford LE (1983) The influence of calcium on shortening velocity of skinned frog muscle cells. J Muscle Res Cell Motil 4:263–282

Podolsky RJ, Costantin LL (1964) Regulation by calcium of the contraction and relaxation of muscle fibers. Feb Proc 23:933–939

Potter JD, Gergely J (1974) Troponin, tropomyosin, and actin interactions in the Ca^{2+} regulation of muscle contraction. Biochemistry 13:2697–2703

Ridgway EB, Gordon AM (1984) Muscle calcium transient: effect of post-stimulus length changes in single muscle fibers. J Gen Physiol 83:75–103

Ringer S (1883) A further contribution regarding the influence of the different constituents of the blood on the contraction of the heart. J Physiol (Lond) 4:29–42

Rios E, Brum G (1987) Involvement of dihydropyridine receptors in excitation-contraction coupling in skeletal muscle. Nature (Lond) 325:717–720

Robertson SP, Johnson JD, Potter JD (1981) The time-course of Ca^{2+} exchange with calmodulin, troponin, parvalbumin, and myosin in response to transient increases in Ca^{2+}. Biophys J 34:559–569

Rosenfeld SS, Taylor EW (1985) Kinetic studies of calcium binding to regulatory complexes from skeletal muscle. J Biol Chem 260:252–261

Rüegg JC (1986) Calcium in muscle activation. A comparative approach, Springer, Berlin Heidelberg New York

Sánchez JA, Stefani E (1978) Inward calcium current in twitch muscle fibres of the frog. J Physiol 283:197–209

Schneider MF, Chandler WK (1973) Voltage dependent charge movement in skeletal muscle: a possible step in excitation-contraction coupling. Nature (Lond) 242:244–246

Schneider MF, Simon BJ (1988) Inactivation of calcium release from the sarcoplasmic reticulum in frog skeletal muscle. J Physiol 405:727–745

Scholey JM, Taylor KA, Kendrick-Jones J (1980) Regulation of nonmuscle myosin assembly by calmodulin-dependent light chain kinase. Nature (Lond) 287:233–235

Schulte B, Yates LD, Allen T, Gordon AM (1987) Rigor and active cross-bridges enhance the fluorescence of fiber-incorporated labeled TnC in the presence and absence of calcium. Biophys J 51:25a

Sellers JR, Chantler PD, Szent-Györgyi AG (1980) Hybrid formation between scallop myofibrils and foreign regulatory light chains. J Mol Biol 144:223–245

Sheterline P (1983) Mechanisms of cell motility: molecular aspects of contractility. Academic Press, London

Simmons RM, Szent-Györgyi AG (1978) Reversible loss of calcium control of tension in scallop striated muscle associated with the removal of regulatory light chains. Nature (Lond) 273:62–64

Smith JS, Cornado R, Meissner G (1985) Sarcoplasmic reticulum contains adenine nucleotide-activated calcium channels. Nature (Lond) 316:446–449

Stafford WF III, Szentkiralyi EM, Szent-Györgyi AG (1979) Regulatory properties of single-headed fragments of scallop myosin. Biochem 18:5273–5280

Stein LA, Schwartz R, Chock PB, Eisenberg E (1979) The mechanism of the actomyosin ATPase. Evidence that ATP hydrolysis can occur without dissociation of the actomyosin complex. Biochemistry 18:3895–3909

Stein LA, Chock PB, Eisenberg E (1981) Mechanisms of the actomyosin ATPase: effect of actin on the ATP hydrolysis step. Proc Natl Acad Sci USA 78:1346–1350

Sundaralingam M, Bergstrom R, Strasburg G, Rao ST, Greaser M, Wang BC (1985) Molecular structure of troponin C from chicken skeletal muscle at 3 Å resolution. Science 227: 945–948

Sweeney HL, Stull JT (1990) Alteration of cross-bridge kinetics by myosin light chain phosphorylation in rabbit skeletal muscle: implications for regulation of actomyosin interaction. Proc Natl Acad Sci USA 87:414–418

Szent-Györgyi AG (1987) Regulation of molluscan myosin by light chains. Acta Biochim Biophys Hung 22:377–389

Szent-Györgyi AG, Szentkiralyi EM (1973) The light chains of scallop myosin as regulatory subunits. J Mol Biol 74:179–203

Takeshima H, Nishimura S, Matsumoto T, Ishida H, Kangawa K, Minamino N, Matsuo H, Ueda M, Hanaoka M, Hirose T, Kangawa K, Minamion N, Matsuo H, Ueda M, Hanaoka M, Hirose T, Numa S (1989) Primary structure and expression from complementary DNA of skeletal muscle ryanodine receptor. Nature 339:439–445

Tanabe T, Beam, KG, Powell JA, Numa S (1988) Restoration of excitation-contraction coupling and slow calcium current in dysgenic muscle by dihydropyridine receptor complementary DNA. Nature 336:134–139

Tobacman LS, Sawyer D (1990) Calcium binds cooperatively to the regulatory sites of the cardiac thin filament. J Biol Chem 265:931–939

Trybus KM, Taylor EW (1980) Kinetic studies of the cooperative binding of subfragment 1 to regulated actin. Proc Natl Acad Sci USA 77:7209–7213

Tsien RY (1980) New calcium indicators and buffers with high selectivity against magnesium and protons: design, synthesis, and properties of prototype structures. Biochemistry 19:2396–2404

Vergara J, Tsien RY, Delay M (1985) Inositol 1,4,5-trisphosphate: a possible chemical link in excitation-contraction coupling in muscle. Proc Natl Acad Sci USA 82:6352–6356

Vibert P, Craig R (1982) Three-dimensional reconstruction of thin filaments decorated with a Ca^{2+}-regulated myosin. J Mol Biol 157:299–319

Volpe P, Salviati G, De Virgilio F, Pozzan T (1985) Inositol 1,4,5-trisphosphate induces calcium release from sarcoplasmic reticulum of skeletal muscle. Nature (Lond) 316: 347–349

Wagenknecht T, Grassucci R, Frank J, Saito A, Inui M, Fleischer S (1989) Three-dimensional architecture of the calcium channel/foot structure of sarcoplasmic reticulum. Nature 338:167–170

Walker JW, Somlyo AV, Goldman YE, Somlyo AP, Trentham DR (1987) Kinetics of smooth and skeletal muscle activation by laser pulse photolysis of caged inositol 1,4,5-trisphosphate. Nature 327:249–252

Weber A, Winicur S (1961) The role of calcium in the superprecipitation of actomyosin. J Biol Chem 236:3198–3202

Winegrad S, Shanes AM (1962) Calcium flux and contractility in guinea pig atria. J Gen Physiol 45:371–394

Yanagida T, Nakase M, Nishiyama K, Oosawa F (1984) Direct observation of motion of single F-actin filaments in the presence of myosin. Nature (Lond) 307:58–60

Yanagida T, Oosawa F (1980) Conformational changes of F-actin-e-ADP in thin filaments in myosin-free muscle fibers induced by Ca^{2+}. J Mol Biol 140:313–320

Yates LD (1985) Calcium activation of fibers. Rec Meat Conf Proc 38:9–25

Yates LD, Greaser ML (1983) Quantitative determination of myosin and actin in rabbit skeletal muscle. J Mol Biol 168:123–141

Zot HG, Potter JD (1982) A structural role for the Ca^{2+}-Mg^{2+} sites on troponin C in the regulation of muscle contraction. J Biol Chem 257:7678–7683

Chapter 2

Regulation of Contractility in Cardiac Muscle

S. Winegrad

Contents

1 Introduction

The function of the heart is to pump blood to all of the organs of the body in sufficient quantity to satisfy the metabolic requirements of the tissues at any given time. The total metabolic needs of the body can vary over a wide range depending on the environment and the level of activity of the organism. In the average healthy human, the amount of blood flow required can vary over a range of five- to sixfold between the requirements during a basal level of activity and intense exercise. There are only two mechanisms by which the heart can pump more blood per unit time: an increase in the amount ejected during a single contraction, stroke volume, and an increase in the frequency of contraction. An augmentation of stroke volume can be produced by an increased resting or end diastolic volume, a more complete emptying of the ventricles during contraction, or a combination of the two. More complete emptying results from neuro-endocrine stimulation of the heart. In the resting human, the ventricles normally eject about 60–65% of their end diastolic volume, but this amount can rise to approximately 90–95% under intense adrenergic stimulation. End diastolic volume does not increase very much during normal function of a healthy human, even during severe exercise, so the contribution to an increased stroke volume from cardiac dilatation, the so-called Frank-Starling Law of the Heart, is essentially restricted to maintaining an equal output from the left and right sides of the heart over any significant period of time. Therefore, the contri-

Department of Physiology, School of Medicine, University of Pennsylvania, Philadelphia, PA 19104-6085, USA

Advances in Comparative and Environmental Physiology
Vol. 12. Edited by H. Sugi

bution to increased cardiac output that can be made by an increase in stroke volume is only 50–60%, which is a minor fraction of the increment of 400% or more required during severe exercise. The major reason for enhanced cardiac output is the rise in frequency of contraction, which may be as much as threefold in the young, healthy human. An important consequence of the augmented frequency of contraction is a decrease in available time for filling. Since the organ can only eject the volume of blood it receives, adequate filling of the heart becomes an important consideration and may be the limiting factor in the function of the organ at very high levels of cardiac output. The kinetics of contraction are, therefore, extremely important. The speed with which the heart contracts and the length of time during which it can fill are important features of the adaptation of the organ to functional demands placed on it. Kinetic factors may even be more important than strength of contraction or force development, a consideration not always sufficiently recognized in analyses of cardiac function.

Consideration of energetics is extremely important in analyzing the functions of the heart. Of all the tissues in the body, the heart removes the largest fraction of arterial oxygen, and it cannot develop a significant oxygen debt, although its metabolic requirements can vary over a range of five- to sixfold. A match between cardiac work and energy supply is crucial, and in diseased states, it can become limiting. The metabolic requirements of the heart are not simply equivalent to the hemodynamic work performed. The basis for this apparent discrepancy becomes clear when one examines the mechanical features of the cardiac contractile cycle. Before the heart can eject blood during a contraction, it must raise the intraventricular pressure to a level that exceeds the pressure in the outflow vessel, the aorta in the case of the left ventricle. This so-called isometric, or more accurately isovolumic, phase of contraction involves the generation of substantial force and performance of considerable work in changing the shape of the heart even though no blood is pumped or external work done. The energy expended in this process is not entirely wasted or degraded as heat because during early diastole, restoration to the resting shape contributes to filling of the ventricle. Consequently, energy utilization varies more closely with the organ's tension-time integral than with the hemodynamics work performed (Alpert and Mulieri 1986). The rate of energy utilization during the isometric or isovolumic phase of contraction is not fixed, but instead can be modified by changing the properties of the contractile proteins (Holubarsch et al. 1990). Since the isovolumic period contributes to the inefficiency of energy utilization in the heart, increase in the frequency of contraction adds to the energy requirements of the organ.

2 The Basic Contractile Process

The fundamental contractile process in cardiac cells is very similar to that in other forms of striated muscle. Contractile proteins are organized into sets of overlapping thick and thin filaments, in which myosin and actin are the major but not exclusive components of the respective filament types (Huxley 1974). Force and motion are generated by an interaction between actin in the thin filament and the head of the myosin molecule, which extends from the body of the thick filament and contains an enzymatic ATPase site (Huxley 1969). The energy for the contraction comes from the hydrolysis of ATP and the multiple reactions involving the hydrolytic products and the buffering of pH (Cain et al. 1962). Each interaction of a myosin molecule with actin is generally considered to be associated with the hydrolysis of a molecule of ATP, although there is some evidence that the energy in one phosphate can be utilized over a distance that would be equivalent to several actin-myosin interactions (Kishino and Yanagida 1988).

The thin filament contains the regulatory proteins, troponin and tropomyosin, which are crucial in the transition between the relaxed and the contracted state (Ebashi 1980). When the concentration of ionic Ca is lower than approximately $1\,\mu$M, as it is in the cytoplasm of relaxed heart cells, the regulatory proteins inhibit the interaction between actin and myosin, and contraction is prevented (Fabiato and Fabiato 1975a). As the concentration of ionic calcium increases above $1\,\mu$M, Ca is bound by one of the three components of troponin (TNC), and the inhibitory effect of the regulatory proteins is overcome. Each regulatory unit, consisting of one molecule each of troponin and tropomyosin, regulates the ability of seven actin molecules to interact with myosin (Murray and Weber 1974). In a highly nonlinear fashion, force rises from zero to a maximum as the Ca concentration in the cytosol rises from about 1 to about $20\,\mu$M.

Myosin, the major component of the thick filament, is a complex molecule consisting of a pair of heavy chains, each about 200 000 daltons, and two pairs of light chains (Harrington 1979). A portion of each heavy chain has a rod-like portion that forms the backbone of the thick filament and a globular region that contains the separate ATPase and actin-binding sites. The two different light chains form part of the globular head. One is called the essential light chain (LC_1), although it does not appear to be essential for contraction, and the other, the regulatory light chain (LC_2) although its regulatory function is not well understood at all.

Another prominent constituent of the thick filament, the C-protein, has a regular distribution, but its function is not known (Offer 1972). Although it becomes heavily phosphorylated when the cells are treated with beta adrenergic agonists, there is, at present, no indication whether its function is immediately concerned with the contractile process or with some other myofibrillar activity, such as filament or protein turnover (Hartzell 1984; Margossian 1985).

The cytosolic concentration of Ca is the trigger for the contraction. It is normally regulated in the cardiac cell by the combined effects of an active transport system in the sarcoplasmic reticulum that concentrates Ca within the lumen of the reticulum, a Na-Ca exchange system located in the sarcolemma and membrane channels that transiently undergo a large increase in Ca conductance during the actin potential. In the resting cell, cystolic Ca is below the level that activates contraction (Allen and Blinks 1978). Membrane depolarization during the action potential increases cytosolic Ca as a result of two main processes: an opening of calcium channels in the sarcolemma, producing a transient inward flow of Ca in response to the existing electrochemical gradient

(Reuter 1983; Tsien et al. 1986) and a release of Ca from stores within the sarcoplasmic reticulum (Fabiato 1985). Relaxation occurs as the Ca concentration in the immediate vicinity of the myofibrils is reduced by re-uptake of calcium by the sarcoplasmic reticulum and the extrusion of Ca from the cell by the Na-Ca exchanger in the sarcolemma. The intensity of the contraction is determined by the amount of Ca bound by troponin and this is in turn a function of the concentration reached in the cytosol (Holroyde et al. 1979a). The duration of the contraction is determined by the time course of rise in cytosolic Ca; the more slowly Ca is reaccumulated, the longer is the contraction. The concentration of Ca achieved in the cytosol is a function of the amount released, the rate of re-uptake and extrusion from the cell and the binding by intracellular proteins. During the normal contraction under most or perhaps even all circumstances, the amount of Ca released to the cytosol is insufficient to produce a maximum contraction (Fabiato 1981).

Since the myosin heads or cross-bridges are distributed along and around each thick filament, the amount of force that can be generated is limited by the degree to which thick and thin filaments overlap, which is, in turn, a function of the resting length of the tissue (Gordon et al. 1966). It is important, however, to appreciate that changes in the length of the tissue are not identical with changes in the degree of overlap of the thick and thin filaments (Katz 1977). Because of the complex structure of the myocardium, cells within the heart may be reoriented without major changes in length.

The velocity with which a cell contracts is determined by the kinetics of the cross-bridge cycle as it passes from the detached state to force generating back to the detached state and by the number of cycling cross-bridges relative to the load being born by the cell (Barany 1967; Brutsaert and Sonnenblick 1971). The basic relation between force and velocity was originally described in skeletal muscle by Hill (1938). The cross-bridge cycle in cardiac muscle involves a series of reactions that appear to be basically very similar to those in skeletal muscle but have not yet been fully described. In the relaxed muscle, myosin binds ATP in the form of its hydrolytic products, ADP and inorganic phosphate (P_i) (Lymn and Taylor 1971; Eisenberg and Hill 1985). The release of these products occurs very slowly until myosin binds to actin after the inhibition of this interaction by the regulatory proteins has been released by Ca. The transition from relaxation to contraction is probably not between bound and unbound states of myosin to actin, but rather between weakly and strongly binding states, the former incapable of generating much force and the latter generating substantial force (Schoenberg 1988). In the strongly bound state myosin serially releases ADP and P_i in association with the development of force and motion. Separation of myosin from actin, the molecular basis of relaxation, occurs as myosin binds a new molecule of ATP because myosin·ATP in contrast to·myosin·ADP·Pi has a low affinity for actin. Important features of the contraction are determined by the time involved in the complete cycle, the time constants for each of the reactions within the cycle, and the relative fraction of the cycle time spent in the force-producing state (Brenner 1988). The force developed by a cardiac cell is a function of the number of cycling cross-bridges and the relative fraction of the cross-bridge cycle during which force is generated. Economy of conversion of chemical energy to mechanical work is also determined by the latter. Velocity is governed by the rate at which the cross-bridge moves during the part of the cross-bridge cycle when translocation is occurring.

The rate of cycling of the cross-bridge is influenced by mechanical factors, presumably because at least some of the reactions in the cross-bridge cycle have rate constants that are sensitive to the position and possibly the conformation of the cross-bridge (Huxley 1957). An important consequence of this sensitivity is the ability of cycling cross-bridges to influence the kinetics of each other. If the load per cross-bridge is modified by a change in the number of cycling cross-bridges or the size of the load, cycling rate is changed. Reduced loads increase cycling rates. This is the basis for the force-velocity relation. When the kinetic properties of individual cross-bridges are not uniform because of the existence of multiple isoforms of the contractile proteins, as is normally true in the hearts of mammals, in particular small mammals, there is a form of mechanical coupling among the cycling cross-bridges (Schwartz et al. 1981). The cycling rates of individual cross-bridges influence each other, and the overall velocity of shortening of the tissue is somewhere between the values that would exist if cross-bridges were homogeneously of the faster or the slower variety.

3 Regulation of the Contraction

3.1 The Contractile Proteins

The contractile proteins may have different responses to the same calcium signal as a result of posttranslational covalent modification of the proteins themselves or as a result of changes in the composition of the cytosol (Winegrad 1984). In order to detect these changes, intracellular indicators, such as aequorin and fura-2, whose optical properties rapidly change with calcium concentration, can be used to track cytosol Ca during a contractile cycle. Alternatively, the cell membrane can be removed as a diffusion barrier, either manually or chemically, allowing the entry of a Ca-buffer system that can control the concentration of Ca in the immediate environment of the contractile proteins.

Modification in the response of the contractile system to calcium has been produced in isolated and skinned preparations. In cardiac cells exposed to α- or β-adrenergic agonists, the same isometric force may be developed at different concentrations of cytosolic Ca, and major changes in isometric tension can occur with little or no alteration in cytosolic Ca (Mope et al. 1980; Winegrad 1984; Blinks and Endoh 1986; Kurihara et al. 1987; Morgan et al. 1988). Force or ATPase activity has been dissociated from the concentration of Ca ions in the immediate environment of the contractile proteins in skinned fibers (McClellan and Winegrad 1978; Horowits and Winegrad 1987), quickly frozen tissue (Winegrad et al. 1986), isolated myofibrils (Ray and England 1976), and isolated contractile proteins (Holroyde et al. 1979b).

Two different types of changes in the relation between Ca concentration and force have been observed. In the first type, no change in maximum Ca-activated force results – the relation between force and Ca concentration is moved to higher or lower Ca concentrations (Ray and England 1976; McClellan and Winegrad 1978; Fig. 1), while in the second, maximum Ca-activated force is modified without any change in the range of Ca concentrations over which tension is developed (McClellan and Winegrad 1980). Here, the first type of change will be called a change in Ca-sensitivity, the second, a change in contractility (Fig. 1).

Since the maximum concentration of calcium achieved during a cardiac contraction probably rarely if ever rises to a level sufficient to produce maximum force, an increase in Ca-sensitivity will generally increase force. Inasmuch as the contraction generally does not last long enough for the cytosolic calcium concentration and calcium binding by troponin to come to equilibrium, changes in the rate of Ca-binding and release can modify the amount of force generated at a given peak concentration of cytosolic calcium.

Altered affinity of the Ca-binding site on troponin can come from a change in either the rate of binding or release of calcium or a combination of both. The effect of a faster rate of release of calcium by troponin is shown in Fig. 2 (Robertson et al. 1982). Although the kinetics of calcium uptake have changed very little, there are major alterations in the maximum level of saturation, the duration of the period during which some sites contain calcium, and the rate at which the proteins lose their calcium. A decrease in calcium affinity from a faster off-rate would produce a lower level of maximum force, a shorter contraction, and a faster rate of relaxation would be expected. Experimental data support these inferences. With the appropriate changes in on- and off-rates of calcium binding by troponin, increases in maximum force and slower rates of rise and fall in tension can also be produced.

Six different, potentially important mechanisms of change in calcium sensitivity in intact cardiac cells have been identified so far. These include

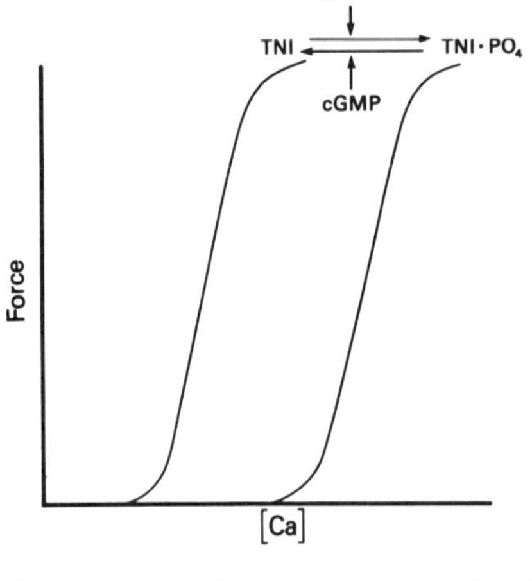

Fig. 1. Diagram of the mechanism by which Ca sensitivity of the contractile system is regulated by phosphorylation of the inhibitory subunit of troponin (*TNI*)

phosphorylation of the inhibitory subunit of troponin (Ray and England 1976; Mope et al. 1980; Robertson et al. 1982), decrease in the concentration of adenosine triphosphate (ATP; McClellan and Winegrad 1980), increase in the resting length of the tissue (Hibberd and Jewell 1982), increase in the concentration of inorganic phosphate (England 1975), a change in the concentration of hydrogen ions (Winegrad 1971; Fabiato and Fabiato 1978), and a change in resting length.

When isolated myofibrils of cardiac muscle are exposed to cAMP and protein kinase, several sites on the contractile proteins are phosphorylated, but the primary site is the inhibitory subunit of the regulatory protein troponin (TNI) (England 1975; Garvey et al. 1988). Associated with this phosphorylation of TNI is an increase in the concentration of Ca^{2+} necessary to activate the actin-dependent ATPase activity of myosin (Fig. 1). A similar change in TNI and calcium sensitivity occurs without destroying the integrity of the membrane or the function of sarcoplasmic reticulum (England 1975) when isolated perfused hearts are exposed to β-adrenergic agonists.

Exposure of EGTA permeabilized or hyperpermeable fibers, which have reduced diffusion barriers to Ca buffers but retain functional adrenergic and cholinergic receptors, to cAMP or to β-agonists produces an increase in the concentration of Ca^{2+} required to activate tension (McClellan and Winegrad 1978; Mope et al. 1980). The shape of the tension-calcium concentration relation is not changed, and the maximum force developed is not changed, but the whole curve is shifted to higher concentrations of Ca^{2+}. Removal of cAMP or the β-agonist reverses the change in calcium sensitivity of the contractile system. Addition of cyclic guanosine monophosphate (cGMP) produces an increase in Ca^{2+} sensitivity, again with no change in maximum Ca^{2+}-activated force. The

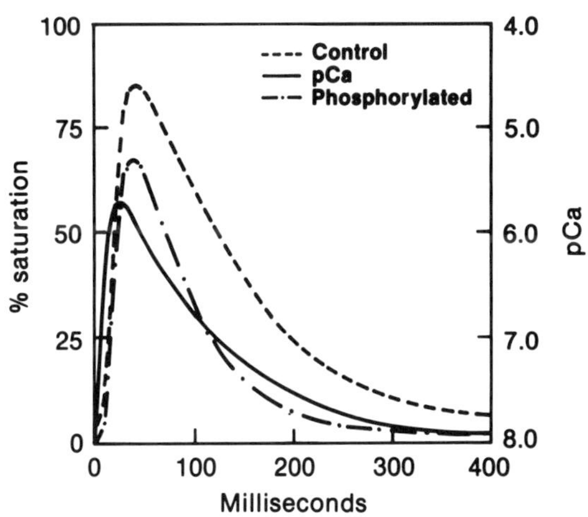

Fig. 2. The effects of a change in Ca sensitivity. The *upper* and *middle curves* show the calculated changes in Ca binding (% saturation) by troponin with a change in Ca affinity from phosphorylation of TNI. *Solid curve* shows the change in ionic Ca in the cytosol during a contraction (Robertson et al. 1982)

amplitude of the shift in calcium sensitivity is closely related to the amount of phosphorylation of TNI, the maximum extent being approximately fivefold. In bundles of hyperpermeable fibers exposed to radioactively labeled ATP, there is a very good correlation between the percentage of TNI phosphorylated and the Ca^{2+} sensitivity; the greater the degree of phosphorylation of TNI, the lower the Ca^{2+} sensitivity. The protein kinase responsible for this control mechanism is located in the surface membrane inasmuch as removal of the sarcolemma inhibits the ability of cAMP to alter calcium sensitivity. Interestingly enough, cGMP is still effective, indicating that its target molecule is not located in the surface membrane.

Under certain circumstances, cholinergic stimulation can alter calcium sensitivity of hyperpermeable fibers (Horowits and Winegrad 1987). In fibers with the highest calcium sensitivity, achieved in the total absence of TNI phosphorylation, cholinergic stimulation does not alter Ca^{2+} sensitivity, but it does raise the sensitivity of fibers with reduced levels of Ca^{2+} sensitivity. Cholinergic agents act by dephosphorylating already phosphorylated TNI, making the extent to which cholinergic stimulation alters Ca^{2+} sensitivity dependent upon the degree of TNI phosphorylation and the level of β-adrenergic activity. In view of the dephosphorylating effect of cGMP and the known increase in cGMP produced by cholinergic stimulation, it is reasonable to assume that at least some of the effect of cholinergic stimulation occurs through a change in the concentration of cGMP. The effect of cholinergic activity on Ca sensitivity is completely inhibited by atropine, indicating that the mechanism of action involves the muscarinic receptors.

Careful studies of the conformation of troponin, using a dansylated form of the molecule, have shown that phosphorylated troponin has a lower affinity for Ca^{2+} entirely as a result of a faster rate of release of bound Ca^{2+} (Holroyde et al. 1980). No significant change occurs in the rate at which Ca^{2+} is bound by troponin. The major effects of TNI phosphorylation should be on both the maximum force developed during a normal contraction and the rate of relaxation, with the little change in the rate of rise of force. In the intact heart, however, phosphorylation of TNI is normally accompanied by phosphorylation of the slow Ca channels leading to an increase in the slow inward Ca^{2+} current during the action potential and a greater rise in the cytosolic calcium concentration. The greater Ca influx will compensate the negative effects of the decrease in Ca^{2+} sensitivity on the level of developed force without altering the more rapid relaxation during β-adrenergic stimulation.

The calcium sensitivity of the contractile system can also be altered by a sufficient decrease in the concentration of ATP (Fabiato and Fabiato 1975b). Maximum Ca^{2+}-activated force is unchanged by reducing ATP concentration from its physiological level of 4–5 to 1 mM. Further reduction of the concentration of ATP raises both maximum Ca^{2+}-activated force and calcium sensitivity. At an ATP concentration of 200 μM, maximum Ca^{2+}-activated force is about one-third greater, and maximum Ca^{2+} activation requires only about one-third or one-fourth of the concentration of Ca^{2+} needed at 5 mM ATP. The effects of decreasing ATP concentration on maximum Ca^{2+} activated force and Ca^{2+} sensitivity reach a maximum at about 50 μM ATP. With further decrease in concentration of ATP, force begins to decline. Phosphorylation of TNI is not involved in this effect of ATP.

In the presence of a low concentration of ATP, the affinity of myosin for actin is so high – this is the basis of rigor interaction – that some myosin molecules bind to actin and shift tropomyosin from its blocking position in which one tropomyosin molecule blocks seven actin molecules (Murray and Weber 1974). One or two rigor interactions may be sufficient to shift a tropomyosin molecule and produce a calcium-insensitive portion of the contractile system. Rigor interactions from low ATP also increase the affinity of troponin for calcium through an allosteric effect. The reason for the increase in maximum Ca^{2+}-activated force is not clear, but it may be related to a state of super contractility of force generators. Unlike the changes in calcium sensitivity that occur under normal, physiological conditions from β-adrenergic or cholinergic stimulation, change from low ATP concentration will not occur in the normal, healthy heart because of the magnitude of the decline in ATP concentration that is necessary.

The affinity of troponin for calcium is sensitive to the resting sarcomere length (Babu et al. 1988; Fig. 3). In skinned or hyperpermeable cardiac cells, an increase in sarcomere length leads to greater Ca sensitivity of contraction. Unlike changes that result from TNI phosphorylation, the Ca^{2+} concentration required for maximum activation is not altered and TNI phosphorylation is not involved. The calcium-binding subunit of cardiac troponin (TNC) is length- or tension-dependent. The skeletal subunit is much less sensitive to changes in length.

This length dependence occurs within the physiological range of sarcomere lengths. Length-dependent Ca^{2+} sensitivity could provide a mechanism for

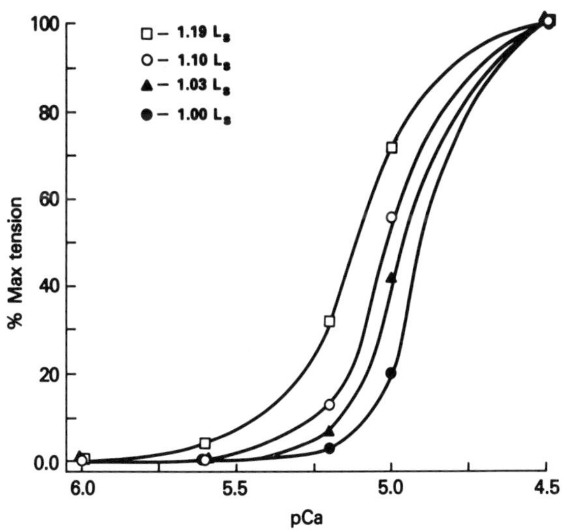

Fig. 3. The increase in Ca sensitivity of the contractile proteins with increase in resting length of the tissue

resisting enlargement of the heart during diastole. A small number of force generators might be activated in the "resting", dilated heart leading to increased stiffness and a smaller change in size of the heart with a given increase in filling pressure.

The calcium sensitivity of the contractile system is decreased by increasing acidity (Fabiato and Fabiato 1978). Within the range that might exist inside cells during normal function and during metabolic stress, such as ischemia or hypoxia, a decrease in pH of 0.5 units will cause an approximately threefold increase in the concentration of Ca^{2+} required for 50% of maximum activation and possibly alter resting tension, active tension, or both.

Both maximum force and Ca-sensitivity of the cardiac contractile system are sensitive to the concentration of inorganic phosphate (Kentish 1986). A rise in the concentration of phosphate from zero to 5 mM causes a reduction in maximum Ca-activated force of about 45%, and a further elevation of the phosphate concentration causes a progressively larger drop in force. In 20 mM phosphate, maximum Ca-activated force is reduced by almost 75%. This influence is independent of any effect of pH or ionic strength. Raising the inorganic phosphate ions from zero to 20 mM doubles the concentration of calcium required to produce 50% of maximum Ca^{2+}-activated force resting. There is an approximately symmetrical shift of the curve relating force to calcium concentration to higher concentrations of the activating ion.

Another form of regulation of calcium sensitivity results from changes in the degree of phosphorylation of the regulating light chain of myosin (LC_2). In this case, the shape of the relation between Ca and developed force changes without alteration in either maximum force or the concentration of Ca necessary to produce maximum force. Phosphorylation of the regulatory light chain of myosin by the Ca-calmodulin-dependent myosin light chain kinase chain increases the rate of rise of force without changing the maximum velocity of unloaded shortening resulting in an apparent increase in Ca-sensitivity (Moreno et al. 1988). In a simple two-state model of cross-bridge cycling that includes only one nonforce-generating and one force-generating step, the changes of mechanical properties produced by phosphorylation of LC_2 would be due to a specific change in the rate of attachment of the cross-bridge.

Exposure of bundles of hyperpermeable cells to either β-adrenergic agonists or micromolar concentrations of cAMP in the presence of a low concentration of nonionic detergent will increase the maximum Ca-activated force by an average of 150% (McClellan and Winegrad 1980). Once the capability for maximum force production has been increased by this combination, it remains at that level. Pretreatment of the tissue with detergent prevents the increase with a subsequent exposure to the β-agonist or cAMP and detergent. Apparently some detergent-sensitive component of the cell, presumably a membrane, is necessary for the increase in contractility to occur. The response to the β-adrenergic agonist, but not to cAMP, can be blocked by including a β-antagonist.

The contractile response of the cells to β-adrenergic activation in the presence of detergent depends on the isoform of myosin that is present. In mammals there are different isoforms of both the heavy and light chains of myosin (Hoh et al. 1977; Lompre et al. 1981). In rat and rabbit, the two animals in which these studies have been conducted, ventricular myosin can exist as three different isozymes (V_1, V_2, V_3) due to the homogeneous and heterogeneous combinations of two α and β isoforms of the heavy chain of myosin. Because of the different maximum velocities, rates of ATPase and economies of energy transductions associated with each of the isozymes, the mechanical properties of the contractile system differ depending upon the relative amount of the three isoforms of myosin. V_1, V_2 and V_3 have decreasing ATPase activities, increasing economies of energy transduction and decreasing maximum velocities of shortening. β-Adrenergic agonists increase the maximum Ca-activated force of bundles of hyperpermeable cardiac cells in accordance with the relative amount of α-myosin heavy chain or V_1 that is present. An approximately linear relation exists between the increment in force and the relative amount of V_1.

These data do not allow distinction between two general types of mechanisms in the regulation of the contractile system: modification of the force of each cross-bridge without change in the total number of active cross-bridges and a change in the number of cross-bridges responding to Ca. The former mechanism would be the equivalent of a rheostat, the latter, of an on-off switch.

Regulation of the contractile proteins has also been studied by using the enzymatic properties as a measure of function in isolated proteins, myofibrils, and more organized models of cardiac cells. The preparation with the greatest degree of structural preservation is the cryostatic section of quickly frozen heart (Winegrad et al. 1986). Because the tissue section is about one-third as thick as a normal ventricular cell and the cell interior is open to a bathing medium, the preparation is a partially skinned fiber with residual, functioning membrane receptors and enzymes. Serial sections can be used for control and experimental samples of very similar populations of cells as well as for separating the relative contributions of V_1 and V_3 force generators to the total enzymatic activity of the myofibrils. The latter is possible because V_3 is alkaline-labile.

Studies with cryostatic tissue sections have demonstrated that even under conditions in which the concentration of Ca ions is not limiting, the ATPase activity of actomyosin can be modified by both adrenergic activity (Winegrad et al. 1986) and by the mechanical state of the heart at the time of quick freezing (Kato et al. 1988). When the level of β-adrenergic activity is raised either in the intact animal by injection of tyramine, in the isolated perfused heart by addition of isoproterenol to the perfusion medium, or the cryostatic section by addition of the drug to the bathing medium, the ATPase activity of myosin is increased. As with the response of force of hyperpermeable fibers to cAMP, the size of the increase in enzymatic activity is related to the relative concentration of V_1.

It is possible to suppress almost totally the enzymatic activity of actomyosin containing a specific isoform of myosin, either V_1 or V_3, in cells with a mixture. Following thyroidectomy in a young rat, the myosin heavy chain in the heart shifts almost totally from the α to the β isoform over a period of 3–4 weeks. The administration of replacement thyroid hormone reverses that change over 7–10 days (Winegrad and Weisberg 1987). During the transitions from α to β and back to α myosin heavy chain, there are periods of time when a mixture of isozymes is present, as determined by nondenaturing gel electrophoresis, but

ATPase activity is due only to myosin containing one of the myosin isoforms. Myosin containing β heavy chain is enzymatically silent 10 days after thyroidectomy, and conversely, 2–3 days after the commencement of replacement therapy the reappearing α heavy chain does not contribute to the total ATPase activity unless the cell is stimulated by β-adrenergic activity. These observations favor an on-off regulation of cross-bridge activity rather than transition between higher and lower levels of activity, a distinction that could not be made from the studies with hyperpermeable fibers.

The mechanical state of the heart also exerts an important influence on the regulatory state of the cross-bridges. In isolated, perfused hearts from older rats, in which there is a substantial mixture of both isoforms of the heavy chain of myosin, the relative contributions of V_1 and V_3 to the total ATPase activity is dependent on filling pressure of the heart (Kato et al. 1990). At low filling pressures, ATPase activity is due almost entirely to V_1 myosin. As filling pressure is increased and cardiac work increases, the relative contribution from V_3 increases, and at the peak of the volume-developed pressure curve, V_3 contributes more to total ATPase activity than V_1. This change occurs within a few minutes and is completely reversible with a lowering of the filling pressure. Interestingly, in spite of a change in the relative contributions of V_1 and V_3, total myosin ATPase activity remains approximately constant. Since V_3 has an enzymatic activity that is about 30% of that of V_1, the constant total ATPase activity with a greater contribution from V_3 means that more enzymatic sites have become active. This conclusion is consistent with the increase in cardiac work performed as the filling pressure has been increased.

The mechanisms for regulating ATPase activity by β-adrenergic activity and by changes in filling pressure are interactive (Kato et al. 1990). As filling pressure is gradually elevated, the ability of β-adrenergic activity to select V_1 force generators is progressively inhibited. Selection of V_3 over V_1 force generators with increased filling pressure is useful for the heart inasmuch as the more energetically economical V_3 force generators are favored for activity. On the other hand, when β-adrenergic stimulation occurs and the heart rate increases, V_1 force generators, which contract more rapidly, are preferentially selected for activity.

In addition to an on-off regulation, the kinetics of the cross-bridge cycle appear to be under adrenergic regulation. Using either oscillatory length changes at different frequencies or perturbations of length in a pseudorandom fashion with a mixture of frequencies, two laboratories have shown changes in stiffness of cardiac tissue that has been placed in contracture by Ba and high frequency stimulation (Berman et al. 1988a; Hoh et al. 1988). The principle of the techniques is that the closer the frequency of the perturbation is to the cycling frequency of the cross-bridge, the lower will be the stiffness of the tissue. With this approach, it has been possible to demonstrate a change in the frequency for minimum stiffness in response to β-adrenergic agonists. Isoproterenol, at a concentration that normally alters the rate of rise of tension of a cyclical contraction, increases the frequency of perturbation at which stiffness is at its minimum by about 50%, indicating an increase in the rate of cycling of the cross-bridges. In rabbit ventricles, which contain primarily V_3, but may have as much as 20% V_1, the frequency minimum observed was the value associated with V_3. After the addition of $10\,\mu$M isoproterenol a second minimum stiffness appeared in the amplitude spectrum at a frequency consistent with the reported for V_1 (Berman et al. 1988b).

Following the exposure to β-adrenergic stimulation (Saeki et al. 1990), myocardial tissue in Ba contracture also shows an acceleration in the rate of length change that occurs during a step decrease in tension. This has been interpreted as an adrenergic effect on the cycling rate of the cross-bridge independent of the level of activation.

Further evidence for a change in the cycling of cross-bridges in response to β-adrenergic stimulation comes from measurements of the economy of energy transduction during contraction. Isoproterenol increased the activation heat by over 150% but also increased the tension-dependent heart normalized to the tension time integral by over 100%. Therefore, isoproterenol not only increases Ca release but also alters the cross-bridge kinetics to produce a less economical cross-bridge cycle. The latter is most easily explained by a decrease in the percentage of the cycle time spent in the force generating state, as would occur with a more rapid detachment of the cross-bridge.

Althought it is not yet possible to bring all of these observations of regulatory changes in the performance of the contractile proteins into a single coherent molecular model, some information already available from biochemical studies may be germane. The prominence of cAMP in the regulatory mechanisms places a focus on phosphorylation. Perfusion of the isolated heart or exposure of isolated cells to β-adrenergic agonists causes the phosphorylation of several proteins, two of which are TNI and C-protein. Phosphorylation of TNI alters Ca-sensitivity but not the maximum force of the contractile system, and the changes in contractility from β-adrenergic stimulation can be dissociated from phosphorylation of TNI (Herring and England 1986). Cardiac C-protein may influence relaxation (Hartzell 1984), and it does seem to have some effect on actomyosin ATPase activity as long as the light chains of myosin are present (Margossian 1985), but these effects have not yet been clearly defined nor has their possible role in the physiological function of the cell been rigorously explored. Although cardiac C-protein becomes heavily phosphorylated with β-adrenergic stimulation, 5 mol of phosphate per mole of protein, in contrast to a much smaller ratio in skeletal muscle, the function of the phosphorylation is not known.

The other phosphorylation reaction of a contractile protein that may be important in physiological regulation is that of the regulatory light chain of myosin. Ca-calmodulin-regulated myosin light chain kinase causes phosphorylation of two residues of LC_2 and produces a change in the shape of the Ca-tension relation as described above (High and Stull 1980; Moreno et al. 1988). Although this reaction appears to be necessary for the generation of force of smooth muscle (Ikebe et al. 1987), there is no evidence of such a role in cardiac muscle.

Experimental data from studies of LC_2 phosphorylation have not produced a consistent story. Although it has been possible to dissociate changes in force from β-adrenergic stimulation by phosphorylation of LC_2 and changes in LC_2

phosphorylation from the amount of force developed (Silver et al. 1986), removal of the regulatory light chain raises actin-activated, but not Ca-activated myosin ATPase, and its replacement reverses the effect (Bhan et al. 1981). If the regulatory light chain is phosphorylated before it is returned to the myofibril, ATPase is not reduced to its previous lower level. The fact that only actin-activated and not Ca-activated ATPase is altered implicates the interaction of myosin with actin rather than the catalytic site. Similar conclusions have been drawn from mechanical studies of skinned fibers, where LC_2 phosphorylation appears to be associated with a modification of the attachment step in the cross-bridge cycle.

X-ray diffraction data show that the cross-bridges can exist in the immediate vicinity of the thin filament in either a tension-producing or a nontension-producing state (Matsubara et al. 1989). When the positions of the cross-bridge mass in the quiescent rat heart and in the heart contracting rhythmically during diastole are compared, differences appear. In the latter case many of the cross-bridges are as close to the thin filament as they are during either systole or rigor when force or enhanced stiffness is present. Apparently, cross-bridge attachment or a transition from a weakly binding to a strongly binding state is rate limiting under isometric rhythmically contracting conditions.

3.2 Genetic Regulation

The performance of the contractile system of cardiac muscle can be altered on three different time scales. A rise in the cytosolic concentration of calcium associated with the activation of contraction will produce force in fractions of a second. Modification of existing contractile proteins will alter the contractile response to a given concentration of Ca in fractions of a minute. Changes in the myofibril from altered protein synthesis and degradation require hours (Everett et al. 1984).

The relative mass for a given body weight may change under pathological conditions, in particular an increase with a greater load on the heart. A considerable amount of work has been carried out in the attempt to identify the necessary and sufficient signals for hypertrophy. While it is clear that an increase in preload or afterload will cause cardiac hypertrophy, the specific stimulus at the cellular level and the chain of events that follow the signals are not yet completely elucidated. Certain aspects are clear at this time. An increase in cardiac work is not necessary for hypertrophy to occur (Xenophontos et al. 1989). The mechanical component of the signal appears to be an increase in diastolic length and/or tension (Komuro et al. 1988; Mann et al. 1989). This has been shown in isolated perfused hearts and in isolated cells suspended on an elastic material that can be used to vary resting tension. In each case, increase in genetic expression and protein synthesis as detected by increase in mRNA, aggregated ribosomal RNA, and the actual synthesis of proteins has been observed (Mahdavi et al. 1982; Long et al. 1989). The presence of Ca ions in the bathing medium appears to be necessary for the effect of greater resting tension to occur. The increase in resting tension does not have to be maintained for the enhancement of protein synthesis. An increase in resting tension lasting only a few minutes is sufficient to produce an increase in protein synthesis that can be detected 2 h later, and changes in genetic activity from brief signals of only a few minutes duration can be detected for at least 6 h after the end of the signaling (Xenophontos et al. 1989). An increase in cell size and contractile protein content has been produced in isolated cells by exposure to α-adrenergic agonist (Simpson 1985). This does not require detectable changes in resting tension.

While the steps following the increase in diastolic tension are not clear, some possible pieces to the puzzle have been identified. Within 30 min of the initiation of the greater diastolic tension or the addition of α-adrenergic agonist (Izumo et al. 1987), increased transcription of protooncogenes, in particular C-fos, can be observed (Kumuro et al. 1988). The increase in transcription appears to be preceded by a transient increase in activity of protein kinase C, and a translocation of the kinase to the nucleus from the cytoplasm. An attractive, but unproven conclusion is that the activation of the gene for C-fos is produced by a protein kinase C-induced phosphorylation, and that the gene product of C-fos, in turn, is involved in the activation of the genes for the contractile proteins. It must be emphasized, however, that this is entirely speculative at the moment.

Although the mechanism by which an increase in resting tension (or length) can initiate a change in the regulation of genes is still obscure, two observations may indicate the general direction. Stretch-activated sodium channels exist in the sarcolemma of cardiac cells as well as other cells, and removal of sodium ions from the bathing medium seems to interfere with the effects of increased resting tension on activating protein synthesis (Kent et al. 1989). The increase in resting tension that produces increased protein synthesis is accompanied by an increase in the intracellular concentration of cAMP (Xenophontos et al. 1989). Modification of the cAMP response appears to be associated with modification of the protein synthetic response.

The increase in contractile proteins that occurs during hypertrophy does not simply produce an increase in myofibrillar protein of the same composition as the existing protein. In most instances of hypertrophy, a change in the relative distribution of the isoforms of myosin and possibly of other contractile proteins occurs (Izumo et al. 1987; Schwartz et al. 1986). In hypertrophy due to increased afterload, there is preferential increase in the synthesis of β-myosin heavy chain and the formation of greater relative amount of V_3. In the case of mild cardiac enlargement from regular periods of exercise, the increase in myocardial mass consists of a greater proportion of V_1 myosin. Cardiac enlargement due to an increase in thyroid hormone is accompanied by a shift to a much larger fraction of V_1.

3.3 The Action Potential

The contraction of cardiac muscle is triggered by a transient increase in the concentration of Ca ions in the immediate vicinity of the myofibrils. The evidence to support this conclusion is overwhelming. In skinned fiber preparations, in which the surface membrane has been removed by manual dissection, an increase in the concentration of Ca ions from approximately 1 to 20 μM causes a transition from complete relaxation to maximum activation of force. A similar change takes place in the ATPase activity of isolated myofibrils or purified systems of contractile proteins that include the regulatory proteins, troponin and tropomyosin. Using different kinds of indicators capable of tracking intracellular Ca-concentrations such as aequorin and fura 2, several laboratories have shown that following membrane depolarization there is a rise in intracellular Ca from a resting level below 1 μM, which is too low to activate tension, to a value sufficiently high for the development of a substantial amount of tension (e.g., Kurihara et al. 1987). Relaxation is preceded by a lowering of the concentration of Ca. From these observations it is clear not only that Ca is the trigger for contraction, but also that the amplitude of the contraction is dependent on the extent to which the concentration of Ca rises.

Extracellular Ca is essential for the maintenance of contractility. Withdrawal of Ca from the extracellular space causes a rapid decline in contractility (Kitazawa 1976). The extent of the decline depends on the size of the reduction of extracellular Ca, total withdrawal producing total loss of contractility in fractions of a minute. Extracellular Ca enters the cell during a contraction. This has been shown by isotopic measurement (Winegrad and Shanes 1962; Niedergerke 1963), by electrical measurement of an increased Ca conductance during the action potential (Reuter 1983) and by measurement with Ca electrodes of a decline in the concentration of extracellular Ca during depolarization of the intact cardiac cell (Bers 1983). All three techniques have demonstrated a quantitative but alinear relation between the amount of Ca entering the cell and the amplitude of contraction. In mammalian heart, all three techniques have also shown that the amount of Ca entering the cell during the contraction is insufficient to produce the level of developed tension (Winegrad and Shanes 1962).

A second reservoir of Ca important in the activation and control of contraction is in the sarcoplasmic reticulum. This internal membrane system surrounds an intracellular compartment in which Ca is heavily concentrated, the equivalent of 1–10 mM in contrast to less than 1 μM concentration in the cytosol of the resting heart cell. Three proteins found in the sarcoplasmic reticulum are critical in its role as an intracellular Ca store: (1) a Ca-transport ATPase; (2) an accessory protein for the Ca transporter called phospholamban; and (3) a Ca-binding protein called calsequestrin (Fleisher and Inui 1989). In isolated preparations, the Ca-transport system is able to lower the Ca concentration in the surrounding medium to less than 1 μM and to establish a gradient of over 10^3 in Ca across its membrane.

Activation of contraction is initiated by the changes in voltage and Ca conductance of the membrane during the action potential. In mammals the relation between the membrane voltage and tension is bell-shaped and passes through a maximum that is at +10 to +20 mV (Morad and Goldman 1973). The amplitude of the associated contraction depends on the characteristics of the first 150 ms of the contraction. Lowering the voltage or shortening the action potential to less than 150 ms produces a weaker contraction. Changing the voltage and/or the duration of the action potential has an effect on the subsequent several contractions (Morad and Trautwein 1968; Wood et al. 1969; Fig. 4). A longer and/or higher plateau increases the amplitude of the subsequent contractions.

During the action potential there is a slow, inward current of Ca ions through Ca channels having a threshold greater than 40 mV (Reuter 1983; Tsien et al. 1986). A greater Ca current occurs with increased depolarization. The inward Ca current appears to be critical for triggering contractions and maintaining contractility (Nabauer et al. 1989). Membrane voltage is thought to be crucial in determining the size of the intracellular Ca transient inasmuch as the amplitude of the Ca transient inside the cell has the same bell-shaped relation to membrane voltage as tension.

An important question is how much the voltage change per se and how much the inward Ca current that is controlled by the voltage each contribute to activation of contraction. Inhibition of the Ca current, either by removal of extracellular Ca or by the use of Ca channel blockers, inhibits contraction (Nabauer et al. 1989) extremely rapidly. Reversal of the blockade of Ca channels between two contractile cycles by inactivating the channel blocker with a flash of light initiates a reversal of the inhibition of contraction in the next contraction. The complete reversal requires six to eight contractions. Inhibition of the Ca conductance even in the presence of depolarization inhibits the contraction, and removal of the inhibition of Ca current immediately initiates a recovery process. The requirement of several contractile cycles for complete reversal of inhibition by the Ca channel blocker is very similar to the response to altering the duration of the plateau of the action potential and the associated Ca conductance. It is probably related to the need of transmembrane Ca movements for reequilibration of the Ca store in the sarcoplasmic reticulum.

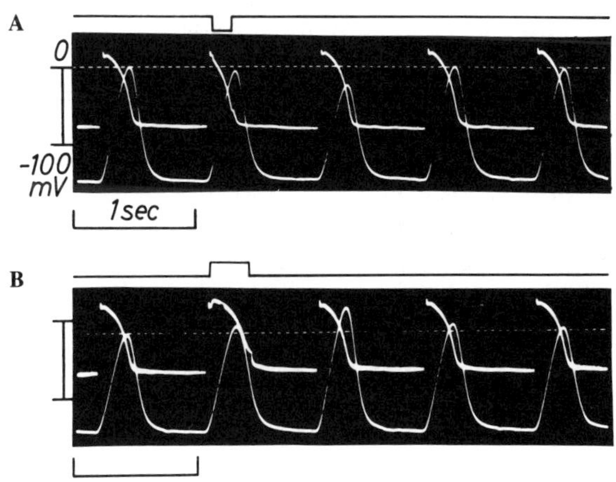

Fig. 4A,B. Simultaneous recording of transmembrane potential (*upper trace*) and force (*lower trace*) in response to changes in the duration of the plateau of the action potential by an applied current. In **A**, the second action potential is shortened and in **B**, the second action potential is lengthened (Wood et al. 1969)

Is the change in membrane voltage required for anything besides opening Ca channels and increasing Ca conductance? On the basis of rapid changes in the extracellular medium during controlled membrane voltage, Morad and colleagues (see Nabauer et al. 1989) have argued that the voltage per se plays no other role. They have shown that the Ca transient declines with the very fast reduction in extracellular Ca in spite of identical depolarization. On the other hand, Berlin et al. (1987) have observed a reduction in the Ca transient by abrupt early repolarization. Whether the latter observation is due to a change in the Ca conductance remains to be seen.

The mechanism by which the membrane events produce a release of Ca from the sarcoplasmic reticulum has been studied most effectively in mechanically skinned single cardiac cells, in which Ca concentration in the myofibrillar space can be changed with a speed approaching that produced in the intact cell by the action potential (Fabiato 1981, 1985). If the rate of rise of Ca concentration is sufficiently rapid, of the order that occurs in the intact cell, there is a release of Ca from the sarcoplasmic reticulum, the so-called Ca-induced Ca release. The amount of Ca released and the amplitude of force produced depend on the rate and extent of rise of the initial or trigger Ca. Ca release from the sarcoplasmic reticulum undergoes inactivation, which can prevent Ca-induced Ca release when the rate of Ca rise is insufficiently fast. With periodic rises in intracellular Ca from Ca pulses, contractile force rises over several cycles just as an intact cell does with abrupt prolongation of the action potential. From these studies it is clear that the early portion of the action potential provokes the release of Ca from the sarcoplasmic reticulum and the plateau maintains the Ca stores of the cell.

The properties of the Ca channel are sensitive to the action of cAMP-dependent protein kinase (Reuter and Scholz 1977; Reuter 1983; Tsien et al. 1986; Reuter et al. 1988; Yatani and Brown 1989). Inward Ca current is increased by β-adrenergic agonists, intracellular injections of cAMP, and exposure to membrane-permeant forms of cAMP. In studies using whole tissue in a double sucrose gap, a concentration jump in cAMP produced a prolongation of the action potential and an increase in the inward calcium current. With fluctuation analysis of recordings made with patch clamps, it has been possible to show that the increase in calcium current is a result of two different changes that can be distinguished by their time courses: (1) an increase in the mean open time of a calcium channel; and (2) increase in the number of functional conductance channels. These changes are generally attributed to phosphorylation of the channel protein by cAMP-dependent protein kinase. A change in the conductance of the slow Ca channel can also occur as a result of direct interaction with a G_s protein without cAMP synthesis following β-adrenergic stimulation.

A second class of compounds, the dihydropyridine Ca agonists such as Bay K 8644, increases Ca channel activity without changing the concentration of cAMP. The major effect of this class of drugs is to increase the channel open time and decrease its closing time, presumably by an effect on channel gating. The mechanism of action is different from that produced by β-adrenergic agonists. Even in the presence of a maximum effect from Bay K 8644, the addition of the β-agonist has an effect by increasing peak current mildly and slowing inactivation.

The dihydrophyridine drugs with Ca-channel blocking activity, such as nifedipine and nitrendipine, increase the rate of decay of the Ca-channel activity and accelerate the rate of inactivation. The blocking effect is very dependent on the membrane potential; it increases with depolarization.

3.4 Sarcoplasmic Reticulum

The sarcoplasmic reticulum is the major source of calcium that activates the contraction (Fleischer and Inui 1988, 1989). Different techniques have been used to estimate the percentage of the activator calcium supplied by the sarcoplasmic reticulum. Although the estimates vary, they tend to indicate that about two-thirds of activator calcium comes from the reticulum and about one-third from the extracellular space via the L-type Ca channels. These values have been inferred from mechanical studies as well as from measurements.

The amount of calcium released by the reticulum depends primarily on two factors: the amount of calcium in the reticulum and the amplitude of the signal for the release of the stored calcium. Calcium in the reticulum has been actively transported into its lumen by a Ca-ATPase pump distributed along the entire surface of the reticulum, both cisternal and tubular. When the concentration in the cytoplasm approachs the range where contraction is stimulated, there is net transport of calcium into the reticulum. The cytosolic calcium comes from the extracellular space through sarcolemmal calcium channels and calcium released from the reticulum earlier. In removing calcium from the cytoplasm, the reticulum competes with sarcolemmal Na-Ca exchange (see below) and the sarcolemmal Ca ATPase pump. Since the two sources of cytoplasmic calcium are the reticulum itself and the extracellular space, the latter through the sarcolemmal Ca channels, the amount of calcium in the reticulum depends on the size of the transmembrane calcium flux, especially during the period of high conductance during the plateau of the action potential. This dependency explains why the strength of contractions is related to the duration and voltage of the plateau of the action potential and why a change in the properties of the plateau disturbs the strength of contraction for several contractile cycles until a new steady state has been established.

Most of the calcium in the reticulum is bound to a calcium-binding protein called calsequestrin (Maclennan et al. 1972), which is bound to the cisternal portion of the reticulum in the vicinity of the junctional reticulum. Consequently, in mammals at least, most of the reticulum calcium is near either the sarcolemma or the transverse tubules, both of which are depolarized during a contraction. Calsequestrin has 40–50 calcium-binding sites with a Michaelis constant (Km) of about 1 mM on the luminal side of the reticulum.

The amount of calcium stored in the reticulum is influenced by the third major protein in the reticulum membrane, phospholamban (Tada et al. 1975). Phospholamban exists as a pentomer throughout the reticulum, apparently in close proximity to the Ca-ATPase pump molecule. Both cAMP-dependent protein kinase and calmodulin-dependent protein kinase produce phosphorylation of phospholamban although at different sites, and dephosphorylation can be produced by a cholinergically controlled phosphatase. There is some question about whether the calmodulin-dependent phosphorylation normally occurs in the intact cell. In the dephosphorylated form, phospholamban appears to raise the Km of the pump (Fleischer and Inui 1988). Phosphorylation of phospholamban removes this inhibition by effectively decreasing the Km for the binding of calcium to the pump molecules. As a result both the rate of calcium transport into the reticulum increases, and the concentration of calcium in the cytoplasm decreases. Ca ATPase reconstituted in membranes containing unphosphorylated phospholamban has the low Km for Ca that is normally seen in the intact reticulum only in the presence of phosphorylated phospholamban. Phosphorylation of phospholamban in this preparation does not alter the Km of the Ca ATPase. It seems that to be inhibitory, phospholamban must have the same steric relation to Ca ATPase that it does in the intact reticulum.

The channel through which calcium is released from the reticulum has recently been isolated and characterized (Inui et al. 1987; Lai et al. 1988; Wagenknecht et al. 1989). It is identical with the foot structures that have been identified electron microscopically between the transverse tubules (or sarcolemma) and the junctional reticulum (Leung et al. 1988). The isolation of the calcium-release channel became possible when it was established that the protein was the ryanodine receptor. In the presence of nanamolar concentrations of ryanodine, the channel is locked in the open position and with micromolar concentrations of ryanodine the channel is locked closed. When the isolated protein is reconstituted in an artificial membrane, the protein acts like a calcium channel that is opened by nanamolar ryanodine, calcium ions, and by ATP and is closed by ruthenium red. Nitrendipine, diltiazem, or D-600, all drugs that modulate the slow calcium channel of the sarcolemma, do not alter Ca conductance of the resuspended Ca-release channel.

The three-dimensional structure of the calcium-release channel, generated from electron micrographs, includes a central pore with four perpendicular channels leading into the pore (Wagenknecht et al. 1989). The amino acid sequence has been deduced from cDNA analysis, and it indicates the presence of two intramembranous regions at either end of the molecule and a large portion of the molecule between these two regions that could act as a regulatory region by binding calcium or ATP (Mikami et al. 1989).

The amount of calcium released by a given signal to the reticulum in intact cells is sensitive to the resting length of the cell. With an increase in length in the range of 70–100% L_{max}, the amount of calcium that is released by a given depolarization is increased (Lakatta and Jewell 1977). The length range over which this increase occurs is very similar to the length range over which the isometric force increases with elongation.

3.5 Na-Ca Exchanger

In addition to the sarcolemma and the Ca pump in the sarcoplasmic reticulum, there is a third mechanism for regulating the intracellular content of calcium, an Na-Ca exchanger (Reuter and Seitz 1968; Bers et al. 1988; Blaustein 1988; Hilgemann 1990). Its presence in cardiac muscle was first inferred from the dependence of Ca fluxes on the concentration of extra- and intracellular Na. Similar observations were made in other tissues. In the squid axon, the failure of cyanide to block the exchange indicated that the energy for the exchange came from the electrochemical gradient of Na and Ca across the surface membrane.

The stoichiometry of the exchange is 3 Na for 1 Ca, making the exchanger electrogenic and the rate of exchange dependent on the difference between the membrane potential and the reversal potential of the exchanger (Blaustein 1988).

The exchanger can transport Ca in either direction, but both entry and exit of Ca are activated by intracellular Ca. The direction of net movement is determined primarily by the intracellular concentrations of Ca and Na as well the relation of the reversal potential for the exchanger to the transmembrane potential.

The major question concerning the exchanger has been its role in the normal contractile cycle. In view of the intracellular concentrations of Na and Ca in the resting heart and the changes induced by the action potential, it is theoretically possible for the exchanger to contribute to Ca influx during depolarization and Ca extrusion following repolarization. At rest, the reversal potential for the exchanger is slightly positive with respect to the resting potential, so a small amount of Ca extrusion occurs. During rapid depolarization, the reversal potential of the exchanger is less than membrane potential favoring Ca entry. The rapid entry of Na enhances the exchange by raising intracellular Na. Ca entry may continue during the plateau, but it should decline with accumulation of intracellular Ca. With repolarization $V_m - E_{Na/Ca} < 0$ and Ca extrusion begins. These are, however, all qualitative considerations and the extent of the contributions of the exchanger to Ca movements during a contractile cycle is dependent upon the kinetics of the exchanger.

Three different preparations have been used in order to determine the basic properties of the exchanger and to deduce its physiological role: (1) isolated vesicles of sarcolemma; (2) isolated cells either internally perfused or loaded with a Ca indicator; and (3) isolated, voltage-clamped patches of sarcolemma. With whole cell voltage clamp and internal perfusion, it has been possible to measure membrane currents and describe the current-voltage relation (Ehara at al. 1989). Following the application of Ca to the outside, there was an outward current that decayed as Ca was accumulated inside the cell. The decay could be inhibited by increasing the intracellular concentraion of EGTA. Certain divalent cations, including Ni, blocked the exchanger current.

With the voltage-clamped sarcolemmal patch (Hilgemann 1990), the exchanger current could be measured as long as nonspecific current leaks were kept small and the Na and Ca pumps blocked. Under these conditions, outward current was activated by cytosolic Na and Ca. An increase in cytosolic Na produced a step increase in membrane current with kinetics that were insensitive to changes in Ca buffering on either side of the membrane. Long-term enhancement of exchange-current transients was produced by the application of 5 mM MgATP to the cytosolic side of the membrane. Most of the ATP-induced increment in current was sensitive to Ca, but a small Ca-insensitive component existed. Steady state outward current increased with cytosolic Ca over the range of 0.1 to 20 μM and then decreased as the Ca concentration increased further and the driving force decreased.

With an intracellular Ca indicator, voltage-clamped individual cardiac cells can be studied to determine the relations among activity of the exchanger, voltage, and intracellular Ca concentration (Hilgemann 1990). The exchanger appears to be too slow to make a significant contribution to Ca entry during the development of the Ca transient following depolarization. On the other hand, it does contribute significantly to the fall in cytosolic Ca during repolarization. The time constant for Ca extrusion through the exchanger is 0.5 s at a membrane potential of −80 mV. The rate of decline of Ca decreases by e-fold with a 77-mV depolarization. These data indicate that the exchanger accounts for about 15% of the decline of the Ca transient following a depolarization. In the absence of extracellular Na, cells relax more slowly. The exchanger should contribute to the regulation of resting cytosolic Ca at transmembrane potentials that are more negative than the equilibrium potential for Ca movement. Twitches can be observed when the Ca channels have been blocked and the concentration of intracellular Na raised by inhibition of the Na pump.

The Na-Ca exchanger should play an important role in pharmacological regulation of contraction. Cardiac glycosides raise intracellular Na by blocking the Na-K ATPase in the sarcolemma. The resultant elevation of intracellular Na should enhance Ca influx through the exchanger by changing the reversal potential of the exchanger.

An additional mechanism for regulating intracellular Ca is Ca-ATPase in the sarcolemma. The affinity of the pump for Ca is about 20 μM in the absence of calmodulin and about 0.4 μM in its presence. Its turnover rate is about one-tenth of that of the Na-Ca exchanger (Blaustein 1988). The affinity is sufficiently high so that under normal physiological conditions the pump is well suited for an important role in setting resting Ca concentration, but a very minor role for restoring Ca after a contraction.

Acknowledgment. This work was supported by N.I.H. grants HL-16010 and HL-33294 and a grant from the W.W. Smith Charitable Trust.

References

Allen D, Blinks JR (1978) Calcium transients in aequorin injected frog cardiac muscle. Nature 273:509–513

Alpert NR, Mulieri LA (1986) Intrinsic determinants of myocardial energetics in normal and hypertrophied hearts. In: Rupp H (ed) The regulation of heart function. Thieme, New York, pp 292–304

Babu A, Sonnenblick E, Gulati J (1988) Molecular basis for the influence of muscle length on myocardial performance. Science 240:74–76

Barany M (1967) The contractile process. Little Brown, Boston, pp 197–220

Berlin J, Cannell M, Lederer W (1987) Regulation of twitch tension in sheep cardiac Purkinje fibers during calcium overload. Am J Physiol 253:H1540

Berman M, Peterson J, Hunter W (1988a) Effect of isoproterenol on force transient time course and on stiffness spectra in rabbit papillary muscle in barium contracture. J Mol Cell Cardiol 20:415–426

Berman M, Peterson J, Hunter W (1988b) Additional evidence that β-adrenergic stimulation enhances V_1 isomyosin in myocardium with mixed isomyosin profile. Biophys J 53:107a

Bers D (1983) Early transient depletion of extracellular Ca during individual cardiac muscle contractions. Am J Physiol 244:H462–468

Bers D, Christensen D, Nguyen T (1988) Can Ca entry via Na-Ca exchange directly activate cardiac muscle contraction. J Mol Cell Cardiol 20:405–414

Bhan A, Malholtra A, Schever J, Conti M, Adelstein R (1981) Subunit function in cardiac myosin. Effects of binding phosphorylated and unphosphorylated myosin light chain 2 to light chain 2 deficient myosin. J Biol Chem 256:7741–7743

Blaustein M (1988) Sodium/calcium exchange and the control of contractility in cardiac muscle and vascular smooth muscle. J Cardiovasc Pharm 12:(Suppl 5) S56–68

Blinks J, Endoh M (1986) Modification of myofibrillar responsiveness to Ca as an inotropic mechanism. Circulation 73:(Suppl III) 85–98

Brenner B (1988) Effect of Ca on cross-bridge turnover kinetics in skinned single rabbit psoas fibers: implication for regulation of muscle contraction. Proc Natl Acad Sci USA 85: 3265–3269

Brutsaert D, Sonnenblick E (1971) Nature of the force-velocity relation in heart muscle. Cardiovasc Res Suppl 1:18–33

Cain DF, Infante A, Davies RE (1962) Chemistry of muscle contraction. Adenosine triphosphate and phosphorylcreatine as energy supplies for single contractions of working muscle. Nature 196:214–217

Ebashi S (1980) Regulation of muscle contraction. Proc R Soc Lond B207:259–286

Ehara T, Matsuoka S, Noma A (1989) Measurement of reversal potential of Na-Ca exchanges current in single guinea-pig ventricular cells. J Physiol 410:227–249

Eisenberg E, Hill T (1985) Muscle contraction and free energy transduction in biological systems. Science 227:999–1006

England P (1975) Correlation between contraction and phosphorylation of the inhibitory subunit of troponin in perfused rat heart. FEBS Lett 50:57–60

Everett A, Simha A, Umeda P, Jakovicic S, Rabinowitz M, Zak R (1984) Regulation of myosin synthesis by thyroid hormone: relative change in the α and β myosin heavy chain mRNA levels in rabbit heart. Biochemistry 23:1596–1599

Fabiato A (1981) Myoplasmic free calcium concentration reached during the twitch of an intact isolated cardiac cell and during calcium induced release of calcium from the sarcoplasmic reticulum of a skinned cardiac cell from the adult rat or rabbit ventricle. J Gen Physiol 78:457–497

Fabiato A (1985) Calcium-induced release of calcium from the sarcoplasmic reticulum. J Gen Physiol 85:189–320

Fabiato A, Fabiato F (1975a) Contractions induced by a calcium triggered release of calcium from the sarcoplasmic reticulum of single skinned cardiac cells. J Physiol Lond 249: 469–495

Fabiato A, Fabiato F (1975b) Effects of magnesium on contractile activation of skinned cardiac cells. J Physiol 249:497–517

Fabiato A, Fabiato F (1978) Effect of pH on the myofilaments and the sarcoplasmic reticulum of skinned cells from cardiac and skeletal muscle. J Physiol 276:233–255

Fleischer S, Inui M (1988) Regulation of muscle contraction and relaxation in heart. The ion pumps: structure, function and regulation. Alan Liss, New York, pp 35-450

Fleischer S, Inui M (1989) Biochemistry and biophysics of excitation-contraction coupling. Annu Rev Biophys Chem 18:333–364

Garvey J, Kranias E, Solaro J (1988) Phosphorylation of C-protein, troponin I and phospholamban in isolated rabbit hearts. Biochem J 249:709–714

Gordon A, Huxley AF, Julian F (1966) The variation in isometric tension with sarcomeric length in vertebrate muscle fibers. J Physiol 184:170–192

Harrington W (1979) Contractile proteins of muscle in the proteins. In: Neurath H, Hill R (eds) Academic Press, New York, pp 246–411

Hartzell C (1984) Phosphorylation of C-protein in intact amphibian cardiac muscle. J Gen Physiol 83:563–588

Herring BP, England PJ (1986) The turnover of phosphate bound to myosin light chain −2 in perfused rat heart. Biochem J 240:205–214
Hibberd, M, Jewell B (1982) Calcium and length dependent force production in rat ventricular muscle. J Physiol 329:527–540
High, C, Stull J (1980) Phosphorylation of myosin in perfused rabbit and rat hearts. Am J Physiol 239:H756–764
Hilgemann D (1990) Regulation and deregulation of cardiac Na-Ca exchange in giant excised sarcolemmal membrane patches. Nature 344:242–245
Hill AV (1938) The heart of shortening and the dynamic constants of muscle. Proc R Soc Lond B126:136–195
Hoh J, McGrath P, Hale P (1977) Electrophoretic analysis of multiple forms of rat cardiac myosin effects of hypophysectomy and thyroxine replacement. J Mol Cell Cardiol 10:1053–1076
Hoh J, Rossmanith G, Kwan L, Hamilton A (1988) Adrenaline increases the rate of cycling of cross-bridges in rat cardiac muscle as measured by pseudo-random binary noise modulated perturbations analysis. Circ Res 62:452–461
Holroyde, M, Howe E, Solaro R (1979a) Modification of calcium requirements for activation of cardiac myofibrillar ATPase by cyclic ATP dependent phosphorylation. Biophys Acta 586:63–69
Holroyde M, Potter J, Solaro R (1979b) The calcium binding properties of phosphorylated and unphosphorylated cardiac and skeletal myosin. J Biol Chem 254:6478-6482
Holroyde M, Robertson S, Johnson D, Solaro J, Potter J (1980) The calcium and magnesium binding sites on cardiac troponin and their role in the regulation of myofibrillar adenosine triphosphatase. J Biol Chem 255:11688–11693
Holubarsch CH, Hasenfuss G, Just H, Blanchard EM, Mulieri LA, Alpert NR (1990) Modulation of myothermal economy of isometric force generation by positive inotropic interventions the guinea pig myocardium. Cardioscience 1:33–41
Horowits R, Winegrad S (1987) cAMP regulation of myosin ATPase activity in the maturing rat heart. Circ Res 61:914–924
Huxley AF (1957) Muscle structure and theories of contraction. Prog Biophys Chem 7:255–318
Huxley AF (1974) Muscular contraction. J Physiol (Lond) 243:1–43
Huxley HE (1969) The mechanism of muscular contraction. Science 164:1356–1366
Ikebe M, Hartshorne D, Elzinga M (1987) Phosphorylation of the 20 000 dalton light chain of smooth muscle myosin by the calcium-activated, phospholipid-dependent protein kinase. J Biol Chem 262:9569–9573
Inui I, Saito A, Fleischer S (1987) Purification of the ryanodine receptor and identity with the feet structures of junctional terminal cisternal of sarcoplasmic reticulum from fast skeletal muscle. J Biol Chem 262:1740–1747
Izumo F, Lompre A, Matsuoka R, Koren G, Schwartz K, Nadal-Ginard B, Mahdavi V (1987) Myosin heavy chain messenger RNA and protein isoform transitions during cardiac hypertrophy. J Clin Invest 79:970–977
Kato NS, Weisberg A, Winegrad S. (1991) Effect of left atrial filling pressure on the activity of specific myosin isozymes in rat heart. Circulation Research 68:1582–1590
Katz A (1977) Physiology of the heart. Raven, New York
Kent R, Hoober K, Cooper G (1989) Load responsiveness of protein synthesis in adult mammalian myocardium: role of cardiac deformation linked to sodium influx. Circ Res 64:74–85
Kentish J (1986) The effects of inorganic phosphate and creatine phosphate on force production in skinned muscle from rat ventricle. J Physiol 370:585-604
Kishino A, Yanagida T (1988) Force measurements by micromanipulation of a single actin filament by glass needles. Nature 334:74–76
Kitazawa T (1976) Physiological significance of Ca uptake by mitochondria in the heart in comparison with that by cardiac sarcoplasmic reticulum. J Biochem (Tokyo) 80:1129–1147
Komuro I, Kurabayashi M, Takaku F, Yazaki Y (1988) Expression of cellular oncogenes in the myocardium during the developmental stage and pressure-overloaded hypertrophy of the rat heart. Circ Res 62:1075–1079

Kurihara S, Konishi M (1987) Effects of B-adrenoreceptor stimulation on intracellular Ca transients and tension in rat ventricular muscle. Pflüger's Arch 409:427–437

Lai F, Erickson H, Rousseau E, Liu Q, Meissner G (1988) Purification and reconstitution of the calcium release channel from skeletal muscle. Nature 331:315–320

Lakatta E, Jewell B (1977) Length dependent activation: its effect on the length tension relation in cat ventricular muscle. Circ Res 40:251–257

Leung A, Imagawa T, Block B, Armstrong CF, Campbell KP (1988) Biochemical and ultrastructural characterization of the dihydropyridine receptor from rabbit skeletal muscle. Evidence for a 52 kilodalton subunit. J Biol Chem 263:994–1001

Lompre AM, Mercadier JJ, Wisnewsky C, Bouveret P, Pantaloni C, D'Albis A, Schwartz K (1981) Species and age-dependent changes in the relative amounts of cardiac myosin isozymes in mammals. Dev Biol 84:286–290

Long C, Ordahl C, Simpson P (1989) Adrenergic receptor stimulation of sarcomeric actin isogene transcription in hypertrophy of cultured rat heart muscle cells. J Clin Invest 83: 1078–1082

Lymn R Taylor E (1971) Mechanism of adenosine triphosphate hydrolysis by actomyosin. Biochemistry 10:4617–4624

Maclennan D, Yep C, Iles G, Seeman P (1972) Isolation of sarcoplasmic reticulum proteins. Cold Spring Harbor Symp Quant Biol 37:469–478

Mahdavi V, Perasamy M, Nadal-Ginard B (1982) Molecular characterization of two myosin heavy chain genes expressed in the adult heart Nature 297:659–665

Mann D, Kent R, Cooper G (1989) Load regulation of the properties of adult feline cardiocytes: growth induction by cellular deformation. Circ Res 64:1079–1090

Margossian S (1985) Reversible dissociation of dog cardiac myosin regulatory light chain 2 and its influence on ATP hydrolysis. J Biol Chem 280:13747–13754

Matsubara I, Maughan D, Saeki Y, Yagi N (1989) Cross-bridge movement in rat cardiac muscle as a function of calcium concentration. J Physiol 417:555–565

McClellan G, Winegrad S (1978) The regulation of the calcium sensitivity of the contractile system in mammalian cardiac muscle. J Gen Physiol 72:737–764

McClellan G, Winegrad S (1980) Cyclic nucleotide regulation of the contractile proteins in mammalian cardiac muscle. J Gen Physiol 75:283–295

Mikami A, Imoto K, Tanabe T, Nudome T, Mori Y, Takeshima H, Narumiya S, Numa S (1989) Primary structure and functional expression of the cardiac dihydropyridine-sensitive calcium channel. Nature 340:230–233

Mope L, McClellan G, Winegrad S (1980) Calcium sensitivity of the contractile system and phosphorylation of troponin in hyperpermeable cardiac cells. J Gen Physiol 75:271–282

Morad M, Goldman Y (1973) Excitation-contraction coupling in heart muscle: membrane control of development of tension. Prog Biophys Mol Biol 27:259–313

Morad M, Trautwein W (1968) The effect of the duration of the action potential on contraction in the mammalian heart muscle. Pflüger's Arch 299:66-82

Moreno I, Arndt H, Gartner C, Ruegg C (1988) Skinned fibers by atrium and ventricle myosin isozymes and contractility. Circ Res 62:632–639

Morgan J, MacKinnon R, Briggs M, Gwathmey J (1988) Calcium and cardiac relaxation in diastolic relaxation of the heart. In: Grossman W, Lorell B (eds) Martinus Nijhoff, The Hague, pp 17–26

Murray J, Weber A (1974) The cooperative action of muscle proteins. Sci Am 230:58–71

Nabauer M, Callewaert G, Cleeman L, Morad M (1989) Regulation of calcium release is gated by calcium current not gating current in cardiac myocytes. science 244:800–803

Niedergerke R (1963) Movements of Ca in beating ventricles of the frog heart. J Physiol 167:551–580

Offer G (1972) C-protein and the periodicity in the thick filaments of vertebrate skeletal muscle. Cold Spring Harb Symp 37:87–93

Ray K, England P (1976) Phosphorylation of the inhibitory subunit of troponin and its effect on calcium dependence of cardiac myofibril adenosine triphosphatase. FEBS Lett 70:11–17

Reuter H (1983) Calcium channel modulation by neurotransmitters, enzymes and drugs. Nature 301:569–574

Reuter H, Scholz H (1977) The regulation of the Ca conductance of cardiac muscle by adrenaline. J Physiol 264:49–62
Reuter H, Seitz H (1968) The dependence of calcium efflux from cardiac muscle on teperature and external ion composition. J Physiol 195:451–470
Reuter H, Porzig H, Kokubun S, Prud'hom B (1988) Calcium channels in the heart. Properties and modulation of dehydropyridine enantiomers. Ann NY Acad Sci 522:16–24
Robertson S, Johnson D, Holroyde M, Kranias M, Potter J, Solaro J (1982) The effect of troponin I phosphorylation on the Ca-binding properties of Ca-regulatory site of bovine cardiac troponin. J Biol Chem 257:260–263
Saeki Y, Shiozawa K, Yanagisawa K, Shibata T (1990) Adenaline increases the rate of cross-bridge cycling in rat cardiac muscle. J Mol Cell Cardiol 22:453–460
Schoenberg M (1988) The kinetics of weakly and strongly binding cross-bridges: implications for contraction and relaxation in molecular mechanism of muscle contraction. In: Sugi H, Pollack G (eds) Molecular mechanism of muscle contraction. Plenum, New York, pp 189-202
Schwartz K, Lecarpentier Y, Martin J, Lompre A, Mercadier J, Swyngedauw B (1981) Myosin enzymatic distribution correlates with speed of myocardial contraction. J Mol Cell Cardiol 13:1071–1075
Schwartz K, de la Bastie D, Bouveret P, Oliviero P, Alonso S, Buckingham M (1986) α-skeletal muscle actin mRNA's accumulate in hypertrophied adult rat hearts. Circ Res 59:551–555
Silver P, Buja M, Stull J (1986) Frequency dependent myosin light chain phosphorylation in isolated myocardium. J Mol Cell Cardiol 18:31–37
Simpson P (1985) Stimulation of hypertrophy of cultured neonatal rat heart cells through an α adrenergic receptor and induction of beating through an α and β adrenergic receptor interaction: evidence for independent regulation of growth and beating. Circ Res 56:884–894
Tada M, Kirchberger M, Katz A (1975) Phosphorylation of a 22000 dalton component of the cardiac sarcoplasmic reticulum by adenosine 3′5′ monophosphate-dependent protein kinase. J Biol Chem 250:2640–2647
Tsien RW, Bean B, Hess P, Lansman J, Nilius B, Nowycky M (1986) Mechanisms of calcium channel modulation of β-adrenergic agents and dihydropyridine calcium agonists. J Mol Cell Cardiol 18:691–710
Wagenknecht T, Grassucci R, Frank J, Saito A, Inui M, Fleischer S (1989) Three dimensional architecture of the calcium channel foot structure of sarcoplasmic reticulum. Nature 338:187–170
Winegrad S (1971) Studies of cardiac muscle with a high permeability to calcium produced by treatment with ethylenediaminetetraacetic acid. J Gen Physiol 58:71–93
Winegrad S (1984) Regulation of cardiac contractile proteins. Cric Res 55:565–574
Winegrad S, Shanes AM (1962) Calcium flux and contractility in Guinea pigs. J Gen Physiol 45:371–394
Winegrad S, Weisberg A (1987) Isozyme specific modification of myosin ATPase by cAMP in rat heart. Circ Res 60:384–392
Winegrad S, Weisberg A, Lin L-E, McClellan G (1986) Adrenergic regulation of myosin adenosine triphosphatase activity. Circ Res 58:83–95
Wood E, Heppner R, Weidmann S (1969) Inotropic effects of electric currents. Circ Res 24:409–445
Yatani A, Brown AM (1989) Rapid β-adrenergic modulation of cardiac calcium currents by a fast G protein pathway. Science 245:71–75
Yatani A, Codina J, Imoto Y, Reeves JP, Birnbaumer L, Brown AM (1987) A G protein directly regulates cardiac calcium channels. Science 238:1288–1292
Xenophontos X, Watson P, Chua B, Haneda T, Morgan H (1989) Increased cyclic AMP content accelerates protein synthesis in rat heart. Circ Res 65:647–656

Chapter 3

Smooth Muscle Activation

G. Pfitzer and *J.C. Rüegg*

Contents

1 Introduction

The contractile structure of vertebrate smooth muscle is extensively discussed by Squire in Chapter 4. Here, we focus on the problem of contractile activation. As in all muscle, calcium ions are essential in the regulation of contraction (for review, see Rüegg 1988). This has been known for many years ever since Filo et al. (1965) showed that the contractile force of permeabilized smooth muscle cells depends on the level of free calcium in contact with the myofilaments. In the following, we shall address first the mechanisms regulating the intracellular calcium ion concentration. Then we shall discuss the relationship between the

Department of Physiology II, University of Heidelberg, Im Neuenheimer Feld 326, 6900 Heidelberg, FRG

Advances in Comparative and Environmental Physiology
Vol. 12. Edited by H. Sugi

intracellular free calcium and contractile force and the mechanisms by which calcium ions activate the contractile machinery. By stimulating the calmodulin-dependent enzyme myosin light chain kinase, calcium increases the extent of myosin phosphorylation which, in smooth muscle, appears to be a prerequisite for the formation of contractile linkages between actin and myosin filaments. However, other mechanisms may also come into play, since there is increasing evidence showing that myosin phosphorylation and contraction may often be dissociated. In discussing smooth muscle regulation, we begin with a description of the transport mechanisms that increase or decrease the level of Ca^{2+} in the cell.

2 Calcium Transport

The contractile machinery of smooth muscle cells, consisting of actin and myosin filaments, is turned off when the calcium ion concentration around the myofilements is less than about $0.1\,\mu$M (DeFeo and Morgan 1985; Yamaguchi 1986; Williams et al. 1989). This down-regulation of the free calcium ion concentration is achieved by different calcium transport mechanisms located at the cell membrane or in the sarcoplasmic reticulum.

2.1 Calcium Handling Structures

The smooth muscle cell membrane contains innumerous transport proteins, receptor proteins, and ion channels. Among these are chloride, sodium, potassium, and calcium channels as well as calcium pumps, sodium-potassium pumps, and sodium-calcium exchangers (Fig. 1). The latter is of particular interest since it "exchanges" intracellular calcium for extracelular sodium thus lowering myoplasmic free calcium. The sodium-potassium ATPase (= sodium-potassium pump), in turn, pumps the intruded sodium out of the cell, thus facilitating also the calcium extrusion by the sodium-calcium exchanger. Because of its high content in voltage-dependent calcium channels, the cell membrane also is often excitable, i.e., it can generate and propagate calcium-dependent action potentials.

However, calcium may also be released into the myoplasm from intracellular calcium stores, the sarcoplasmic reticulum (SR), which also sequesters calcium (for review, see Somlyo and Himpens 1989). Some of this SR is lying directly under the sarcolemma and is in close contact with it (subsarcolemmal SR), while another portion of the reticulum is situated deep inside the cell (deep SR). The SR is somewhat less abundant than in skeletal muscle, but quite variable. Thus, its volume ranges from 1 to 7% of the cell volume in comparison with, for instance, 15% in frog skeletal muscle. Unlike skeletal muscle, T-tubular structures are absent, so that the excitable cell membrane cannot communicate directly with the deep SR. The mitochondria are involved in calcium homeostasis only under conditions of calcium overload. In the normal

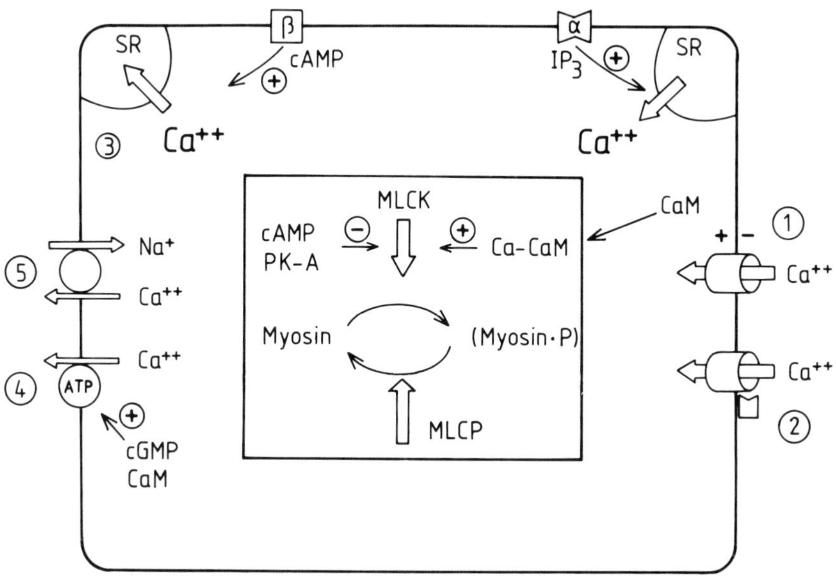

Fig. 1. Calcium transport in smooth muscle activation and relaxation. Calcium release into myoplasm occurs via voltage- and receptor-operated membrane channels (*1*,*2*) and/or by release from the sarcoplasmic reticulum (*SR*) after α-receptor stimulation and release of inositol trisphosphate (IP_3). In response to an elevated calcium concentration, the calcium-calmodulin (*CaM*) complex will be formed which, in turn, will activate myosin light chain kinase (*MLCK*). MLCK will phosphorylate myosin (*myosin·P*), leading to interaction with actin and, finally, contraction of smooth muscle. Calcium removal from the myoplasm is effected by the calcium pump of the sarcoplasmic reticulum (*3*) and the sarcolemma (*4*) or by the Na-Ca exchanger (*5*). Then MLCK is inhibited and the myosin light chain is dephosphorylated by the myosin phosphatase (*MLCP*). β-Receptor stimulation causes an increase in the level of cAMP. This second messenger stimulates Ca uptake into the SR and inhibition of MLCK

contraction-relaxation cycle, however, mitochondria have no calcium-regulatory function (Somlyo et al. 1979).

2.2 Calcium Influx Through Calcium Channels of the Cell Membrane

During action potentials or during maintained membrane depolarization, extracellular calcium is entering the smooth muscle cell through calcium channels of the cell membrane (Johansson and Somlyo 1980). The influx is entirely passive since there is a large concentration gradient as well as an electrical gradient driving the calcium ions into the cell (the actual driving force is the so-called driving potential which is the difference between the calcium-equilibrium potential and the actual membrane potential). It is clear then that the cell would be flooded with calcium if the cell membrane was not practically calcium impermeable in the resting cell. Here, the channels are closed and act as a barrier, preventing high concentrations of extracellular calcium from entering the cell.

This is important, as calcium in high concentrations in the myoplasm normally elicits "spasms" and it would be toxic; for instance, it activates calcium-dependent proteases digesting cellular proteins and, in addition, it would damage the oxidative metabolism by flooding the mitochondria engaged in oxidative phosphorylation. Calcium channels therefore restrict the influx of calcium ions, allowing only the small amount of calcium to enter that is required for contractile activation. Naturally, this amount of calcium depends on the Ca^{2+}-driving potential, on the number of calcium channels available, and on their open time (their probability of being open). The latter is determined by gates within the channels which are either voltage-dependent (voltage-dependent channels) or controlled by the occupancy of receptors with specific agonists (receptor-operated channels). The calcium entry via voltage-dependent channels and thus contraction can be inhibited by drugs which hyperpolarize the cell membrane (such as K-channel agonists), or which block the calcium channels (calcium antagonists). Experimentally, of course, contraction may also be inhibited simply by binding the extracellular calcium with a calcium chelator, such as EGTA.

Often, however, the contractile effect of hormones or drugs cannot be readily blocked by calcium antagonists or by binding extracellular calcium to calcium-chelating agents. Therefore, in this case, contraction seems to be elicited not by calcium entering the cells from the extracellular space, but by calcium released from intracellular stores.

2.3 Calcium Release from the Sarcoplasmic Reticulum

The sarcoplasmic reticulum (SR) plays a major role in the physiological regulation of cytoplasmic free calcium in smooth muscle. Calcium is stored both in the junctional SR that forms surface couplings with the plasma membrane and in the central "deep" SR which is not connected to the outer membrane (Somlyo et al. 1982). It has been shown that noradrenaline causes a calcium release from the internal SR rather than by opening potentially dependent calcium channels. Thus, the signal transmission from the membrane to the calcium stores must be chemical since there is no T-system in smooth muscle. The messenger which releases calcium from intracellular stores in response to extracellular signals has been identified as inositol trisphosphate (IP_3; Suematsu et al. 1984; Somlyo et al. 1985) and is derived from phosphoinositides (minor phospholipids), located in the cell membrane. After occupation of certain receptors by their agonist (e.g., noradrenaline), IP_3 is released together with other isomers of inositol phosphates and diacylglycerol (DG). The formation of IP_3 and DG from phosphoinositol bisphosphate is catalyzed by phospholipase C. Diacylglycerol serves as a second messenger activating protein kinase C that causes the phosphorylation of specific target proteins such as the myosin light chains and other protein components, but the function of various isomers of inositol phosphates is still unclear and controversial (Barnes 1986; Berridge 1987).

On the other hand, it has been established that the release of intracellular IP_3 is sufficient and fast enough to trigger contractions in smooth muscle via calcium release from the SR. This finding suggests that IP_3 can act as a second messenger for pharmacomechanical coupling in smooth muscle (Somlyo et al. 1988). This mechanism is independent of the membrane potential and enables excitatory drugs to release calcium from internal stores and, hence, elicit contraction. A high IP_3-phosphatase activity has been measured in smooth muscle tissue which allows for the necessary rapid breakdown of the inositol trisphosphate.

2.4 Calcium Reuptake into the Sarcoplasmic Reticulum

The calcium ion concentration of the myoplasm depends on the dynamic equilibrium of two opposing processes, the calcium release or influx into the myoplasm, on the one hand, and the calcium sequestration or extrusion, on the other hand. Thus, the calcium concentration and hence contraction can be also increased simply by decreasing the rate of calcium extrusion through the cell membrane or the calcium sequestration by the sarcoplasmic reticulum.

Sequestration is carried out by the calcium-ion pump which transports calcium ions from the myoplasm against the calcium concentration gradient into the inside of the vesicles of the sarcoplasmic reticulum; there it is stored at high concentrations (Somlyo et al. 1982). This active transport process requires energy derived from ATP splitting by a calcium-transport ATPase that is stimulated by calcium ions as well as by the second messenger cyclic AMP. This messenger activates a protein kinase which transfers phosphate from ATP to the phospholamban which activates the transport ATPase. Consequently, an increase in intracellular cyclic AMP levels would increase the rate of calcium sequestration, thus promoting smooth muscle relaxation. Note that the levels of cyclic AMP are precisely regulated by the relative activities of an enzyme which is engaged in the production of cyclic AMP, the adenylate cyclase, and by another enzyme family which decreases the cyclic AMP levels, the various types of phosphodiesterases. Consequently, cyclic AMP levels rise and smooth muscle relaxes, if the phosphodiesterase is inhibited by papaverin or other phosphodiesterase inhibitors or if the adenylate cyclase is stimulated, for instance, by forskolin or, more specifically, when the β-receptors are stimulated by adrenaline or noradrenaline. While the mechanism just described is an important pathway involved in the mediation of β-adrenergic relaxation, it is, however, not the only one. Alternative mechanisms may involve the inhibition of myosin light chain kinase, as will be discussed below. In addition, sodium-calcium exchange may also be promoted, because cyclic AMP-dependent protein kinase may stimulate the sodium-potassium ATPase, thus causing a reduction in the intracellular sodium contents (Scheid et al. 1979). This then facilitates the exchange of intracellular calcium with extracellular sodium, so that calcium is driven out of the cell. Conversely, inhibition of the sodium-potassium pump would reduce the sodium gradient across the cell membrane, thereby inhibiting the sodium-calcium exchange. It has been discussed whether the Na-Ca exchanger also plays a role when the sodium-potassium pump is inhibited with a high concentration of cardiac glycosides and perhaps also in hypertension (Blaustein 1977).

Another cyclic nucleotide which is involved in calcium extrusion from the cell and smooth muscle relaxation is cyclic GMP. This cyclic nucleotide is said to stimulate the calcium pump of the cell membrane (Popescu et al. 1985), but possibly also the Ca pump of the SR. In this way, the intracellular calcium level and thus contractile force is reduced. A number of drugs and relaxing factors have been found recently to increase cyclic-GMP levels, thus causing relaxation. Among these drugs are sodium nitroprusside as well as nitric oxide compounds which are used as vasodilators. A relaxing factor derived from the endothelium has been identified as being nitric oxide (Palmer et al. 1987).

In summary, smooth muscle tone depends on the balance of factors and mechanisms which increase the intracellular calcium ion concentration and others which decrease it. In addition, however, the calcium responsiveness, i.e., the relationship between contractile activity and intracellular free calcium, may also be varied by these cyclic nucleotides.

3 The Dependence of Smooth Muscle Tone on the Intracellular Concentration of Ca^{2+}

In smooth muscle, contraction force is calcium-dependent. Thus, it can be increased both by increasing the intracellular free calcium, i.e., the amplitude of the Ca transient, or the calcium responsiveness of the "contractile proteins".

3.1 Role of Calcium Amplitude Modulation

The intracellular calcium ion concentration in a resting smooth muscle cell is about $0.1\,\mu$M and may be increased without membrane depolarization (pharmacomechanical coupling; cf. Somlyo and Somlyo 1968) or by electromechanical coupling. In the latter case contraction is elicited by membrane depolarization, when the muscle is stimulated electrically by graded membrane depolarization or by spontaneous action potentials. These action potentials are caused by calcium currents through voltage-dependent calcium channels. In spontaneously active smooth muscle, the potentials are generated by pacemaker cells, thus spreading to other smooth muscle cells via gap junctions. A slight but maintained depolarization of the pacemaker cells increases the firing frequency of the action potentials, thus causing a superposition of twitches and eventually their fusion to a maintained tetanic contraction, the myogenic tone. Rythmic fluctuations of myogenic tone for periods of seconds and minutes may then be ascribed to spontaneous fluctuations of the firing frequency of pacemaker cells. These are associated with and possibly due to spontaneous alterations of the resting membrane potential (slow waves) on which spikes (action potentials) are superimposed (Connor et al. 1977). Each spike is caused by an inward current of calcium ions through calcium channels that, in turn, elicit intracellular calcium transients resembling the spikes. At high spike frequency, however, the calcium

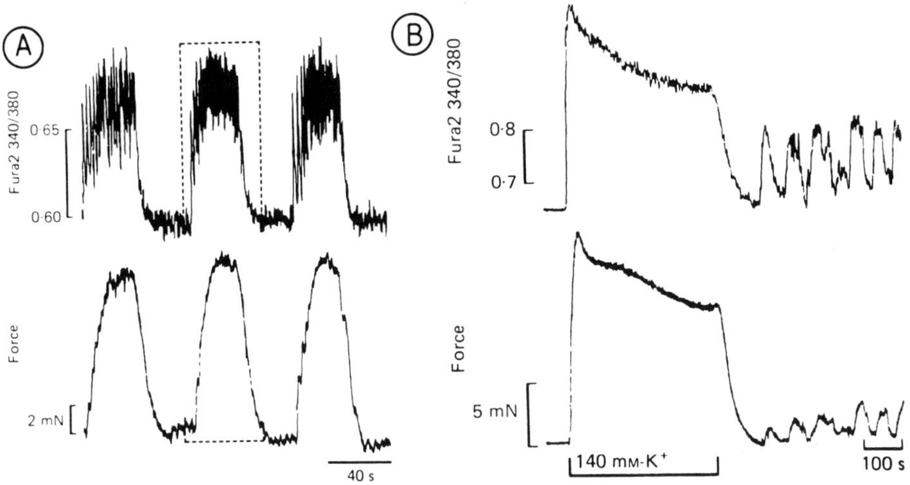

Fig. 2. **A** The spontaneous (myogenic) rythmic contraction in the ileum of guinea pigs (*lower tracing*) is triggered by calcium transients (*upper tracing*). Note "spike-like transients" superimposed on "slow calcium waves". The muscle cells were loaded with the fluorescent calcium indicator Fura2, and calcium was detected by measuring Fura2 fluorescence at 510 nm, excited at 340 and 380 nm, respectively. **B** Depolarization with high K^+ (140 mM) induces an increase in intracellular Ca^{2+} contraction. Force of both the phasic and tonic component of contracture is proportional to the increase in Ca^{2+}. (Himpens and Somlyo 1988)

transients may fuse to cause a maintained elevation of intracellular free calcium ion concentration (Fig. 2A; cf. Himpens and Somlyo 1988). Greater action potential frequency correlates with increased amounts of calcium entering the cells and hence with an increase in intracellular free calcium and force. In this manner, then, the tone is increased when intestinal smooth muscle is stretched or stimulated by the application of stimulatory hormones such as acetylcholine.

If intestinal smooth muscle is depolarized by potassium chloride rather than by action potentials, the increase in calcium ion concentration is smooth and maintained, and related to the extent of depolarization and force production (Himpens et al. 1989). A depolarization by 140 mM potassium chloride, for instance, induces a maximum increase in free calcium from about 100 to 600 nM associated with an increase in force. Subsequently, however, both force and free calcium ion concentration decline again to some extent in spite of continued depolarization (Fig. 2B). In K-depolarized muscle, too, the increase in intracellular free calcium is due to the influx of extracellular calcium via potential-dependent calcium channels that are opened by the depolarization. Thus, after removing external calcium, intracellular free calcium rapidly returns to resting levels, in spite of continuous depolarization by potassium chloride. Intracellular free calcium and force recover again following the readdition of external calcium, at least in vascular smooth muscle. In intestinal smooth muscle, the readdition of external calcium also causes intracellular free calcium to rise again, surprisingly, however, without an increase in force. Obviously, the contractile machinery had been desensitized to calcium. Clearly, the relationship between

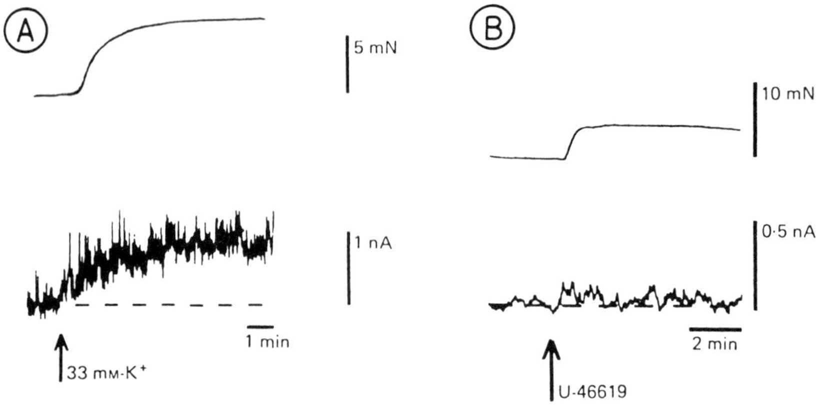

Fig. 3. **A** Contractile activation (force) with K^+ by increasing myoplasmic Ca^{2+} or **B** by increasing calcium responsiveness (stimulation with U 44619) in a vertebrate smooth muscle (coronary artery). Noisy tracing: calcium transient as detected by photomultiplyer output of aequorin light. (Bradley and Morgan 1987)

intracellular free calcium and force is not unique, but may be altered, indicating a change in calcium sensitivity (Himpens et al. 1989).

3.2 Calcium Sensitivity Modulation of Tone

An interesting example of calcium desensitization has been reported by Morgan. Thus, potassium, in a concentration of 46 mM caused a contraction of ferret portal veins maintained by an elevated intracellular free calcium ion concentration, but after addition of isoprenaline, the preparation relaxed completely without any decrease in calcium ion concentration (Morgan and Morgan 1984). These isoprenaline effects are associated with and probably due to a stimulation of adenylate cyclase causing an increase in cyclic-AMP levels and activation of protein kinase A, phosphorylating and inhibiting myosin light chain kinase and contractile activation as discussed below.

When stimulated by phenylephrine, an α-receptor agonist, the portal vein responded with a maintained contraction, though the calcium transient was brief and small. As shown in Fig. 3, however, stimulation by the endoperoxide analog U 44619 even induced a contraction without any increase in intracellular free calcium ion concentration (Bradley and Morgan (1987) in contrast to K^+ depolarization which increases both force and myoplasmic Ca^{2+}. However, many agonists such as prostaglandine $F2_\alpha$ or phenylephrine (Fig. 4) may increase both the intracellular free calcium ion concentration as well as the calcium responsiveness of the contractile machinery or the "calcium sensitivity". This dual effect may be partly due to the fact that such agonists also stimulate phospholipase C and this causes both an increase in intracellular levels of inositol trisphosphate and diacylglycerol. Inositol trisphosphate has been shown to bind to the intracellular calcium channels of the sarcoplasmic reticulum and

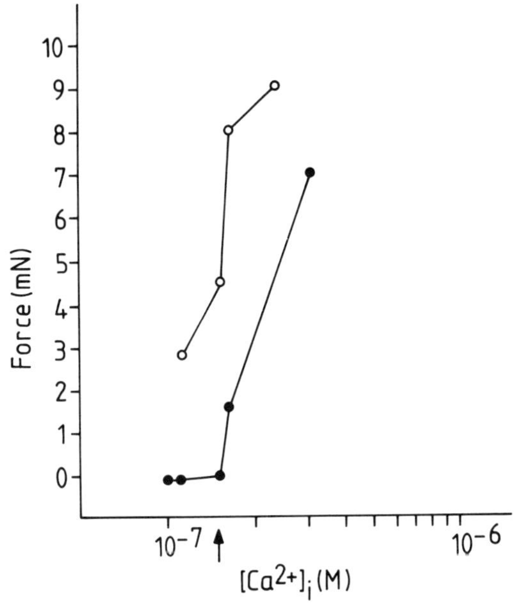

Fig. 4. Relation between intracellular free Ca^{2+} concentration and contractile response during stimulation with phenylephrine (*open symbols*) and K^+ depolarization (*closed symbols*). Note increased calcium sensitivity with phenyleephrine. (After De Feo and Morgan 1985)

cause release from the internal calcium stores followed by a rapid increase in intracellular calcium. Diacylglycerol, on the other hand, is a powerful activator of protein kinase C and this enzyme has been implicated in causing an increase in calcium sensitivity mediated by small G-proteins that may influence (perhaps are not yet understood. Perhaps, as discussed below, they may involve a change in calcium sensitivity mediated by small g-proteins that may influence (perhaps indirectly) myosin phosphatase activity. These effects will be discussed below in relation to the phosphorylation theory of smooth muscle contraction.

From a comparative point of view, it is interesting to note that a dissociation of force and intracellular free calcium may also be found in certain invertebrate smooth muscles. Thus, in the byssus retractor of the bivalve *Mytilus*, the edible clam, a tonic contraction can be induced by acetylcholine, which elicits an initial rise in intracellular free calcium. Subsequently, the intracellular calcium then rapidly declines to basal levels, whereas force is maintained for prolonged periods with little energy consumption (catch state). Catch can, however, be abolished by neurohormones, such as serotonine, without any change in the level of intracellular free calcium (Ishii et al. 1989; cf. also Atsumi and Sugi 1976). Since serotonin stimulates adenylate cyclase and increases intracellular levels of cyclic AMP, relaxation from catch may also involve calcium desensitization by cAMP-dependent protein kinase, analogous perhaps to the cAMP-induced desensitization in the portal vein described above. Incidentally, vascular smooth muscle also exhibit catch-like states (often called latch), which will be described in more detail below. In these smooth muscles, a tonic contraction is often initiated by a transient increase of myoplasmic free calcium, but is then maintained (in latch) at low levels of Ca^{2+} and myosin light chain phosphorylation (Rembold 1991). In this state, tension is also maintained very economically, i.e., at a low energetic tension cost, while the shortening velocity of the contractile mechanism is "geared down" (cf. Paul 1990 for a review).

In summary then, smooth muscle contraction may be initiated by an initial rise in intracellular free calcium, but its relationship to force of contraction is not unique, but quite variable. For instance, calcium sensitivity may differ in phasic and tonic mammalian smooth muscle (Himpens et al. 1988) and it may also depend on the kind of muscle stimulation and on the duration of contraction. Clearly, the variable relation of Ca^{2+} and contractile activity suggests that the establishment of a certain Ca^{2+} concentration cannot be the ultimate step in the activation of a contractile machinery. It is, on the contrary, the beginning of a cascade reaction involving calmodulin as calcium receptor, activation of myosin light chain kinase, phosphorylation of myosin, and actin-myosin interaction as will be discussed next.

4 Calcium Activation of Smooth Muscle Contractile Proteins

Any model of smooth muscle regulation has to accomodate several special features of smooth muscle contraction: namely, the variable force-calcium relation and the fact that in the course of a tonic contraction, the energy cost of isometric tension and the speed with which the muscle may shorten are reduced while force is maintained. This state is called "latch". As mentioned previously, calcium activation is mediated via phosphorylation of myosin light chains, but alterations in the thin filament proteins may also be important. Before discussing these mechanisms in more detail, it is necessary, therefore, to summarize our knowledge on the contractile and regulatory proteins within thick and thin myofilaments.

4.1 Smooth Muscle Contractile and Regulatory Proteins

In smooth muscle cells, two filamentous domains can be distinguished: actin-myosin and actin-intermediate filament domains. It is generally assumed that the MgATP-dependent cyclic interaction of myosin cross-bridges with actin is the molecular basis of force generation and shortening. While the number of cross-bridges interacting with actin determines force, their cycling frequency determines the speed with which the muscle may shorten. Smooth muscle is remarkable in that both the number and frequency of the cross-bridges may be regulated independently.

It has recently been suggested that the actin-intermediate filament domain may contribute to force maintenance of smooth muscle (Small et al. 1986; Rasmussen et al. 1987).

Actin, the major component of the thin filaments, is a globular protein (G-actin) with a molecular weight of 42000 daltons, existing in different isoforms, which under physiological conditions polymerizes into filaments (F-actin). The distribution of the isoforms (a- and γ-smooth muscle and β- and γ-nonmuscle actin) is specific for the different smooth muscle tissues (Fatigati and Murphy 1984). In addition, some 30% of the actin exists as monomeric G-actin allowing for the possibility that a dynamic equilibrium between G- and F-actin may not only be important for nonmuscle motility, but also for smooth muscle contraction (Boels and Pfitzer 1990; Mauss et al. 1990).

Additional proteins of the thin filaments of smooth muscle are tropomyosin and caldesmon, an elongated molecule with a molecular weight of 93 000 daltons, which is localized solely in the actin-myosin domain (Fürst et al. 1986). Recently, a further actin-binding protein, calponin, has been purified which is immunologically related to cardiac troponin-T (Takahashi et al. 1988).

The thin filaments insert into membrane-associated dense bodies containing the actin-binding protein vinculin and into cytoplasmic dense bodies having α-actinin as a major component. The structural organization of the thin filaments has been reviewed in detail by Marston and Smith (1985).

The enzyme myosin is the primary protein of the thick filaments and is composed of two high molecular weight subunits, or heavy chains, and two types of low molecular-weight subunits, or light chains. The molecular weight of each heavy chain subunit is 200 000 daltons, whereas the light chains are 20 000 and 17 000 daltons, respectively. The native molecule is composed of a tail region embedded in the thick filament and a globular head region (S1 subfragment) that contains the actin-binding domain and the catalytic site for ATP hydrolysis. In in vitro motility assays, the S1 fragment is sufficient to move actin filaments (Toyoshima et al. 1987). The light chains are also associated with the globular head region. The 20 000 dalton or regulatory light chain is phosphorylated by the Ca^{2+}/calmodulin-dependent myosin light chain kinase (MLCK) which is thought to be the dominant mechanism of the activation of the actin-activated myosin MgATPase (Kamm and Stull 1985). Phosphorylation of the regulatory light chain affects the conformational state of myosin, i.e., it promotes the transition from the folded (sedimentation coefficient 11S) to the extended conformation (6S) of myosin which is then able to assemble into filaments (Craig et al. 1983). This change in the conformation may also underlie the activation of the myosin MgATPase (Ito and Hartshorne 1990).

4.2 The Phosphorylation Theory of Smooth Muscle Activation

The reversible phosphorylation and dephosphorylation of the regulatory light chains of myosin (LC_{20}) catalyzed by the calcium- and calmodulin-dependent myosin light chain kinase and specific phosphatases is generally believed to be the primary regulatory mechanism of smooth muscle activation (Fig. 5). Stimulation of smooth muscle causes a rise in the intracellular free $[Ca^{2+}]$ which then binds to calmodulin, a protein that is related to troponin C. When 3–4 mol of calcium are bound to 1 mol of calmodulin, it can activate the enzyme myosin light chain kinase (MLCK) which results in the phosphorylation of the serine-19 residue in the amino terminus of the 20 000 dalton light chain of

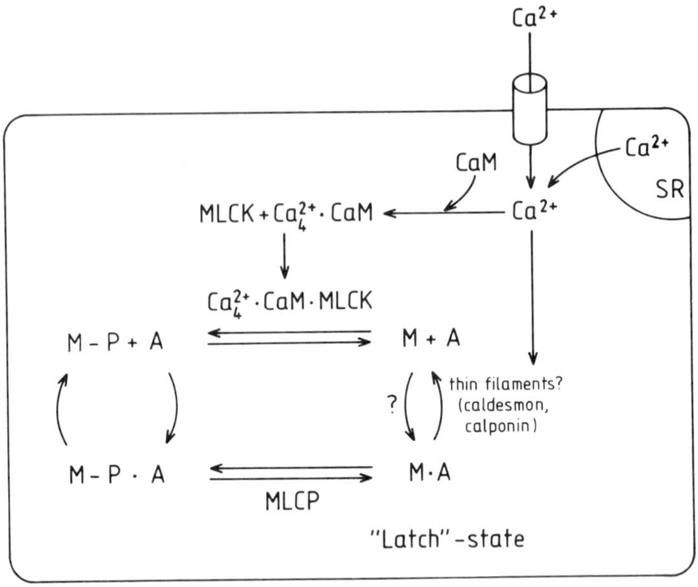

Fig. 5. Scheme of activation of smooth muscle proteins. A rise in the cytoplasmic free calcium concentration binds to calmodulin (*CaM*) which then activates myosin light chain kinase (*MLCK*), thus inducing phosphorylation of the regulatory light chains of myosin (*M-P*). Phosphorylated myosin attaches to actin (*A*) and generates force. Dephosphorylation of myosin induces either relaxation or may allow cross-bridges to enter the latch state. This state may, in addition, be regulated via a thin filament linked regulatory mechanism in a calcium-dependent manner

myosin (LC_{20}) and in the activation of the myosin MgATPase activity by actin (Hartshorne and Mrwa 1982; Kamm and Stull 1985). Under certain conditions the LC_{20} may also be phosphorylated at a second site (threonine-18) by MLCK (reviewed in Kamm and Stull 1989). Dephosphorylation of the LC_{20} by a specific phosphatase results in the inhibition of the actin-activated MgATPase of myosin (Sellers et al. 1981). While MLCK is a well-characterized enzyme, the identity of the myosin phosphatase (MLCP) is not known (Takai et al. 1989). Although little is known about the regulation of MLCP, it is assumed that MLCP is equally effective in the presence and absence of calcium. This, therefore, creates a futile cycle in the presence of calcium, which at the expense of a "pseudo-ATPase" (due to the phosphorylation/dephosphorylation cycle) allows for large responses with only minor changes in the activity of MLCK (Rüegg and Pfitzer 1985).

How phosphorylation regulates the interaction between myosin and actin is still a matter of debate. In gizzard myosin, the effect of phosphorylation of LC_{20} appears to be similar to the effect of calcium binding to the troponin-tropomyosin system in striated muscle, in that it accelerates a process which occurs after binding of myosin to actin, while little affecting the affinity of myosin for actin (Adelstein and Sellers 1987). In contrast, the ability of aortic smooth muscle myosin to bind to actin is increased by phosphorylation of

LC_{20} (Wagner and Vu 1986). In skeletal muscle, phosphorylation of the regulatory light chains of myosin increases the rate of transition (f_{app}) from weakly, nonforce-generating to strongly, force-generating binding states (Brenner and Morano 1990). The effect of phosphorylation on gizzard myosin would be in line with a phosphorylation-dependent increase in f_{app} in smooth muscle as well.

In a number of smooth muscle preparations from different sources, both skinned and intact, it has been demonstrated that the increase in force is typically preceded by an increase in myosin phosphorylation (reviewed in Kamm and Stull 1985). This is even true for the very rapid tension development, occurring within seconds in tissues undergoing electrical field stimulation (Kamm and Stull 1986; Fischer and Pfitzer 1989). The following experiments underline the dominant role of myosin phosphorylation in the regulation of smooth muscle contraction. Limited digestion of MLCK with trypsin results in an enzyme that is active in the absence of calcium and calmodulin. This enzyme elicits a contraction in skinned smooth muscle fibers in the absence of calcium (Mrwa et al. 1985), and after injection into single smooth muscle cells from the stomach of *Bufo marinus*, it induces shortening of these cells at basal concentrations of intracellular free Ca^{2+} (Itoh et al. 1989). Similarly, skinned fibers permanently phosphorylated with the ATP analog ATP-γ-S contract in the absence of calcium. Thus, phosphorylation of LC_{20} in the absence of calcium is sufficient to induce a contraction in smooth muscle.

Relaxation of smooth muscle has been correlated with dephosphorylation of myosin. In fact, it appears that dephosphorylation of myosin is a prerequisite of relaxation. This is suggested by the finding that skinned fibers from chicken gizzard deficient in phosphatase activity relax only after reconstitution with purified phosphatase (Bialojan et al. 1985).

However, although dephosphorylation of myosin reduces actomyosin MgATPase activity and can be correlated with relaxation in intact and skinned preparations, other processes may be involved in the regulation of smooth muscle relaxation. In intact smooth muscle from chicken gizzard, relaxation is regulated at apparently basal levels of LC_{20} phosphorylation (Fischer and Pfitzer 1989). Moreover, tension may develop slowly in tissues stimulated with phorbolesters (Chatterjee and Tejada 1986) or the potassium channel antagonist 3,4-diaminopyridine (Pfitzer and Fischer 1990) with no or only minor increases in LC_{20} phosphorylation. Furthermore, force may be maintained in a calcium-dependent manner (Aksoy et al. 1983) in the presence of light chain dephosphorylation to nearly basal levels (reviewed in Kamm and Stull 1985). The dephosphorylation of LC_{20}, while force is maintained, is also associated with a decline in the cytosolic free Ca^{2+} (review Hai and Murphy 1989), suggesting that the regulatory mechanism responsible for force maintenance has a high calcium sensitivity.

During maintained force, both the shortening velocity and the rate of ATP consumption decline with time (Siegman et al. 1980, 1984). In fact, in some preparations, phosphorylation seems to correlate better with unloaded shortening velocity than with force, suggesting that phosphorylation regulates shortening velocity rather than force. Tension maintenance at low levels of phosphorylation was attributed to so-called latch-bridges which cycle slowly or

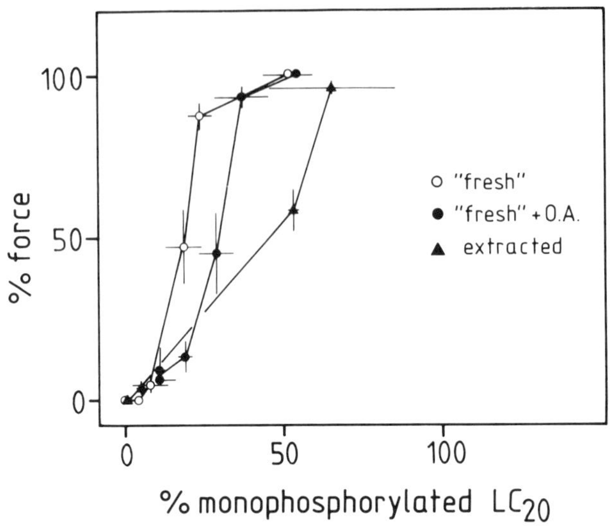

Fig. 6. Relation between isometric tension and myosin phosphorylation in skinned chicken gizzard. Smooth muscle strips from chicken gizzard, skinned with 1% Triton-X-100 and stored in a glycerol buffer at −20 °C until use, were used within 4 days after skinning (denoted as "*fresh*"). They were stimulated by elevating Ca^{2+} with or without okadaic acid (*O.A.*). Note that inhibition of phosphatase by okadaic acid produces a rightward shift of the force-phosphorylation relation, i.e., less force is produced for a given amount of phosphorylation. Extraction of skinned gizzard in the glycerol buffer for 10–12 h at 37 °C also reduced the activity of the phosphatase, while MLCK activity was not affected. However, the force phosphorylation shifted even further to the right, suggesting that an additional protein has been extracted (Pfitzer and Schmidt unpubl.)

not at all but are able to maintain force via a second calcium-dependent regulatory mechanism (Dillon et al. 1981).

In a recent modification of their initial "latch" hypothesis, Murphy and coworkers (Hai and Murphy 1989) have proposed a model in which myosin phosphorylation and dephosphorylation can account for tension development and maintenance at low levels of phosphorylation. The central features of their model are: cross-bridges may only attach in the phosphorylated state; if they are dephosphorylated before detachment from actin occurs, they are transformed into latch-bridges, which in contrast to phosphorylated cross-bridges have a low rate of detachment. Since according to Murphy and coworkers, detached cross-bridges would not reattach unless they were phosphorylated by MLCK, the maintenance of the latch state would depend on the continuous supply of latch-bridges from previously phosphorylated ones. This model allows one to simulate time courses of force and phosphorylation in intact smooth muscles and also accounts for the steep and nonlinear force-phosphorylation relation in skinned chicken gizzard (Fig. 6). As predicted by the model, the force-phosphorylation relation is shifted to the right toward higher levels of phosphorylation in the presence of the phosphatase inhibitor okadaic acid. However, the force-phosphorylation relation remains nonlinear even with high concentrations of okadaic acid, which may be attributed to the fact that okadaic acid only inhibits some 90 to 95% of the phosphatase activity. In fact, the model predicts that only complete inhibition of phosphatase results in a linear relation between force and phosphorylation. A nonlinear relation between force and phosphorylation would also result if, as in skeletal muscle, phosphorylation of LC_{20} increased f_{app} and the rate of the transition from force-generating into nonforce-generating states, g_{app}, remained constant (Rüegg and Pfitzer 1991). Thus, the shape of the force-phosphorylation relation may be determined by more than one factor and is therefore not very sensitive in discriminating between different models. Clearly, more work is needed to understand the regulation of the "latch state". In particular, methods are needed which distinguish the tension-dependent isometric ATPase, which is a related to g_{app} (Brenner 1988), from the "pseudo-ATPase" due to the phosphorylation-dephosphorylation cycle. This is essential to determine whether g_{app} of dephosphorylated cross-bridges is lower than that of phosphorylated cross-bridges.

Work in intact and skinned smooth muscle fibers and in isolated biochemical systems (Marston and Smith 1985) suggest additional calcium-dependent regulatory mechanisms: prolonged extraction of the skinned gizzard preparations in a glycerol/ATP buffer also reduces phosphatase activity to about 10% of the initial activity, while MLCK activity is barely affected. However, in these preparations (cf. Fig. 6), the relation between force and phosphorylation is linear, suggesting that additional factors regulate smooth muscle force. The model of Hai and Murphy (1989) also does not account for the regulation of the relaxation rate (Fischer and Pfitzer 1989), as has also been pointed out by Marston (1989), or some energetic aspects (Paul 1990). The model further predicts that unloaded shortening velocity is governed by the level of phosphorylation. However, shortening velocity does not depend on phosphorylation in a unique manner (Siegman et al. 1984; Haeberle et al. 1985).

4.3 Regulation by Binding of Calcium to Myosin

As in the case of molluscan smooth muscle, direct binding of calcium to myosin has also been implicated in vertebrate smooth muscle regulation (Chacko and Rosenfeld 1982). What is the evidence?

Smooth muscle actomyosin ATPase activity (Schirmer 1965) and force in skinned smooth muscle (Filo et al. 1965) are affected by the concentration of free Mg^{2+}, which affects the conformational state of myosin independent of LC_{20} phosphorylation (Ikebe et al. 1984). Under low but not under high concentrations of Mg^{2+}, there is a direct effect of calcium on shortening velocity of skinned fibers independent of phosphorylation (Barsotti et al. 1987), which may be due to calcium binding to the nonspecific Ca^{2+}-Mg^{2+} sites of myosin. In intact smooth muscle, a phosphorylation-independent effect of calcium on shortening velocity is also observed (Siegman et al. 1984). In intact smooth muscle, the free Mg^{2+} is probably 0.1 to 0.3 μM (Dillon et al. 1982), suggesting that under normal physiological conditions, the shortening velocity may be influenced by the interaction of calcium with myosin.

As force maintenance occurs at low levels of myosin phosphorylation, the regulatory site controlling this effect must have a higher affinity for calcium than does calmodulin. Since the affinity of calcium for myosin is weaker than calcium for calmodulin, it is unlikely that the interaction of calcium with myosin is involved in the regulation of "latch" (Barsotti et al. 1987).

4.4 Thin Filament Linked Regulatory Mechanisms

Considerable biochemical evidence, e.g., from the "competition assay", suggests a thin filament linked regulatory mechanism (Marston and Smith 1985) in addition to myosin phosphorylation. Two proteins of the thin filaments, namely, caldesmon and calponin, have been purified and biochemically characterized. However, both inhibit the ATPase of unphosphorylated and dephosphorylated myosin which is in contrast to the activation of ATPase of phosphorylated myosin by native thin filaments (Marston and Smith 1985).

4.4.1 Caldesmon

Caldesmon is a widely distributed (Clark et al. 1986) actin-linked and calmodulin-binding protein with a molecular weight of 93 000. In gizzard smooth muscle, it is the major calmodulin-binding protein (8–11 μM caldesmon as opposed to about 30 μM calmodulin). Caldesmon is a constituent of native thin filaments of smooth muscle which are composed of actin, tropomyosin, and caldesmon in molar ratios of 28:4:1 (Marston and Smith 1985). Since caldesmon may be as long as 150 nm, it may span over 28 actin molecules (Fürst et al. 1986). Sobue and coworkers (reviewed in Sobue et al. 1988) found that caldesmon binds to actin in the absence of calcium and to calmodulin in the presence of calcium. Thus, caldesmon may regulate actin-myosin interaction in a calcium- and calmodulin-dependent flip-flop mechanism. Indeed, many investigators found inhibition of actin activated ATPase of phosphorylated and dephosphorylated myosin by caldesmon which was overcome by the addition of calcium in the presence of calmodulin (Clark et al. 1986; Sobue et al. 1988). The disinhibition does not necessarily result in the dissociation of caldesmon from actin-tropomyosin filaments (Marston et al. 1988). It has, therefore, been suggested that, together with tropomyosin and calmodulin or another calcium-binding protein, caldesmon may function similar to the troponin-tropomyosin regulatory system of striated muscle. There is some controversy regarding the stoichiometry and mechanism of inhibition of actin-activated myosin MgATPase by caldesmon. According to Marston and coworkers (Marston and Smith 1985), the stoichiometry of caldesmon binding to actin is 1 caldesmon to 23 actin monomers, while Chalovich and coworkers (see Velaz et al. 1989) reported a stoichiometry between 1:7 and 1:10. In intact cells, a 1:7 stoichiometry would be possible, if caldesmon binds to only some of the actin. Indeed, within smooth muscle cells, caldesmon is only located in the actomyosin domain and not in the actin-intermediate filament domain (Fürst et al. 1986).

Caldesmon also binds to smooth muscle myosin (Ikebe and Reardon 1988) and enhances the binding of smooth muscle heavy meromyosin and myosin subfragment-1 to actin in the presence of ATP (Hemric and Chalovich 1988). While myosin, therefore, may associate with the caldesmon-actin complex, the binding of myosin to the productive site on actin seems to be inhibited by caldesmon, resulting in the inhibition of the actin-activated MgATPase (Hemric and Chalovic 1988). This contrasts with the initial hypothesis that caldesmon might inhibit product release, e.g., ADP (Lash et al. 1986).

There are few functional studies on caldesmon so that its role in the regulation of smooth muscle contraction is not clear. The enhanced binding of myosin to actin and simultaneous inhibition of ATPase by caldesmon allows for a role as "latch protein". Yet, it is unclear whether the binding of caldesmon to myosin has the potential to support tension (Velaz et al. 1989). Skinned smooth muscle strips from chicken gizzard are relaxed by caldesmon without altering the level of myosin light chain phosphorylation (Spazenko et al. 1985). Caldesmon inhibits the movement of myosin-coated beads along actin cables of *Nitella* (Sellers and Shirinsky 1987). Caldesmon inhibits tension in skeletal skinned fibers compatible with a steric blocking mechanism (Brenner et al. 1990). In all cases, the inhibition is reversed by calcium/calmodulin. Any model of caldesmon function has to accomodate the fact that phosphorylation of myosin is sufficient to induce a contraction at basal levels of calcium (cf. Sect. 3.2).

In addition to calcium, caldesmon may be regulated by phosphorylation which also reverses the inhibition of ATPase and the binding to myosin (Sutherland and Walsh 1989). Gizzard caldesmon has intrinsic kinase activity, and may also be phosphorylated by a Ca^{2+}/calmodulin-dependent kinase, distinct from MLCK, and by protein kinase C (reviewed in Kamm and Stull 1989). However, additional studies are required to understand the functional role of caldesmon phosphorylation.

4.4.2 Calponin

Calponin, a heat-stable, basic, 34-kD protein, which has been purified from chicken gizzard and bovine aorta, interacts with F-actin and tropomyosin in a calcium-independent manner and with calmodulin in a calcium-dependent manner (Takahasi et al. 1986, 1988). The protein, which is immunologically related to troponin-T, inhibits actin-activated MgATPase activity of phosphorylated and unphosphorylated smooth muscle myosin (Winder and Walsh 1990). The inhibitory effect is calcium-independent but is reversed by phosphorylation of calponin, which is a substrate for protein kinase C and Ca^{2+}/calmodulin-dependent protein kinase II. Since calponin also has sequence homologies with G-binding proteins, there is also the possibility that it may confer some of the effects observed with GTP-γ-S (Takahashi et al. 1991)

4.5 The Variable Calcium Responsiveness of the Contractile System

As discussed in Section 2.2, the relationship between force development and intracellular free Ca^{2+} is quite variable in smooth muscle. Such modulation of calcium sensitivity may be due to an alteration of the dependence of myosin light chain phosphorylation on free calcium. Alternatively, however, the coupling between calcium-induced phosphorylation and force may also be variable. There is now much evidence for the physiological modulation of calcium sensitivity, and in many cases this has been analyzed at the level of the calcium regulatory proteins.

4.5.1 GTP-γ-S Increases Calcium Sensitivity

As pointed out in Section 2.2, the sensitivity to the activator calcium may increase in smooth muscle stimulated by agonists. This apparent increase in the calcium sensitivity of force may, in fact, be due to an increase in the sensitivity of LC_{20} phosphorylation to calcium (Rembold 1991). An increase in the calcium sensitivity of LC_{20}-phosphorylation occurs when the ratio of phosphorylating to dephosphorylating reactions at constant Ca^{2+} is shifted toward phosphorylation. The exact mechanism regarding how the calcium sensitivity of force and phosphorylation are increased is not yet known and it should be noted that the calcium sensitivity of force and phosphorylation might be regulated independently. However, a putative mechanism has been suggested using α-toxin or β-escin permeabilized smooth muscle (Nishimura et al. 1988; Kitazawa et al. 1989). In these preparations, the agonist-receptor-messenger-response chain is still intact (Pfitzer and Boels 1990) and agonists induce an increase in the calcium sensitivity of force in the presence of GTP. This response is mimicked by GTP-γ-S, suggesting the involvement of a GTP-binding protein. GTP-γ-S also increases phosphorylation at clamped calcium (Fujiwara et al. 1989) and it has been suggested that this is due to G-protein-mediated inhibition of myosin phosphatase (Kamm and Stull 1990).

Interestingly, during the sustained phase of an agonist-induced contraction, there is not only a decrease in $[Ca^{2+}]$ but also in LC_{20} phosphorylation. This indicates that the above mentioned mechanism cannot account for the high force at low calcium and low LC_{20} phosphorylation. It may, however, be involved in the increased Ca sensitivity of myosin phosphorylation in agonist-stimulated tissues as opposed to depolarized tissues (Rembold 1991).

4.5.2 Down-Regulation of Calcium Sensitivity

Myosin light chain kinase is phosphorylated by cAMP-dependent protein kinase at two specific sites in the absence of calcium and calmodulin. This reduces the affinity of MLCK for the Ca^{2+}/calmodulin complex (reviewed in Adelstein and Sellers 1987) and would therefore decrease the calcium sensitivity of phosphorylation-dependent force development. If MLCK is complexed with calcium and calmodulin, only one site is phosphorylated which has no effect on the activity of the enzyme.

Submaximally activated skinned fibers are relaxed by cAMP or the catalytic subunit of cAMP-dependent protein kinase at clamped concentrations of calcium (Pfitzer et al. 1985), which is associated with dephosphorylation of LC_{20}. Although it has not been demonstrated directly, this suggests, under the assumption of an unregulated phosphatase, that MLCK activity is reduced in the presence of the catalytic subunit of the cAMP-dependent protein kinase at constant calcium, resulting in a net dephosphorylation of LC_{20}.

The physiological role of this mechanism is still under dispute, since measurements of MLCK activity in homogenates of intact tissues have not yielded unequivocal results. Formation of cAMP and relaxation were associated with a decrease in MLCK activity in uterine but not in tracheal smooth muscle (Miller et al. 1983; Nishikori et al. 1983).

As discussed in Section 2.2, there is a decrease in the calcium sensitivity of force in the presence of agents that stimulate adenylate cyclase. Further studies are required to clarify whether cAMP-dependent phosphorylation is involved in the desensitization or whether other substrates of cAMP-dependent protein kinase have to be considered.

Proteins of the intermediate filaments (e.g., desmin) are also substrates of cAMP-dependent protein kinase (Park and Rasmussen 1986). This is of particular interest, since intermediate filaments may contribute to force (see Sect. 3.4.5.3) and may thus be involved in the apparent desensitization. Furthermore, under certain conditions, cGMP-dependent protein kinase also desensitizes skinned and intact preparations to calcium (Pfitzer et al. 1984, 1986; Sato et al. 1988). This desensitization clearly is not mediated through phosphorylation of MLCK (Kamm and Stull 1989).

4.5.3 Role of Protein Kinase C and Intermediate Filaments

Calcium- and diacylglycerol-activated protein kinase C has also been implicated in calcium sensitivity modulation (Itoh et al. 1988). Protein kinase C phosphorylates myosin heavy chains, myosin light chains at serine-1, serine-2, threonine-9, caldesmon, and cytoskeletal proteins (desmin, filamin, vimentin) (reviewed in Kamm and Stull 1989). The physiological significance is far from being clear.

Phorbolesters, activators of protein kinase C, induce slowly developing contractions in intact and skinned fibers with only small increases in LC_{20} phosphorylation (Chatterjee and Tejada 1986; cf. also Kamm and Stull 1989) which may be due to a small increment in $[Ca^{2+}]$ followed by activation of MLCK (Rembold and Murphy 1988). In this regard it is interesting that the slow contraction induced by phorbolesters is associated with ^{32}P incorporation into proteins of the intermediate filaments, namely desmin, and into caldesmon. A similar phosphorylation pattern has been observed during the tonic phase of a contraction elicited with carbachol (Park and Rasmussen 1986).

Intermediate filaments consist of relatively insoluble proteins forming the skeletal framework of the cytoplasm. Cytoskeletal and contractile proteins appear in two filamentous domains: (1) intermediate filaments containing actin, desmin, and filamin; and (2) contractile proteins containing actin, myosin, tropomyosin, and caldesmon. The arrangement of these domains has led to the speculation that the intermediate filament domain may be involved in the regulation of tonic contraction, while the actomyosin domain is responsible for the initial, rapidly rising phase of a contraction (Fürst et al. 1986). One way to regulate the intermediate filament domain is via phosphorylation as desoribed for cAMP-dependent protein kinase and protein kinase C.

4.6 Concluding Remarks and Future Prospect

The smooth muscle contractile machinery is, as in all muscle, calcium-activated. However, the clacium responsiveness may vary over a wide range, both in intact muscle and in isolated contractile structures. As we have seen, vertebrate smooth muscle differs from striated muscle in many ways. Most importantly, calcium regulation is affected not via the troponin-tropomyosin system, but by a plethora of various regulatory proteins including the myosin light chains, calmodulin, caldesmon, and calponin. The functional properties of the latter two proteins are still ill-defined, nor do we know how smooth muscle manages to gear down the kinetic cross-bridge properties during prolonged contraction. This phenomenon, called latch (or catch), is most intriguing and has been a problem for nearly a century, ever since it was realized that smooth muscles are unique in their capacity to maintain tension for prolonged periods with little energy expenditure. Given the new tools of analyzing intracellular regulatory mechanisms, rapid progress in this field may, however, be forthcoming.

References

Adelstein RS, Sellers JR (1987) Effects of calcium on vascular smooth muscle contraction. Am J Cardiol 59:4B–10B

Aksoy MO, Mras S, Kamm KE, Murphy RA (1983) Ca^{2+}, cAMP, and changes in myosin phosphorylation during contraction of smooth muscle. Am J Physiol 245:C255–C270

Atsumi S, Sugi H (1976) Localization of calcium-accumulating structures in the anterior byssal retractor muscle of *Mytilus edulis* and their role in the regulation of active and catch contraction. J Physiol (Lond) 257:549–560

Barnes DM (1986) Inositol phospholipid metabolism and calcium regulation. Science 234:287–288

Barsotti RJ, Ikebe M, Hartshorne DJ (1987) Effects of Ca^{2+}, Mg^{2+}, and myosin phosphorylation on skinned smooth muscle fibres. Am J Physiol 252:C543–C554

Berridge MJ (1987) Inositol trisphosphate and diacylglycerol: two interacting second messengers. Annu Rev Biochem 56:159–193

Bialojan C, Merkel L, Rüegg JC, Gifford D, Disalvo J (1985) Prolonged relaxation of detergent-skinned smooth muscle involves decreased endogenous phosphatase activity. Proc Soc Exp Biol Med 178:648–652

Blaustein MP (1977) Sodium ions, calcium ions, blood pressure regulation, and hypertension: a reassessment and a hypothesis. Am J Physiol 232:C165–C173

Boels PJ, Pfitzer G (1990) Phalloidin (PH) and cytochalasin D (CYD) relax skinned mesenteric arterioles (MA). Biophys J 57:327a

Bradley AB, Morgan KG (1987) Alterations in cytoplasmic calcium sensitivity during porcine coronary artery contractions as detected by aequorin. J Physiol (Lond) 385:437–448

Brenner B (1988) Effect of Ca^{2+} on cross-bridge turnover kinetics in skinned single rabbit psoas fibers: implications for regulation of muscle contraction. Proc Natl Acad Sci USA 85:3265–3269

Brenner B, Morano I (1990) Effects of myosin light chain phosphorylation on isometric force and cross-bridge turnover kinetics. Pflügers Arch 415:R73

Brenner B, Yu LC, Chalovich JM (1990) Caldesmon inhibits generation of active tension in skinned rabbit psoas fibres by blocking attachment of weak binding crossbridges to actin. Biophys J 57:397a

Chacko S, Rosenfeld A (1982) Regulation of actin-activated ATP hydrolysis by arterial myosin. Proc Natl Acad Sci USA 74:129–133

Chatterjee M, Tejada M (1986) Phorbol ester-induced contraction in chemically skinned vascular smooth muscle. Am J Physiol 251:C356–C361

Clark T, Ngai PK, Sutherland C, Gröschel-Stewart U, Walsh MP (1986) Vascular smooth muscle caldesmon. J Biol Chem 261:8028–8035

Connor JA, Kreulen D, Prosser CL, Weigel R (1977) Interaction between longitudinal and circular muscle in intestine of cat. J Physiol (Lond) 273:665–689

Craig R, Smith R, Kendrick-Jones J (1983) Light-chain phosphorylation controls the conformation of vertebrate non-muscle and smooth muscle myosin molecules. Nature 302:436–439

De Feo TT, Morgan KG (1985) Calcium-force relationships as detected with aequorin in two different vascular smooth muscles of the ferret. J Physiol (Lond) 369:269–282

Dillon PF, Aksoy MO, Driska SP, Murphy RA (1981) Myosin phosphorylation and the cross-bridge cycle in arterial smooth muscle. Science 211:495–497

Dillon PF, Meyer RA, Kushmerick MJ (1982) Cytoplasmic Mg^{2+} in smooth muscle by ^{31}P-NMR. Fed Proc 41:987

Fatigati V, Murphy RA (1984) Actin and tropomyosin variants in smooth muscles. J Biol Chem 259:14383–14388

Filo RS, Rüegg JC, Bohr DF (1965) Glycerinated skeletal and smooth muscle. Calcium and magnesium dependence. Science 147:1581–1583

Fischer W, Pfitzer G (1989) Rapid myosin phosphorylation transients in phasic contractions in chicken gizzard smooth muscle. FEBS Lett 258:59–62

Fujiwara T, Itoh T, Kubota Y, Kuriyama H (1989) Effects of guanosine nucleotides on skinned smooth muscle tissue of the rabbit mesenteric artery. J Physiol 408:535–547

Fürst DO, Cross RA, Demey J, Small JV (1986) Caldesmon is an elongated, flexible molecule localized in the actomyosin domains of smooth muscle. EMBO J 5:251–257

Haeberle JR, Hott JW, Hathaway DR (1985) Regulation of isometric force and isotonic shortening velocity by phosphorylation of the 20000 dalton myosin light chain of rat uterine smooth muscle. Pflügers Arch 403:215–219

Hai C-M, Murphy RA (1989) Ca^{2+}, crossbridge phosphorylation, and contraction. Annu Rev Physiol 51:285–298

Hartshorne DJ, Mrwa U (1982) Regulation of smooth muscle actomyosin. Blood Vessels 19:1–18

Hemric ME, Chalovich JM (1988) Effect of caldesmon on the ATPase activity and the binding of smooth and skeletal myosin subfragments to actin. J Biol Chem 263:1878–1885

Himpens B, Somlyo AP (1988) Free-calcium and force transients during depolarization and pharmacomechanical coupling in guinea-pig smooth muscle. J Physiol (Lond) 395:507–530

Himpens B, Matthijs G, Somlyo AP (1989) Desensitization to cytoplasmic Ca^{2+} and Ca^{2+} sensitivities of guinea-pig ileum and rabbit pulmonary artery smooth muscle. J Physiol (Lond) 413:489–503

Ikebe M, Reardon S (1988) Binding of caldesmon to smooth muscle myosin. J Biol Chem 263:3055–3058

Ikebe M, Barsotti RJ, Hinkins S, Hartshorne DJ (1984) Effects of magnesium chloride an smooth muscle actomyosin adenosine-5′-triphosphatase activity, myosin conformation, and tension development in glycerinated smooth muscle fibers. Biochemistry 23:5062–5068

Ishii N, Mitsumori F, Takahashi K, Simpson AW, Ashley CC (1989) Intracellular metabolite and free calcium concentrations during the "catch" contraction and relaxation in a molluscan smooth muscle. Prog Clin Biol Res 315:463–464

Ito M, Hartshorne DJ (1990) Phosphorylation of myosin as a regulatory mechanism in smooth muscle. In: Sperelakis N, Wood JD (eds) Frontiers in smooth muscle research. Wiley-Liss, New York, pp 57–72

Itoh T, Kubota Y, Kuriyama H (1988) Effects of a phorbol ester on acetylcholine-induced Ca^{2+} mobilization and contraction in the porcine coronary artery. J Physiol (Lond) 240:917–920

Itoh T, Ikebe M, Kargacin GJ, Hartshorne DJ, Kemp BE, Fay FS (1989) Effects of modulators of myosin light-chain kinase activity in single smooth muscle cells. Nature 338:164–167

Johansson B, Somlyo AP (1980) Electrophysiology and excitation-contraction coupling. In: Bohr DF, Somlyo AP, Sparks HV (eds) The cardiovascular system. Handbook of Physiology, Sect 2, vol II. Am Physiol Soc Bethesda, Maryland, pp 301–323

Kamm KE, Stull JT (1985) The function of myosin and myosin light chain kinase phosphorylation in smooth muscle. Annu Rev Pharmacol Toxicol 25:593–620

Kamm KE, Stull JT (1986) Activation of smooth muscle contraction: relation between myosin phosphorylation and stiffness. Science 232:80–82
Kamm KE, Stull JT (1989) Regulation of smooth muscle contractile elements by second messengers. Annu Rev Physiol 51:299–313
Kamm KE, Stull JT (1990) Mechanism of GTP-γ-S-dependent regulation of smooth muscle contraction. Biophys J 57:163a
Kitazawa T, Kobayashi S, Horiuti K, Somlyo AV, Somlyo AP (1989) Receptor coupled, permeabilized smooth muscle: role of the phosphatidylinositol cascade, G-proteins and modulation of the contractile response to Ca^{2+}. J Biol Chem 264:5339
Lash, JA, Sellers JR, Hathaway DR (1986) The effects of caldesmon on smooth muscle heavy actomeromyosin ATPase activity and binding of heavy meromyosin to actin. J Biol Chem 261:16155–16160
Marston S, Lehmann W, Moody C, Pritchard K, Smith C (1988) Caldesmon and Ca^{2+} regulation in smooth muscles. In: Gerday C, Bolis L, Gilles R (eds) Calcium and calcium binding proteins. Springer, Berlin Heidelberg New York, pp 69–81
Marston SB (1989) What is latch? New ideas about tonic contraction in smooth muscle. J Muscle Res Cell Motil 10:97–100
Marston SB, Smith CW (1985) The thin filaments of smooth muscles. J Muscle Res Cell Motil 6:669–708
Mauss S, Koch G, Kreye VAW, Aktories K (1990) Inhibition of the contraction of the isolated longitudinal muscle of the guinea pig ileum by botulinum C2 toxin – evidence for a role of G/F-actin transition in smooth muscle contraction. Naunyn-Schmiedeberg's Arch Pharmacol 340:435–351
Miller JR, Silver PJ, Stull JT (1983) The role of myosin light chain kinase phosphorylation of tracheal smooth muscle. Mol Pharmacol 24:235–242
Morgan JP, Morgan KG (1984) Alteration of cytoplasmic ionized calcium levels in smooth muscle by vasodilators in the ferret. J Physiol (Lond) 357:539–551
Mrwa U, Güth K, Rüegg JC, Paul RJ, Boström S, Barsotti S, Hartshorne D (1984) Mechanical and biochemical characterisation of the contraction elicited by a calcium-independent light chain kinase in chemically skinned smooth muscle. Experientia 41:1002–1006
Nishikori K, Weisbrod NW, Sherwood OD (1983) Effects of relaxin on rat uterine myosin light chain kinase activity and myosin light chain phosphorylation. J Biol Chem 258:2468–2474
Nishimura J, Kolber M, Van Breemen C (1988) Norepinephrine and GTP-y-S increase myofilament Ca^{2+} sensitivity in alpha-toxin permeabilized arterial smooth muscle. Biochem Biophys Res Commun 157:677–683
Palmer RM, Ferrige AG, Moncada S (1987) Nitric oxide release accounts for the biological activity of endothelium-derived relaxing factor. Nature (Lond) 327:524–526
Park S, Rasmussen H (1986) Carbachol-induced protein phosphorylation changes in bovine tracheal smooth muscle. J Biol Chem 261:15734–15739
Paul RJ (1990) Smooth muscle energetics and theories of cross-bridge regulation. Am J Physiol 258 (Cell Physiol 27) C369–C375
Pfitzer G, Boels PJ (1991) Differential skinning of smooth muscle: a new approach to EC-coupling. Blood Vessels 28:262–267
Pfitzer G, Fischer W (1990) Latch-like contraction in intact and skinned smooth muscle strips of chicken gizzard. Pflügers Arch 415:R77
Pfitzer G, Hofmann F, Disalvo J, Rüegg JC (1984) cGMP and cAMP inhibit tension development in skinned coronary arteries. Pflügers Arch 401:277–280
Pfitzer G, Rüegg JC, Zimmer M, Hofmann F (1985) Relaxation of skinned coronary arteries depends on the relative concentrations of Ca^{2+}, calmodulin and active cAMP-dependent protein kinase. Pflügers Arch 405:70–76
Pfitzer G, Merkel L, Rüegg, JC, Hofmann F (1986) Cyclic GMP-dependent protein kinase relaxes skinned fibers from guinea pig taenia coli but not from chicken gizzard. Pflügers Arch 407:87–91
Popescu LM, Panoiu C, Hinescu M, Nutu O (1985) The mechanism of cGMP-induced relaxation in vascular smooth muscle. Eur J Pharmacol 107:393–394
Rasmussen H, Takuwa A, Park S (1987) Protein kinase C in the regulation of smooth muscle contraction. FASEB J 1:177–185

Rembold CM (1991) Modulation of the Ca^{2+}-sensitivity of myosin phosphorylation in intact swine arterial smooth muscle. J Physiol (Lond) 429:77–94

Rembold CM, Murphy RA (1988) Ca^{2+}-dependent myosin phosphorylation in phorboldiester stimulated smooth muscle contraction. Am J Physiol 255:C719–723

Rüegg JC (1988) Calcium in muscle activation, 2nd edn. Springer, Berlin Heidelberg New York

Rüegg JC, Pfitzer G (1985) Excitation-contraction coupling in coronary smooth muscle. In: Rupp H (ed) The regulation of heart function. Thieme, New York

Rüegg JC, Pfitzer G (1991) Contractile protein interactions in smooth muscle. Blood Vessels 28:159–163

Sato K, Ozaki H, Karaki H (1988) Changes in cytosolic calcium levels in vascular smooth muscle strip measured simultaneously with contraction using fluorescent calcium indicator fura 2. J Pharmacol Exp Ther 246:294–300

Scheid CR, Honeyman TW, Fay FS (1979) Mechanism of adrenergic relaxation of smooth muscle. Nature (Lond) 277:32–36

Schirmer RH (1965) Die Besonderheiten des Kontraktilen Proteins der Arterien. Biochem Z 343:269–282

Sellers JR, Shirinsky VP (1987) Caldesmon inhibits movement of myosin-coated beads on Nitella actin cables. J Cell Biol 150:194a

Sellers JR, Pato MD, Adelstein RS (1981) Reversible phosphorylation of smooth muscle myosin, heavy meromyosin, and platelet myosin. J Biol Chem 256:13137–13142

Siegman MJ, Butler TM, Mooers SU, Davies RE (1980) Chemical energetics of force development, force maintenance and relaxation in mammalian smooth muscle. J Gen Physiol 76:609–629

Siegman MJ, Butler TM, Mooers SU, Michalek A, (1984) Ca^{2+} can affect Vmax without changes in myosin light chain phosphorylation in smooth muscle. Pflügers Arch 401:385–390

Small JV, Fürst DO, De Mey J (1986) Localization of filamin in smooth muscle. J Biol Chem 102:210–220

Sobue K, Kanda K, Tanaka T, Ueki N (1988) Caldesmon: a common actin-linked regulatory protein in the smooth muscle and nonmuscle contractile system. J Cell Biol 37:317–325

Somlyo AP, Himpens B (1989) Cell calcium and its regulation in smooth muscle. FASEB J 3:2266–2276

Somlyo AP, Somlyo AV, Shuman H (1979) Electron probe analysis of vascular smooth muscle: composition of mitochondria, nuclei and cytoplasm. J Cell Biol 81:316–335

Somlyo AP, Somlyo AV, Shuman H, Endo M (1982) Calcium and monovalent ions in smooth muscle. Fed Proc 41:2883–2890

Somlyo AP, Walker JW, Goldman YE, Trenthan DR, Kobayashi S, Kitazawa T, Somlyo AV (1988) Inositol trisphosphate, calcium and muscle contraction. Philos Trans R Soc Lond B 320:399–414

Somlyo AV, Somlyo AP (1968) Electromechanical and pharmacomechanical coupling in vascular smooth muscle. J Pharmacol Exp Ther 159:129–145

Somlyo AV, Bond M, Somlyo AP, Scarpa A (1985) Inositol trisphosphate-induced calcium release and contraction in vascular smooth muscle. Proc Natl Acad Sci USA 82:5231–5235

Spazenko A, Wagner J, Dabrowska R, Rüegg JC (1985) Caldesmon-induced inhibition of ATPase activity of actomyosin and contraction of skinned fibres of chicken gizzard smooth muscle. FEBS Lett 192:9–12

Suematsu E, Hirata J, Hasimoto T, Kuriyama H (1984) Inositol 1,4,5-trisphosphate releases Ca^{2+} from intracellular store sites in skinned single cells of porcine coronary artery. Biochem Biophys Res Commun 120:481–485

Sutherland C, Walsh MP (1989) Phosphorylation of caldesmon prevents its interaction with smooth muscle myosin. J Biol Chem 264:578–583

Takahashi K, Hiwada K, Kokubu T (1986) Isolation and characterization of a 34 000-dalton calmodulin- and F-actin-binding protein from chicken gizzard smooth muscle. Biochem Biophys Res Commun 141:20–26

Takahashi K, Hiwada K, Kokubu T (1988) Vascular smooth muscle calponin. A novel troponin T-like protein. Hypertension 11:620–626

Takahashi K, Abe M, Nishida W, Hiwada K, Nadal-Ginard B (1991) Primary structure, tropomyosin binding, and regulatory function of smooth muscle calponin: a new homolog of troponin T. J Muscle Res Cell Motil (in Press)

Takai A, Troschka M, Mieskes G, Somlyo AV (1989) Protein phosphatase composition in the smooth muscle of guinea pig ileum studied with okadaic acid and inhibitor 2. Biochem J 262:617–623

Toyoshima YY, Kron SJ, McNally EM, Niebling KR, Toyoshima, C, Spudich JA (1987) Myosin subfragment-1 is sufficient to move actin filaments in vitro. Nature 328:536–539

Velaz L, Hemric ME, Benson CE, Chalovich JM (1989) The binding of caldesmon to actin and its effect on the ATPase activity of soluble myosin subfragments in the presence and absence of tropomyosin. J Biol Chem 264:6902–9610

Wagner PD, Vu N-D (1986) Regulation of the actin-activated ATPase of aorta smooth muscle myosin. J Biol Chem 261:7778–7783

Williams DA, Beckers PL, Fay FS (1989) Regional changes in calcium underlying contraction of single smooth muscle cells. Science 235:1644–1648

Winder SJ, Walsh MP (1990) Smooth muscle calponin. Inhibition of actomyosin MgATPase and regulation by phosphorylation. J Biol Chem 265:10148–10155

Yamaguchi H (1986) Recording of intracellular Ca^{2+} from smooth muscle cells by sub-micron tip, double-barrelled Ca^{2+}-selective microelectrodes. Cell Calcium 7:203–219

Chapter 4

The Structures of Striated and Smooth Muscles Related to Their Function

J.M. Squire

Contents

1 The Development of Organisation in Muscle Cells

1.1 Introduction

It is a truism to say that without directed movement life as we know it would not exist, but it is notable that the highly sophisticated life forms found today require that this movement should appear in a variety of different roles and guises. Such movements include the flagella-driven swimming of sperm cells, translation of chromosomes towards the centrioles and the later active constriction of the contractile ring during cell division, active migration of cells such as fibroblasts, movement of cell organelles and particles within cells and especially along nerve axons and, of course, the highly developed movements in muscle. Without prior knowledge it might well be thought that these different processes require many fundamentally different molecular mechanisms, but, in fact, nature has been characteristically frugal in the use of its resources. There appear to be only two major motile systems, one based on actin filaments and actin-binding proteins and the other based on microtubules and their associated

Biophysics Section, Blackett Laboratory, Imperial College of Science, Technology and Medicine, London SW7 2BZ, UK

Advances in Comparative and Environmental Physiology
Vol. 12. Edited by H. Sugi

proteins. Even then there may be considerable similarities between the two systems. Many of the actin-based systems involve myosin as the actin-binding ATPase, and relative movement (sliding) between actin and myosin filaments is thought to be due to the altered conformation of the actin-myosin complex (myosin heads swinging on actin) when the products of ATP hydrolysis (i.e. ADP and P_i) are released. Similarly, in the case of microtubules, interactions between tubulin and various microtubule-associated ATPases may be the origin either of microtubule sliding over other microtubules as in cilia and flagella or organelles sliding along microtubules as in neuronal transport. Rather than produce a diversity of fundamentally different motile systems, nature has found an ATPase-driven (protein swinging?) mechanism that works, has evolved what appear to be two different varieties of a similar process and has then adapted these two varieties, by means of altered structural environments or different interactions with modifying, sometimes regulatory, proteins, to give the diversity of contractile systems that is now known. This chapter is about the actin-myosin system, which is found in all cells in the contractile ring and in highly developed forms in different muscle types, and it considers the ways that have evolved for making radically different uses of the same molecular mechanism, the actin-myosin crossbridge cycle. The chapter closes with a brief summary of recent structural evidence about the fundamental molecular movements involved in this cross-bridge cycle itself.

1.2 Evolution of Muscle Structure

Since movement is essential for life, it is natural to suppose that first the microtubule system and then the actin-myosin system must have been key developments in the evolution of even the most primitive eukaryotes. We shall never know what these early forms were like, but some of their features may be inferred from what we know of these motile systems today. In the case of the actin-myosin system it seems that the contractile ring that develops during cytokinesis must be an ever-changing collection of growing and depolymerising actin filaments together with some form of myosin assembly. In order to be able to produce active constriction of the dividing cell, it is also necessary for the actin or myosin assemblies to be suitably anchored to the cell membrane. Furthermore, it is evident, since actin filaments are polar structures, that even the most primitive contractile assemblies must contain organising centres for the actin filaments. Likewise, the highly polar myosin molecules may also need to be suitably attached either to the cell membrane, to other actin filaments or to other myosin molecules, in order to be able to transmit any tension that is produced back to the points where net movement is needed. Although it is conceivable that primitive myosins were of the type found in *Acanthamoeba* myosin I, with one ATPase "head" and no tail (Pollard 1975), or that single, two-headed myosin molecules could themselves produce relative sliding of two oppositely polarised actin filaments (Trinick and Offer 1979), it is evident that an anti-parallel myosin dimer (Fig. 1a,b) would be the most efficient basic unit for producing directed movement of actin filaments. Figure 1 illustrates possible

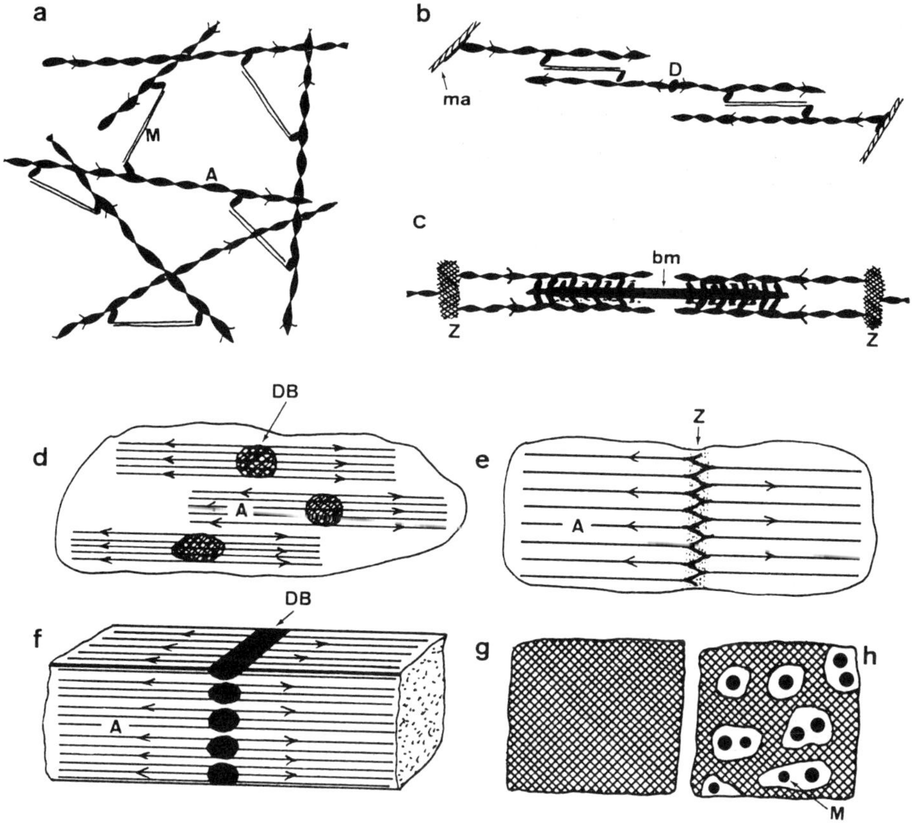

Fig. 1a–c. Illustration of the possible development of the contractile proteins, actin and myosin, into gradually more ordered assemblies. **a** The simplest form of contractile unit; **b** anchorage of actin to the cell membrane, other organelles, or to further contractile units; **c** development of larger myosin aggregates. *A* Actin; *M* myosin dimers; *D, Z* actin anchorage points; *ma* membrane attachment; *bm* bipolar myosin aggregates; *arrowheads* indicate actin polarity. **d** to **g** Stages of development of actin anchorage structures in invertebrate muscles from **d** dense bodies to **f** pseudo Z-bands (aligned dense bodies) to **e** complete Z-band structures. Some arthropods have Z-bands (**g**) transverse view, which become perforated (**h**) to allow penetration of the thick filaments (*M*). *A Actin; DB* dense bodies; *Z* Z-band; *arrows* indicate actin polarity. (Squire 1986b)

evolutionary stages in the organisation of muscle contractile units starting from a rudimentary assembly of actin filaments and myosin dimers (a), moving on to actin filaments with membrane attachments and sites of polarity reversal (b), and then to actin filaments anchored at a polarity reversing Z-line and interacting with a relatively large assembly (filament) of myosin molecules to give a muscle sarcomere in which concerted filament sliding can occur (Huxley and Hanson 1954; Huxley and Niedergerke 1954). The known sites of polarity reversal for the actin filament invariably involve the protein α-actinin. In non-muscle cells the α-actinin can reversibly bind calcium ions and can form tem-

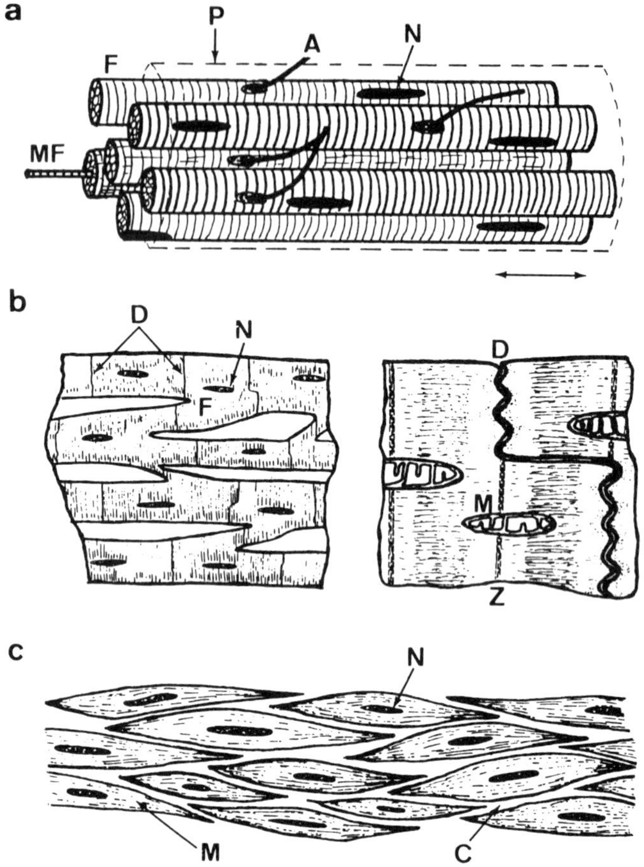

Fig. 2. Schematic representations of the different cell structures in **a** vertebrate skeletal muscle; **b** vertebrate cardiac muscle; and **c** vertebrate smooth muscle. **a** Long, straight fibres (*F*) each containing many nuclei (*N*) and connected to motor nerves (*A*). The fibres are composed of myofibrils (*MF*), and several fibres form a fascicle surrounded by a connective tissue layer, the perimysium (*P*). **b**, *Left* The arrangement of the branched fibres in cardiac muscle, which are linked through intercalated discs (*D*). The single nuclei (*N*) are centrally located. *Right* The enlargement (*right*) shows the dense granular structure on each side of the intercellular boundary. *Z* Z-band; *M* mitochondria. **c** The arrangement of the spindle-shaped cells in vertebrate smooth muscles. The nuclei (*N*) are centrally placed, the region *M* contains the contractile material, and there is much collagenous connective tissue (*C*) in the intracellular space. These cells are often very much longer and thinner than shown. (Squire 1986b)

porary interactions with actin as the needs of the cell require (see review by Blanchard et al. 1989). In muscle cells the α-actinin does not bind calcium, but forms relatively permanent bridging structures between anti-parallel actin filaments, as discussed in Section 2.2.

Some of the characteristic appearances of muscle cells across the animal kingdom are illustrated in Fig. 2. These range from the organised parallel cross-striated fibres (a) in vertebrate skeletal muscles and some invertebrate striated muscles (e.g. in insects, crustaceans, molluscs), to the specialised striated muscles in vertebrate hearts (b), to the elegantly tapered cells (c) in vertebrate smooth muscles. Apart from the muscles illustrated here, the molluscan smooth muscles contain long cylindrical fibres as in Fig. 2a, but these do not show the cross-striation pattern.

All of these muscles contain actin-anchorage structures, invariably containing the protein α-actinin, but these anchorage points show various degrees of organisation. On the whole, the smooth muscles (vertebrate visceral muscles or invertebrate smooth muscles) have parallel bundles of actin filaments anchored through large dense bodies to similar bundles of opposite polarity (Fig. 1d). It is because these anchorage points are irregularly distributed in three dimensions that such muscles do not appear cross-striated. However, as discussed later, they do contain rudimentary contractile units which extend along the cells rather like the myofibrils in cross-striated muscles. In slightly more organised muscles, the dense bodies are laterally elongated and are aligned to produce pseudo Z-bands (Fig. 1f). In some muscles, such as the body wall muscles in annelids and nematodes, adjacent rod-like dense bodies are axially stepped to give an "obliquely striated" appearance (Rosenbluth 1965, 1968). Some molluscan muscles are also obliquely striated. In the cross-striated muscles, there are organised planar networks of actin-anchorage structures, the Z-bands (or Z-lines or Z-discs; Fig. 1e), where opposite faces of the network serve to another actin filaments of opposite polarity. As shown later, the patterns (2-D lattices) of actin-anchoring sites (Z-bands) are highly ordered and have unit cell sizes and shapes which correlate systematically with the organisation of the myosin filaments in the A-band of the muscle sarcomere.

An additional feature of some muscles is that holes can appear in the Z-band (Fig. 1g,h) to allow penetration of myosin filaments as the sarcomere shortens. These perforated Z-bands, found by Hoyle et al. (1965) in the giant fibres of the barnacle *Balanus nubilis* and by Osborne (1967) in the abdominal muscle of the blowfly larva, permit a larger working range of sarcomere lengths than is possible in muscles with impenetrable Z-bands, where there is an abrupt fall off of tension with reducing sarcomere length as the myosin filaments butt into the sarcomere ends (Gordon et al. 1966).

Studies of the evolutionary history of different phyla and subphyla, as summarised in Fig. 3, show that the cross-striated muscles in crustaceans, the obliquely striated muscles in annelids and the smooth muscles in molluscs were all relatively early developments which occurred some 200 million years before the evolution of primitive vertebrates. One of the most notable differences between vertebrate muscles and these earlier contractile systems is the lack of the protein paramyosin. Paramyosin is a two-chain α-helical coiled-coil molecule, rather like the rod part of the myosin molecule without its globular (ATPase) heads (Fig. 4c; see review in Fraser and MacRae 1973). A feature that paramyosin confers on muscles is the ability to make very large myosin-containing filaments. Vertebrate striated muscles, with no paramyosin, contain the smallest of known muscle myosin filaments (Huxley 1963). As illustrated in Fig. 4b, these filaments have the heads of three myosin molecules at the same axial level along the filament (a so-called crown of heads) and similar crowns

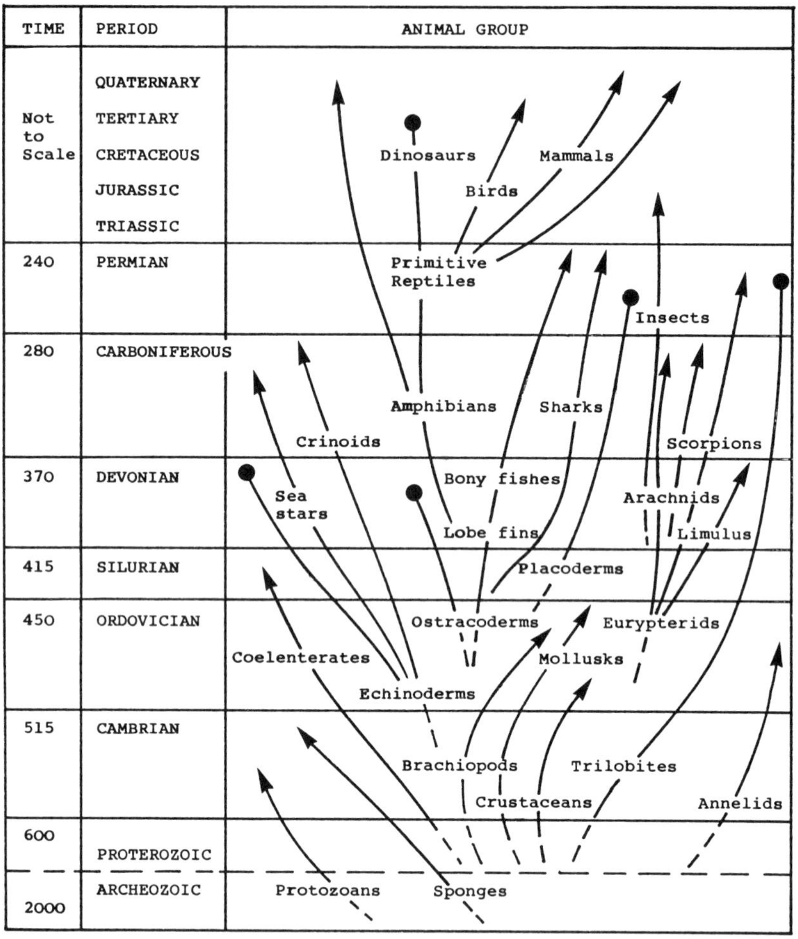

Fig. 3. Summary of the probable evolutionary history of different phyla or subphyla. Time in millions of years: divisions approximate (After Storey and Usinger 1965; Black 1970; Patterson 1978). *Arrows* indicate forms that survive today. (Squire 1986b)

occur at 143 Å axial intervals and with 40° rotations around the filament axis to give the filament a three-stranded helical structure of repeat 429 Å (Squire 1972). The filament backbone, formed by the myosin rods, is about 140 to 150 Å in diameter and the filaments are 1.57 μm long, probably containing 294 myosin molecules (Sjostrom and Squire 1977a,b; Squire 1986a,b; Offer 1987).

Apart from vertebrates, all other animals have muscles containing paramyosin. The smallest paramyosin-containing filaments are those in some insect flight muscles (e.g. *Lethocerus*, *Apis*; see Pringle 1972). They contain up to 11% paramyosin (Levine et al. 1976), have four pairs of myosin heads in each 145 Å-spaced crown (Reedy et al. 1981; Fuller and Squire 1987; Morris et al. 1991), have a helical repeat of 1160 Å (8 × 145 Å), a backbone diameter of 160 to

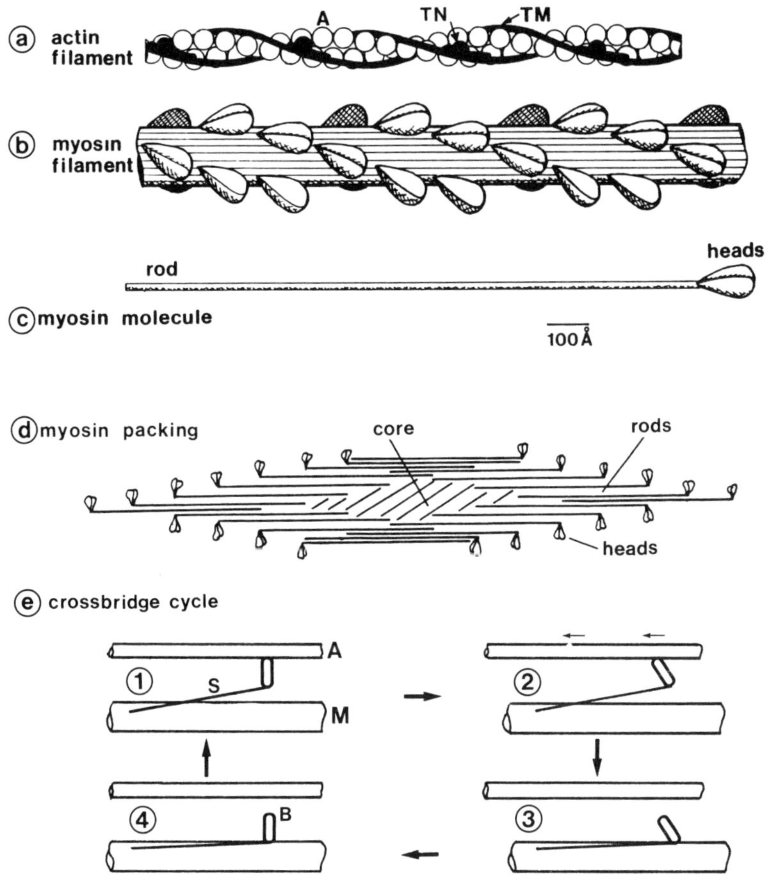

Fig. 4. Illustrations of **a** actin filaments; **b** myosin filaments; and **c** myosin molecules. Actin filaments contain polar, globular (not spherical) actin monomers (*A*), together with tropomyosin (*TM*) and troponin (*TN*). Myosin molecules have a rod and two elongated globular heads. The rods form the filament backbone and the heads project from the filament surface. The scale in **c** also applies to **a** and **b**. **d** Illustration of the anti-parallel packing of myosin molecules halfway along a myosin filament and the parallel packing elsewhere. The whole structure, which is shown very schematically, is bipolar. Myosin filaments in invertebrates contain a core of paramyosin, but there is probably no core in vertebrate striated muscle thick filaments. **e** Summary of the simplest scheme relating events in the acto-myosin ATPase cycle to mechanical events in a possible cross-bridge cycle. *A* Actin; *M* myosin; *B* myosin heads; *S* myosin S2. States indicated are *4* relaxed muscle; *1* initial cross-bridge attachment; *2* rigor attachment which could represent the end of the cross-bridge working stroke; and *3* the initial detached cross-bridge state. (Squire 1987)

200 Å (Reedy 1968; Beinbrech et al. 1988) and are four-stranded helical filaments about 2 μm long. Other invertebrate myosin filaments are also four-stranded, but much longer than 2 μm (e.g. *Limulus*, Levine et al. 1988; see Sect. 3.2) or they may have more myosin heads per crown, greater lengths than 2 μm, larger diameters and contain a substantial amount of paramyosin. For

example, some crustacean myosin filaments may be five-stranded (Wray 1979a), scallop striated muscle myosin filaments are seven-stranded (Vibert and Craig 1983), and some molluscan smooth muscles have myosin-containing filaments up to 2000 Å in diameter and 25 μm in length (Sobiescek 1973). In such filaments myosin is a relatively minor component and they are thus known as paramyosin filaments. It appears likely that the myosin in such filaments forms a surface shell on an extended semi-crystalline core of paramyosin (Elliott and Bennett 1984).

These different muscle characteristics are necessarily related to their different functional requirements. Molluscan smooth muscles, with relatively poor three-dimensional order, very large paramyosin filaments and very long actin filaments, can only contract relatively slowly, but they can sustain very high tensions. In addition, many display the paramyosin-mediated "catch" response (also found in some other invertebrates) in which tension is maintained over long periods (hours) with little expenditure of energy (Lowy and Millman 1963). At the other extreme, where very fast shortening speeds are required, muscles have highly organised short sarcomeres with small myosin filaments, short actin filaments and the minimum number (six) of actin filaments to each myosin filament (see later). It is evident that for a given length of striated muscle fibre, with the same intrinsic actin-myosin ATPase rates, a fibre with a large number of short sarcomeres in series will shorten more quickly than one with a smaller number of longer sarcomeres, but the latter can sustain tension levels which are higher and which can be maintained over a much greater range of muscle lengths. Different muscles and fibre types may contain alternative isoforms of the same protein and some of the differences can be seen as altered structural arrangements within the sarcomere. It is these basic differences in sarcomere design, together with differences in the geometrical arrangements of fibres within the whole muscle (Fig. 5a–e), that allow the same contractile mechanism, the cross-bridge cycle (Fig. 4e; A.F. Huxley 1957; H.E. Huxley 1969; see reviews in Cooke 1986; Squire 1986b, 1990), to be harnessed to fulfil the great diversity of functional needs demanded by the highly varied skeletons, environments and life styles of different animals.

2 Striated Muscle Sarcomeres

2.1 Fibre Ultrastructure and the Cytoskeleton

It was shown in Fig. 2a that striated muscles comprise parallel cylindrical fibres, often up to 100 μm in diameter, each of which is a multi-nucleate cell enclosed by an excitable membrane with which motor nerve terminals are associated. The whole fibre assembly is embedded and surrounded in connective tissue, the outer part of which (the perimyseum) delimits a muscle fascicle. These fibres are themselves composed of smaller cylindrical myofibrils, often about 1 to 2 μm in diameter, between which are mitochondria and the sarcoplasmic reticulum (SR; Fig. 5i), which is a membranous structure associated with the release and uptake of calcium ions during activation and relaxation. The fibre membrane

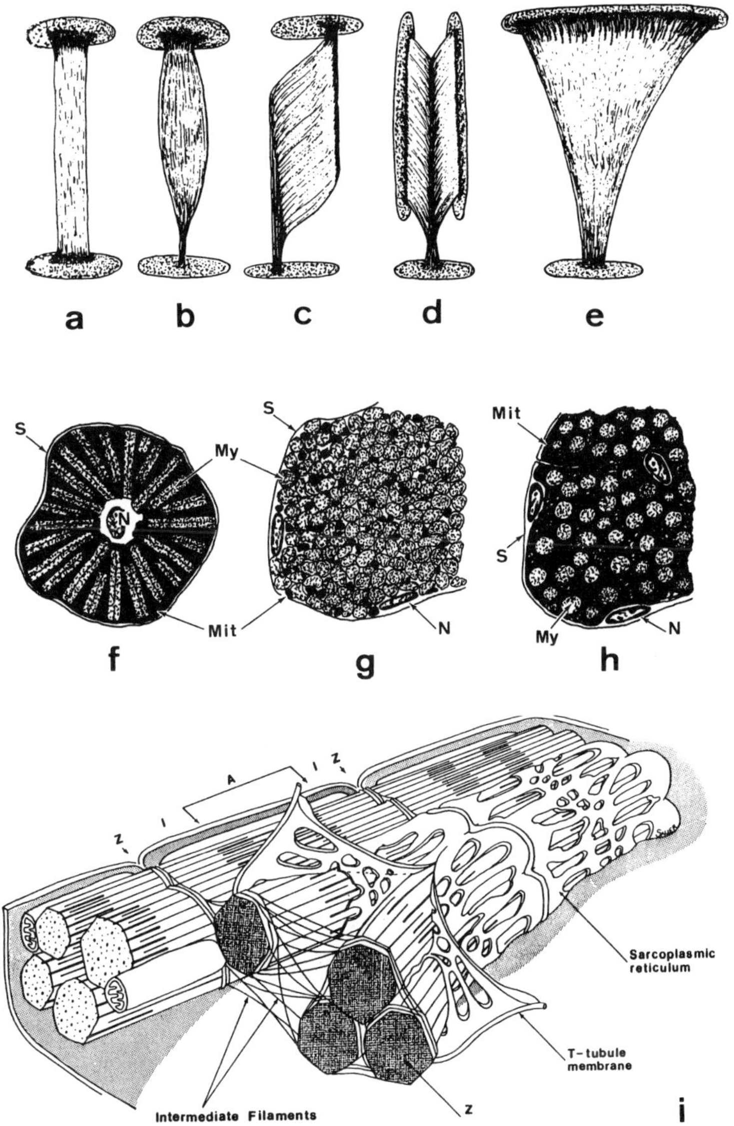

Fig. 5a–e. Various fibre organisations within muscles of different shapes and actions: **a** strap; **b** fusiform; **c** unipennate; **d** bipennate; **e** deltoid. **f** to **h** Different types of organisation of the contractile material in invertebrate muscle fibres: **f** tubular structure; **g** closely packed structure and **h** fibrillar structure. *S* Sarcolemma; *My* myofibrils; *Mit* mitochondria; *N* nuclei. (Squire 1986b, after Chapman 1982). **i** Three-dimensional arrangement of various cytoskeletal components in muscle fibres relative to the myofibril structure and the locations of the SR and T-systems (Lazarides 1980). *A* *A*-band; *I* I-band; *Z* Z-band. (**i** reprinted by permission from Nature *283*, 249–255 Copyright © 1980 MacMillan Magazines)

(sarcolemma) is continuous with excitable tubular invaginations known as T-tubules which run laterally through the muscle fibres and around individual myofibrils. In this way depolarisation of the sarcolemma, stimulated by the motor nerves, is propagated into the interior of the fibres via the T-tubules so that Ca^{2+} release by the SR is rapidly triggered. Simple diffusion of Ca^{2+} from the fibre surface would be much too slow.

Apart from the collagenous connective tissue, which provides a structural framework between fibres, there is also a complex cytoskeletal network within fibres and between myofibrils, which probably serves to hold the myofibrils in place and to coordinate sarcomere length changes between adjacent myofibrils. Major components of this intermyofibrillar cytoskeleton are intermediate filaments composed principally of desmin and skelemin (Fig. 5i). In vertebrate muscles desmin appears to provide lateral connections between adjacent myofibrils at the level of the Z-band (Lazarides 1980), and skelemin serves a similar role at the M-band (Price 1987), although some intermediate filaments are thought to run axially along the myofibril surface between successive Z-band levels. There is additional cytoskeletal material within the muscle myofibrils as discussed later.

The organisation of myofibrils within fibres and the content and distribution of mitochondria, SR and nuclei are very different in different muscle types. Figure 5(f–h) illustrates the diverse fibre structures found in different invertebrate muscles (Chapman 1982). Tubular fibres (f), found in insects and arachnids, have centrally placed nuclei and flat ribbon-like myofibrils between which abundant mitochondria are located. Figure 5g shows a closely packed structure, typical of crustacean fibres and some insects, in which nuclei are peripherally located, the myofibrils are cylindrical and the mitochondria are dispersed between myofibrils. Insect asynchronous flight muscles (Fig. 5h), which use energy at exceedingly high rates, contain a very high proportion of mitochondria (up to 40% of the total volume) between which the cylindrical myofibrils are located.

Vertebrates also contain striated muscle fibres with diverse structures, although, in most vertebrates, the fibres tend to be variations on Fig. 5g. The myofibrils are usually cylindrical and comprise most of the fibre volume, with mitochondria scattered between them. Slow oxidative fibres (Type I) have a relatively large mitochondrial content, whereas faster fibres (Type IIa,b) have fewer mitochondria and more glycogen in varying amounts according to their needs.

The fast white muscles of bony fish, which have proved to be very valuable in studies of sarcomere ultrastructure and the myosin head (cross-bridge) cycle (see later), have fibres in some ways like the illustration in Fig. 5f, with numerous, highly elongated ribbon-shaped myofibrils, about 0.5×10 to $15\,\mu m$ in cross-section. However, unlike Fig. 5f, in these fish fibres the myofibrils radiate out, not from central nuclei, but from more circular or polyhedral central myofibrils. The myofibrils are closely packed with relatively few mitochondria between them and the fibre nuclei are peripherally located. Fish muscle slow red fibres (primarily associated with the lateral line) also contain some ribbon-shaped myofibrils, 1×3 to $6\,\mu m$ in cross-section, especially at the fibre periphery, but most internal myofibrils are circular or polyhedral, giving a fibre appearance not unlike Fig. 5g with relatively numerous mitochondria scattered through the fibre, and with peripherally placed nuclei (see Munro 1986; Munro et al. 1992, for details of fish muscle ultrastructure).

2.2 Vertebrate Muscle Sarcomeres

Figure 6 illustrates the filament organisation in vertebrate striated muscle sarcomeres. The bipolar myosin filaments (see Fig. 4d) are centrally placed to form the A-band of the sarcomere. They are cross-linked halfway along their length by M-bridges which occur in three to five different positions separated axially by multiples of about 220 Å (Sjostrom and Squire 1977a) to form the M-band. On each side of the M-band and before the start of the region on the myosin filaments where myosin heads are located (the bridge region), is a region of the myosin filament that is apparently devoid of either M-band proteins or myosin heads and for this reason it is termed the bare region (Luther and Squire 1980).

Interdigitating with the two polar ends of the myosin filaments are two sets of actin filaments with opposite polarity; the polarity being determined by the actin filament interactions at the Z-band (Fig. 1e). Apart from actin, these filaments contain the regulatory proteins tropomyosin and troponin (see Fig. 4a; Ebashi et al. 1969; Cohen and Vibert 1987). In cross-sections through the overlap region of the A-band, myosin and actin filaments are organised in a hexagonal lattice, with actin filaments placed at the mid-point (the trigonal point) of a triangle formed by three mutually adjacent myosin filaments (Fig. 6b; H.E. Huxley 1957). There are twice as many actin filaments as myosin filaments in such a cross-section and it is often called a 2:1 structure. At the M-band, the M-bridges form a hexagonal cross-linking structure between the myosin filaments (Fig. 7b), although not all adjacent myosin filaments are cross-linked at all M-bridge levels (see Sect. 3.1). At the Z-band the actin filaments, perhaps surprisingly, are not hexagonally arranged but form an almost square array (Knappeis and Carlsen 1962; Pringle 1968), with Z-bridges (probably α-actinin) linking actin filaments in one sarcomere to those in the next. In axial view this sometimes gives rise to what is known as the zig-zag appearance (Fig. 7c,d), whereas in transverse view they often show either a "basket weave" or a "small square lattice" appearance (Fig. 7h,i). A single model to account for some of these appearances and sketched in Fig. 7j was first proposed by Knappeis and Carlsen (1962). Although many features of this model are correct, recent 3-D reconstructions of different vertebrate Z-bands and Z-crystals have shown important differences not only between the Z-bands in different vertebrate muscle types, but also between the observed Z-band structures and the Knappeis and Carlsen model. Some of these features are discussed in Section 4.

In addition to the intermediate filament cytoskeleton around myofibrils, there is also a cytoskeleton within myofibrils comprising the unusual proteins titin (connectin) and nebulin (Maruyama et al. 1977; Wang et al. 1979; Trinick

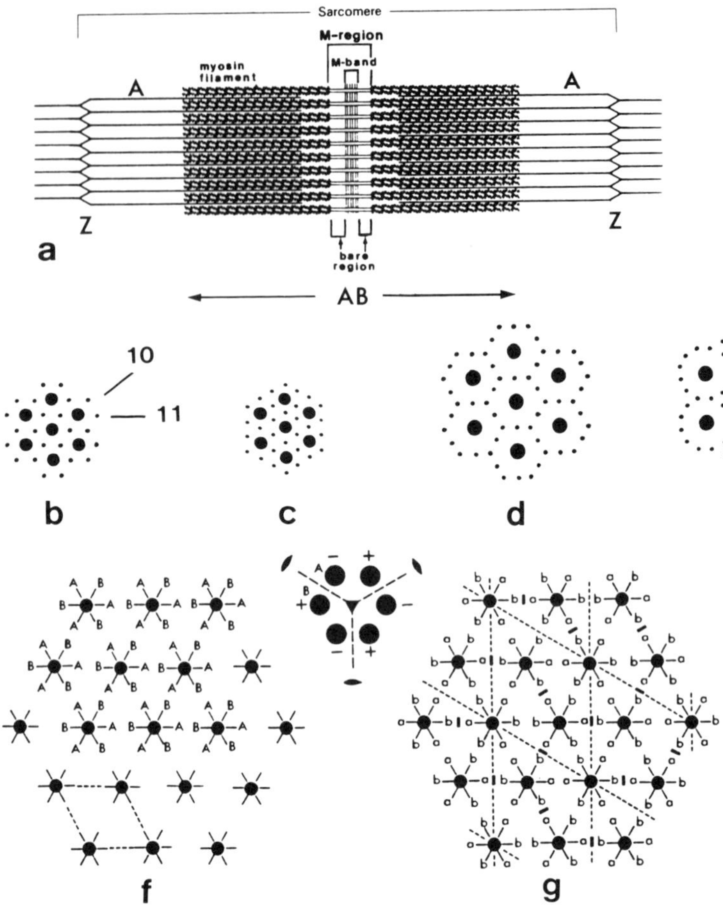

Fig. 6. **a** Sarcomere structure in vertebrate striated muscles. The A-band is the location of bipolar myosin filaments which are cross-linked at the M-band. Actin filaments (*A*) interdigitate with the A-band lattice and extend out to the Z-bands (*Z*) at the two ends of the sarcomere. **b** to **e** Arrangements of thick and thin filaments found in different muscle cross-sections: **b** vertebrate striated muscles (*10* and *11* lattice planes indicated), **c** insect fibrillar flight muscle, **d,e** two possible arrangements of 12 actin filaments around each myosin filament, as in arthropod leg and trunk muscles and some synchronous flight muscles. The ratios of thin to thick filaments are **b** 2:1, **c** 3:1, **d** 5:1, and **e** 6:1 (Toselli and Pepe 1968). **f** to **g** *Inset* Illustration of the structural implications of the 32-point group symmetry of the vertebrate myosin filaments at the level of M1, halfway along the filament, where two-fold axes are located. The *filled circular objects* marked + (and denoting an arbitrary asymmetric unit) are equivalent to those marked − except for their polarity with respect to the filament axis (i.e. + points up out of the page and − points down). If the interaction sites corresponding to − and + are called *A* and *B* respectively, then it is clear that M-bridge connections between adjacent filaments can be from A to A or from A to B. Illustrated in **f** and **g** are the two ways that *A* and *B* sites can be connected systematically to give the M-band. In **f** unlike interactions (A to B) are favoured, and all thick filaments have identical orientations, as in fish muscle. Like interactions (A to A or B to B) are favoured in **g**. They cannot be totally satisfied in this case, but if they are maximized as in **g**, then a superlattice of the kind observed in frog muscle and other higher vertebrate muscle is inevitably produced. (Luther and Squire 1980). **b** to **e** Reproduced from the Journal of Cell Biology 1968, vol 37, pp. 445–461, by copyright permission of the Rockefeller University Press

et al. 1984; see Squire et al. 1987 for review). Titin, a single chain protein of molecular weight about 3 × 10^6 daltons and representing about 15% of the myofibrillar mass, has been shown by antibody labelling studies to extend axially through the sarcomere from close to the M-band to near the Z-band (Trinick et al. 1984; Furst et al. 1988; Itoh et al. 1988; Pierobon-Bormioli et al. 1989). For many years it has been known that myofibrils from which actin and myosin have been exhaustively extracted still retain their structural integrity (Huxley and Hanson 1954) and it has been postulated that elastic "gap filaments" may extend through the sarcomere (Locker and Leet 1975, 1976), probably from the myosin filament tips to the Z-band, and possibly forming a myosin filament core through to the M-band region. The titin molecule fulfils many of the requirements for such gap filaments. For example, it is extensible in the I-band region as shown by systematic changes in epitope separations with changing sarcomere length (Itoh et al. 1988; Pierobon-Bormioli et al. 1989). However, the accessibility of titin to antibody in the A-band region shows that it does not form a myosin filament core, but is probably associated with the myosin filament surface. Pierobon-Bormioli et al. (1989) have shown that antibody labelled rabbit psoas muscle fibres show no relative movement of titin epitopes in the A-band during stretches up to 2.8 μm sarcomere length, but, above that length, "elastic" movements start to occur. There is clearly a specific linkage between titin and the A-band at short sarcomere lengths, but it appears to be one that can be disrupted at large stretches.

The elastic properties of titin have been implicated in the role of stabilising sarcomere structure during sustained contractions (Horowitz and Podolsky 1987). Muscle sarcomeres have an inherently unstable design; slight axial movement of an A-band towards one end of a sarcomere will increase the myosin filament overlap with actin at that end and decrease it at the other end. This imbalance, unless counteracted in some way, will tend to make the A-band move further off centre; a positive feedback effect. Horowitz and Podolsky (1987) showed that fibres with their titin molecules preferentially severed by irradiation were much less stable than sarcomeres with intact titin.

A measure of the total elasticity of fibres, including the effects of titin, intermediate filaments, membranous components, connective tissue, etc., is the so-called passive elasticity. The change in tension with sarcomere length of an unstimulated fibre provides an indication of the cytoskeletal forces available to maintain sarcomere alignment, A-band location and indirect tension transmission along the fibre. It is notable that some muscles, such as rabbit psoas muscle and frog sartorius muscle, have passive tensions that start low at rest length (2.3 μm) but then increase rapidly so that the muscles become relatively inextensible beyond 2.8 to 3 μm sarcomere length. Frog semitendinosus muscle, on the other hand, shows relatively little passive tension even up to the non-overlap sarcomere length of 3.6 μm. The fibres of bony fish muscles (e.g. plaice or turbot fin muscles), cannot be extended beyond about 2.9 μm in the intact muscle, but are easily extended to over 4 μm when skinned and isolated (Brenner and Squire 1987); thus indicating that a major constraint on extensibility in this case is the connective tissue matrix in the muscle, rather than the intermyofibrillar (intermediate filament) or intramyofibrillar (titin) cytoskeleton.

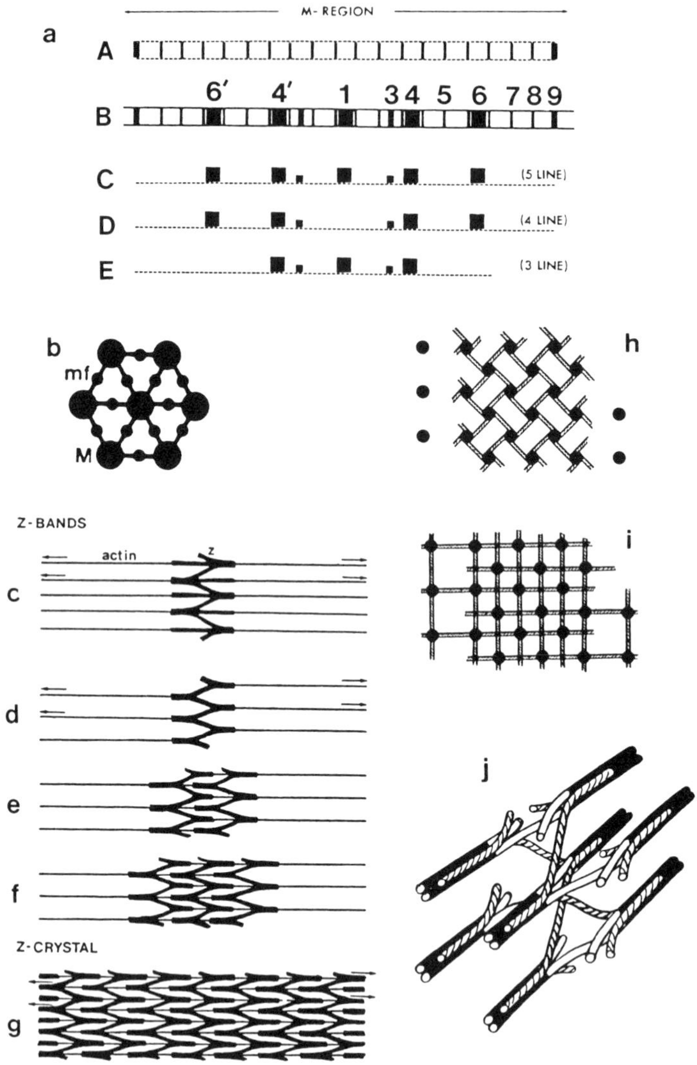

Fig. 7. **a** Schematic diagram illustrating some different observed M-band appearances and their interpretation. *Row B* shows all of the observed lines in the M-region, including five strong M-band lines (M6, M4, M1, M4′, M6′). These strong lines occasionally appear to be split. *Row A* shows the positions of the lines in two overlapping 145 Å arrays that are thought to be caused by stain-excluding features on the negatively stained anti-parallel myosin rods in the cryosectioned M-region. *Rows C, D,* and *E* represent different appearances that are seen if the densities of M1 and M6 (M6′) vary; they give 5-line, 4-line, and 3-line M-band patterns, respectively. The line M3 appears to be a constant feature of these M-bands, but does not fit onto the 145 Å arrays. It may be part of the M-bridge/M-filament assembly as suggested by Luther and Squire (1980). Based on the work of M. Sjostrom, cited in Squire (1981). **b** Detail of the transverse M-band structure in vertebrate striated muscles. Myosin filaments (*M*) are crosslinked by M-bridges (*thin lines*) via M-filaments (*mf*) to other myosin filaments in the hexagonal lattice. **c** and **d** Illustrations of the longitudinal appearance of the Z-band structure in simple single layer Z-bands when viewed in different directions (45° apart). **d** to **f**

The role of the protein nebulin, which has a molecular weight of about 5×10^6 daltons and is located along the I-band, is unknown. It does not appear to have similar elastic properties to titin (Pierobon-Bormioli et al. 1989), but may be closely associated with the thin filaments. A great deal of work on both titin and nebulin is necessary in order to understand properly their properties and interactions with other sarcomere components.

3 Lattice Specialisations of Striated Muscles

3.1 Classes of Vertebrate Striated Muscle

Vertebrate striated muscles can be broadly divided into two types, skeletal and cardiac muscles (see Fig. 2). Apart from the presence of some radically different protein isoforms, particularly myosin, troponin-I and C-protein (England 1986), the basic sarcomere structure of cardiac muscles is similar to that of skeletal muscles and the myosin and actin filaments appear to have the same symmetry. However, a unique characteristic of cardiac muscle fibres is that they do not, like skeletal muscles, extend between tendon connections, but they are themselves linked end to end through intercalated discs (Fig. 2b) which provide structural and electrical connections between successive cells so that concerted contractions are possible.

Vertebrate skeletal muscles are distinguished into fibre types according to their speed, fatigue resistance, energy source and myofibril ultrastructure, as discussed earlier (Brooke and Kaiser 1970; Beckett and Bourne 1972; Dubowitz and Brooke 1973; Dubowitz 1981; Squire 1986b), but there is a fundamental ultrastructural difference which separates the muscles of the teleosts (e.g. bony fish) from all other vertebrate muscles. This difference concerns the way that the myosin and actin filaments are organised in the sarcomere. Both muscle types have sarcomere structures as in Fig. 6a with hexagonally arranged filaments as in Fig. 6b, but the way the filaments are placed on this lattice is different.

With the notable exception of bony fish muscles, all vertebrate striated muscles have their three-stranded myosin filaments placed in one of two orienta-

Corresponding views showing the zig-zag structure of the Z-filaments in Z-bands of different thickness: **d** one-layer, **e** two-layer, **f** three-layer. **g** The same structure extended to many layers to give Z-crystals or anomalous Z-bands (Squire 1981). **h** and **i** Z-band appearances seen in muscle cross-sections and shown schematically: **h** is the basket-weave appearance and **i** is the small square lattice appearance (for detailed references see Squire 1981). At the two sides of each diagram the actin filament positions in the two linking sarcomeres are indicated, as is the relative displacement of these two lattices at the Z-band. **j** Perspective drawing of the possible basic structure of the Z-band. The Z-filaments shown schematically as *shaded* or *white* rods would account for both the zig-zag structure seen in longitudinal sections (see **d** to **g**) and the basket-weave appearance seen in transverse sections (**h**). The *black* rods merely indicate the approximate two-fold symmetry of the actin filaments but clearly not their helical twist. In the Knappeis and Carlsen model (1962), both the shaded and white rods would represent equivalent Z-bridges and the whole structure would appear to have four-fold rotational symmetry around the actin filament position as in **h** and **i** (Squire 1981)

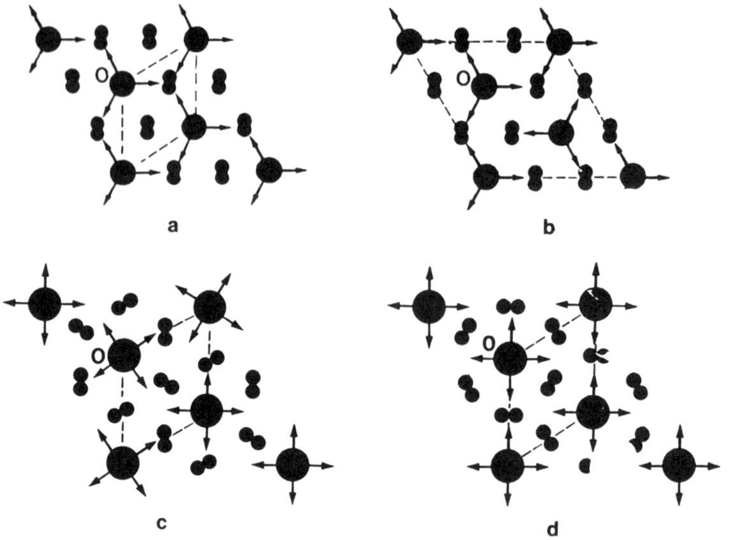

Fig. 8. The filament lattices in the overlap regions of the A-bands in various striated muscles. **a** Teleost skeletal muscle; **b** frog skeletal muscle (typical of non-teleost vertebrates); **c** *Lethocerus* (giant water bug) flight muscle; **d** *Apis* (bee) flight muscle. In each case the *large black circles* represent myosin filaments and the pairs of *small circles* represent actin filaments and indicate their relative orientations. The *dashed outlines* show the unit cells of the hexagonal lattices. In **b** and **c** these are statistical unit cells. The *arrows* indicate the symmetry of the cross-bridge arrangements on the myosin filaments; three-fold in vertebrate skeletal muscles (**a,b**) and four-fold in insect flight muscles (**c,d**). In each diagram the "imaginary observer" discussed in the text and who would "see" the lattices drawn in Figs. 11 and 14 would be on the axis of the myosin filaments marked *O*. (Squire 1992)

tions around their long axes (Fig. 8b). The arrangement is not regular, but two building rules, the "no-three alike rules" described by Luther and Squire (1980), define the distribution of filament orientations. These rules state that (1) no three adjacent filaments in a triangle can all have the same orientation and (2) no three successive filaments along a 100 direction (see Fig. 6b) through the lattice can all have identical orientations. These two rules, when applied systematically, generate the superlattice "unit cell" containing three myosin filaments first described, but with the wrong symmetry, by Huxley and Brown (1967). However, even applying these two rules rigorously, some filament orientations are not defined and there is still sufficient randomness that adjacent unit cells of the superlattice contain different structures.

Unlike all other vertebrate muscles, the muscles of bony fish do not have a superlattice; all the myosin filaments in a particular A-band have identical azimuthal orientations (Fig. 8a; Luther and Squire 1980; Luther et al. 1981). The whole A-band has a regular three-dimensional order. Since the skeletal muscles of higher vertebrates all have the superlattice, one must assume that there is some functional advantage in having disorder rather than order. Likewise one might assume that to bony fish the simple lattice has an advantage.

However, the nature of these different advantages is by no means clear, especially since cartilaginous fish have superlattice A-band structures. Of course, it is possible that both lattices are equally good, and that chance evolution of the simple lattice from the superlattice found in more primitive vertebrates muscles (e.g. hagfish, lampreys; Luther and Squire 1992) represented a neutral change. It is shown later that this change, although possibly not crucial to muscle function, has been of considerable value to muscle researchers involved in structural studies of the contractile mechanism.

It has been shown by Luther and Squire (1980) that different M-bridge interactions can easily account for the differences in the two vertebrate lattice types (Fig. 6f,g). Interactions between non-equivalent M-bridge sites ("unlike" interactions; Fig. 6f) produce a simple lattice A-band, whereas "like" interactions (Fig. 6g), which cannot be systematically sustained across the A-band, will produce the observed superlattices if they are maximised. However, there are other differences between M-bands, especially between M-bands in different fibre types, which seem to be associated with the physiological role of the fibres. As discussed earlier, in longitudinal view (Fig. 7a) there are up to five levels of M-bridge connections, spaced at 220 Å axial intervals. However, sometimes only the central three levels (M4, M1 and M4′) appear dense in electron micrographs of longitudinal sections (three-line M-band), sometimes M1 is missing but M6, M4, M4′ and M6′ are strong (four-line M-band) and sometimes all five levels are strong (five-line M-band). On the whole it has been found in a given animal that the fastest fibres have a three-line M-band, that the slowest fibres have a four-line M-band, and that the five-line M-band represents a halfway house between these two extremes. Reductions are seen in the staining density of lines M6 (M6′) as fibres tend from slow to fast, with corresponding increases in M1 density (Sjöström and Squire 1977a,b; Squire 1986b). The M4 (M4′) bridges are a relatively constant feature of all M-bands (Edman et al. 1988), and may be associated with essential structural proteins involved in defining the A-band lattice. It can be shown that interactions at M4 are sufficient to define the difference between the simple and superlattice A-bands. M-band structure in fish muscles has been determined by three-dimensional reconstruction (Luther et al. 1981; Crowther and Luther 1984; Luther and Crowther 1984; Varriano-Marston et al. 1987). The conclusion that M-band structure and fibre function are related is supported both by the fact that identified M-band proteins include the enzyme creatine kinase (MM-CK; Walliman et al. 1982), together with the structural proteins M-protein and myomesin (Strehler et al. 1983; Grove et al. 1984), and the observation that M-band structure can be changed, along with contractile properties, in cross-innervation experiments between slow and fast muscles (see Squire 1986b).

Vertebrate muscle Z-bands were described briefly in Section 2.2 and illustrated in Fig. 7. The Z-band model of Knappeis and Carlsen (1962; Fig. 7c,d,h), as it was originally drawn, has four-fold rotational symmetry around the actin filament axis and electron microscope evidence for the presence of such symmetry has often been claimed (Reedy 1964; Franzini-Armstrong 1973). However, the actin filaments themselves clearly do not possess four-fold symmetry and only approximately do they have ever a two-fold rotation axis. There is, therefore, an immediate uncertainty about how actin filaments should be placed on the Knappeis and Carlsen structure. Squire (1974) discussed two possible Z-band forms, one in which all the actin filaments on the same side of a given Z-band have identical azimuthal orientations about their long axes and one in which there are 90° rotations between adjacent filaments. Of these, it was argued that the former structure with identical filament azimuths was by far the more plausible, and evidence since then has tended to support this conclusion. Three-D reconstructions from sections of Z-bands in frog sartorius muscle (Lee et al. 1987, 1991) and in bony fish muscle (Luther 1991) and also from cardiac Z-crystals (Morris et al. 1990) have all tended to confirm a single azimuth for actin filaments with the same polarity, as have studies of S-1 labelled rabbit muscle I-bands which show labelled actin filaments with their helix crossover points in axial register (Hirose and Wakabayashi 1988). However, at the resolution of these studies, there is still a 180° ambiguity in actin filament orientation at each lattice point, since at low resolution actin filaments often appear to have a two-fold rotation axis.

One of the striking features of the 3-D reconstructions of frog and fish muscle Z-bands is that the two structures are closely related but different. The essential difference concerns the way in which actin filaments on one side of the Z-band are related to those on the other side. In theory, these two Z-band halves, assumed to be equivalent, could be related either by two-fold rotational symmetry (two-sided plane group C12; this also has 2_1 screw symmetry) or by two-fold screw symmetry alone (two-sided plane group $P12_1$). These lattices are drawn schematically in Fig. 9a,b. The recent 3-D reconstructions show that the C12 lattice occurs in the frog sartorius muscle Z-band and the $P12_1$ structure occurs in the fish muscle Z-band. The nemaline rods (Z-crystals; Stromer et al. 1976) which have been studied by Morris et al. (1990), show a mixture of two-fold rotation and 2_1 screw axes perpendicular to the actin filament axis. At low resolution this gives an apparent space group $P4_22_12$, although due to the lack of true two-fold symmetry in the actin filament, the real symmetry may in fact be $P4_32_12$ (see Fig. 9c).

A key feature of the Knappeis and Carlsen Z-band model is that the Z-bridge connections only occur between anti-parallel actin filaments. However, the frog and fish muscle Z-bands both show the presence of what may be additional bridging material between filaments on the *same* side of the Z-band. In the $P12_1$ structure of fish muscle these form parallel lines of density on one face of the Z-band, and these lines are related by the 2_1 screw symmetry to lines of density running in the *same* direction between the filaments of opposite polarity on the other face of the Z-band (Fig. 9b). On the other hand, the two-fold rotation axis in the plane of the frog muscle Z-band relates parallel tracks on one Z-band face to similar tracks rotated around the muscle long axis by 90° on the other Z-band face. These extra densities are not so immediately apparent in the Z-crystals, although they may still be there. However, it was suggested by Rowe (1973), Yamaguchi et al. (1978), Squire (1981) (see Fig. 7) that single zig-zag Z-bands (c,d), multiple layer Z-bands (e,f) and Z-crystals (g) might be related. The new 3-D reconstructions show that, in the frog and fish muscle Z-bands, the Z-bridge connections (which are probably α-actinin) are in both cases

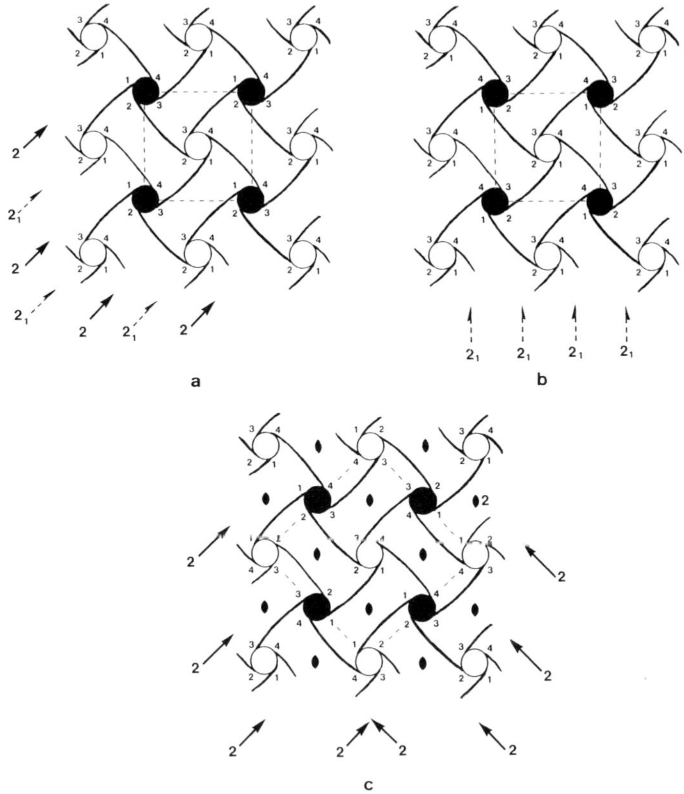

Fig. 9. Illustrations of the symmetries of different Z-band structures. Actin filaments of opposite polarity (from adjacent sarcomeres) are shown as *open* and *filled circles*, bridges between anti-parallel actin filaments are indicated schematically by *curved lines*. The primitive unit cell in each case is indicated by *broken lines*. In each case the four non-equivalent sites of interaction on an individual actin filament (which does not have exact four-fold or even two-fold rotational symmetry) are labelled *1* to *4* and the distributions of these numbers indicate the relationships between adjacent filaments. **a** The symmetry of the frog muscle Z-band with both two-fold rotation and 2_1 screw axes in the plane of the Z-band (Lee et al. 1987, 1991); **b** the $p12_1$ symmetry of the fish white muscle Z-band in which there are two different 2_1 screw axes and no two-fold rotation axes (Luther 1990); **c** one possible structure for nemaline rod Z-crystals in which the true unit cell, *dashed outline*, is a "superlattice" containing two actin filaments of each orientation (Morris et al. 1990). The structure contains three perpendicular two-fold rotation axes, and each actin filament position has 4_3 screw symmetry. When averaged over a simple unit cell as in **a** and **b**, this structure would appear as space group $4_2 2_1 2$ as observed. (Squire et al. 1992)

similar to single zig-zag layers of the Z-crystal structure. Further studies of the relationships between these structures should provide much needed insights into the way that α-actinin molecules and actin filaments interact; insights that will help in understanding membranous actin connections and dense bodies, both of which appear to involve α-actinin.

3.2 Lattice Specialisation in Invertebrate Muscles

The fastest of the invertebrate striated muscles are the asynchronous flight muscles of insects. Here, the working state of the muscles corresponds to a rapid active oscillation of the muscle length at a frequency (the resonant frequency of the muscle/wing/thorax assembly) much greater than the frequency of nervous stimulation. Wing beat frequencies of up of 1000 Hz are found in some cases (e.g. the midge *Formicopymia*). The fibre structures of various types of insect flight muscles (Smith 1961, 1965; Pringle 1972) are illustrated in Fig. 5f–h (see Chapman 1982). Because of the fibrillar appearance of their fibres (Fig. 5h), insect asynchronous flight muscles are often referred to as fibrillar muscles. The abundance of mitochondria between well-separated cylindrical myofibrils in these muscles permits the very high rates of energy consumption involved in sustained flight. The general sarcomere layout of these muscles is as in Fig. 6a, but the A-band length is about 2.2 μm, the actin filament length is about 1 μm, the I-band length is *very* short and the myosin filaments are linked directly to the Z-band by short, elastic C-filaments (White 1967). The myosin filaments are four-stranded (Reedy et al. 1981; Fuller and Squire 1987; Morris et al. 1991), are Ca^{2+} – regulated (Lehman et al. 1973; see Squire 1981, p. 239) and contain different amounts of paramyosin depending on the insect (Levine et al. 1976). Some insect filament backbones appear solid in cross-section and others tubular (Hodge 1955; Auber and Couteaux 1963; Auber 1965; Beinbrech et al. 1988), possibly depending to some extent on the paramyosin content. One of the unique properties of such muscles is that the helix repeat of the myosin cross-bridge array (8 × 145 Å) is exactly three times the crossover repeat of the insect actin filaments (385 Å). This matching of the filament repeats may be linked to the enhanced stretch-activation properties of these muscles (Desherevskii 1971; Pringle 1972; Wray 1979b; see Sect. 4).

One of the most well-studied types of asynchronous flight muscle is that of the giant water bug *Lethocerus*. These muscles have been reported to be very highly ordered and they have therefore been studied extensively by X-ray diffraction (Reedy et al. 1965; Holmes et al. 1980), by mechanical tests (Pringle 1967; Thorsen and White 1969) and using electron microscopy (Ashhurst 1967; Reedy 1968). For many, this muscle type has represented the flagship of insect fight muscle research. But it turns out that in some respects it may not be an ideal muscle and it is certainly not always "typical" of other insect flight muscles. Although in all asynchronous flight muscles the actin and myosin filament repeats, the myosin filament symmetry and the major protein components all seem rather similar, recent studies have confirmed earlier observations that the lattice arrangements of the myosin and actin filaments are very different in different muscles, as are the M- and Z-band structures.

Insect asynchronous flight muscles, which are specialised for speed, contain the kind of A-band filament lattice illustrated in Fig. 6c. Unlike the 2:1 structure in vertebrates, here the actin filaments are located directly between two adjacent myosin filaments, giving a 3:1 filament ratio. Striated muscles other than in vertebrates, in insect fibrillar muscles and in the faster synchronous insect flight

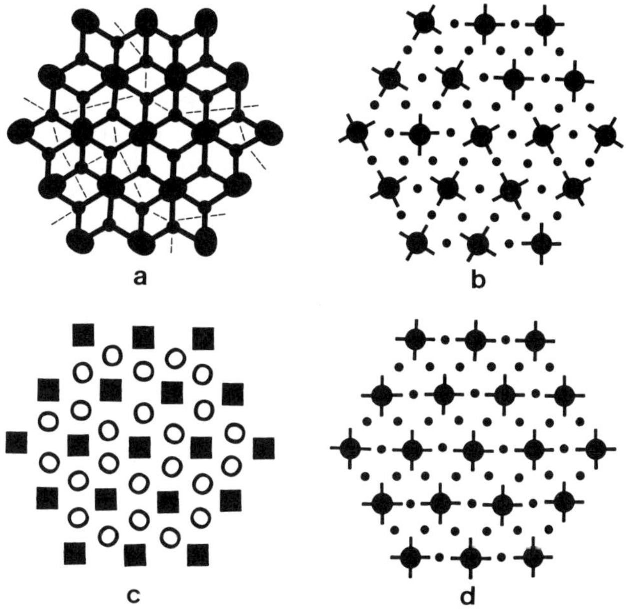

Fig. 10. Filament arrangements in cross-section at the M-bands **a** and **c** and A-band overlap regions **b** and **d** for *Lethocerus* flight muscle (**a,b**) and *Apis* flight muscle (**c,d**). In **a** the myosin filaments have oval cross-sectional profiles and these ovals are oriented at one of three positions, 60° apart. Adjacent myosin filaments are cross-linked by M-bridges which meet at the trigonal points of the hexagonal lattice. The distribution of myosin filament orientations is not totally random but is described by a random-biased structure (Freundlich and Squire 1983). Domains of identically oriented filaments occur, as indicated by the *dashed lines*. In fact, the average domain sizes would be larger than indicated here. **b** The myosin filament arrangement in **a** extended to the cross-bridge region of the A-band where the filaments have four-fold symmetry. One 145 Å spaced level ("crown") of myosin heads is indicated schematically by the lines extending out from the filament backbone, each line representing the location of a myosin head pair. The myosin head lattice is disordered. **c** and **d** Comparable figures to **a** and **b** but for *Apis* muscle. In the M-band the myosin filaments appear square or diamond shaped in profile (Auber 1967), and ring-like structures at the trigonal points of the lattice appear to represent the cross-linking structures in this muscle. All of the square profiles seem to be identically oriented over large regions of a myofibril, and, by extrapolation, in the A-band overlap region the myosin cross-bridges are regularly organised in three dimensions **d**. (Squire et al. 1992b)

muscles tend to have many more actin filaments per myosin filament (see Fig. 6d,e and later discussion).

Figure 10a illustrates the M-band of *Lethocerus* flight muscle in cross-section (Ashhurst 1967; Freundlich and Squire 1983). The myosin filament profiles are oval in shape and these ovals have different relative orientations. In their analysis of this M-band, Freundlich and Squire (1983) identified three different orientations 60° apart, distributed in what they termed a random-biased lattice. The distribution was not quite random; there was a slight preference for adjacent myosin filaments to have the same orientation. This bias meant that the myofibril actually contained domains of identically oriented filaments with a broad range of domain sizes and shapes.

Translating this M-band structure to the cross-bridge region of the A-band, where the myosin filaments have their four-stranded helical arrangement of myosin heads, gives the result in Fig. 10b. There is an irregular cross-bridge distribution across the myofibril. At first sight this appears to contradict the observation of well-sampled myosin layer lines in X-ray diffraction patterns from *Lethocerus* muscle (Reedy et al. 1965) and the apparently beautifully ordered electron microscope images of this muscle in the rigor state (Reedy 1968). However, the well-ordered domains scattered across the myofibril appear to be large enough to account for good sampling of the X-ray diffraction pattern (Squire et al. in preparation). Likewise, the ordered rigor structure may be dominated by the beautifully ordered array of actin filaments in *Lethocerus* (space group $P6_4$; Holmes et al. 1980; Fig. 8c). Unlike vertebrate muscle where the actin filament lattice has to change from square at the Z-band to hexagonal in the overlap region of the A-band, *Lethocerus* actin filaments are anchored into exactly the right hexagonal Z-band structure to allow them to fit directly into the hexagonal myosin filament array. The actin filaments also contain an extra complement of tropomyosin as part of "heavy troponin-I" (troponin-H; Bullard et al. 1988), together with the unusual protein arthrin (Bullard et al. 1985), and these may well confer relatively great rotational rigidity to the actin filaments compared with thin filaments in other muscles. The regular labelling pattern of flared-Xs in rigor *Lethocerus* muscle (Reedy 1968) may well reflect the overwhelming influence of this ordered actin array on the cross-bridge interactions that can occur.

Another insect asynchronous flight muscle that has been studied for many years is that from the bee (*Apis*; Auber 1967; Trombitas and Tigyi-Sebes 1974, 1977; Squire 1992). Recently, a three-dimensional reconstruction of the Z-band of honeybee has been carried out (Cheng and Deatherage 1989; Deatherage et al. 1989). Interestingly, this structure has different symmetry from that in *Lethocerus* muscle. The actin filament arrangement inferred from the bee Z-line is illustrated in Fig. 8d; for comparison the $P6_4$ arrangement in *Lethocerus* is illustrated in Fig. 8c. The M-band appearance in bee muscle is also different from that in *Lethocerus*. The myosin filaments do not appear oval at the M-band but are roughly square in shape and all the square profiles appear to be identically oriented (Fig. 10c; Auber 1967). Between the myosin filaments, where the M-bridges would normally be, there is a ring-like structure which possibly serves the same purpose as the M-bridges, but which may alternatively correspond to membranous extensions of the sarcoplasmic reticulum, as observed in some butterfly muscles (Reger and Cooper 1967). The regular orientation of the myosin filaments at the M-band implies that in the cross-bridge region of the A-band (Fig. 10d) there is intrinsically more regularity than in the *Lethocerus* A-band. In the sense that the A-band contains good three-dimensional order and a true unit cell, it has a similar simplicity of structure to that in fish muscle and it may have similar advantages in ultrastructural studies (Squire et al. 1992a).

A third form of insect flight muscle A-band symmetry is found in the flight muscles of flies. For example, the flesh fly (*Sarcophaga*) probably contains axial shifts between adjacent myosin filaments of multiples of 48A (one-third of the crown spacing of 145 Å; Elliott 1965; Reedy et al. 1973). The details of the Z- and M-band structures in these muscles have yet to be determined, but the actin and myosin filaments appear to have the same axial repeats as in *Lethocerus* and *Apis* (Wray 1979b).

Moving away from the highly specialised and relatively well-ordered asynchronous flight muscles of some insects, other insect muscles, including some of the synchronous flight muscles and all insect leg muscles, which tend to be much slower in contraction speed, have many more actin filaments per myosin filament. Sometimes they have 5:1 or 6:1 filament lattices, as illustrated in Fig. 6d,e, but 4:1 and even 7:1 ratios have also been observed, as have less simple filament ratios (Pringle 1972). The filament ratios in the synchronous flight muscles of insects depend on the wing-beat frequency, with the flight muscles of the cockcroach and locust (Orthoptera), dragon- and damselflies (Odonata) and moths and butterflies (Lepidoptera) having filament ratios ranging between the 3:1 ratio fo the very fast muscles and the 6:1 ratio of the slowest body or visceral muscles (see below). Like the asynchronous flight muscles, the fastest synchronous flight muscles (frequencies more than about 10 Hz and even as high as 200 Hz), such as those in dragonflies (*Aeshna*; Smith 1966), damselflies and some butterflies (*Achalarus lyciades*) tend to have a 3:1 filament ratio (Reger and Cooper 1967). But, for example, the flight muscles of the "painted lady" butterfly (*Vanessa*), which has a low wing-beat frequency, and the cockcroach tergocoxal muscle, both have a 4:1 filament ratio with seven to nine actin filaments surrounding each myosin. The bluebottle (blowfly; *Calliphora*) tergotrochanteral muscle, which is involved both in jumping and in the start of flight, has a 5:1 filament ratio (see Pringle 1972 for detailed references). The tonic muscles involved in the control of flight usually contain tubular fibres (Fig. 5f) with long sarcomeres and long I-bands and the high 5:1 or 6:1 lattices typical of invertebrate slow muscles. Some insect skeletal muscles also show variations in sarcomere and myosin filament lengths. In the case of the flexor (clasper) muscle in the femur of *Lethocerus*, Tregear and Squire (1973) quoted a mean myosin filament length of $3.5\,\mu m$ (5% fiducial limits; 3.1 to $3.8\,\mu m$) and $2\,\mu m$ (5% f.l.; 1.8 to $2.2\,\mu m$) for the actin filaments. The actin to myosin filament ratio was observed to be 5.2:1 (5% f.l.; 4.8 to 5.6).

In many ways striated muscles in other invertebrates (e.g. in crustaceans, molluscs, arachnids) are similar to this with similar filament ratios, and variability in sarcomere length is a common feature (Atwood 1972). For example, Franzini-Armstrong (1970) observed three groups of fibres of average sarcomere length 4, 5 and $7\,\mu m$ in the walking legs of the crab *Portunus depurator*, with variations along individual fibres of 25% and even with variations between the two halves of a given sarcomere. Little or no M-band structure could be seen in such fibres. Unusually, some white fibres of an eyestalk muscle in crab (Hoyle and McNeill 1968) have a 7:1 filament ratio. Other crustacean muscles contain sarcomeres of uniform size, some being relatively long, such as the (slow) tonic abdominal muscles of crayfish and lobsters (s = 6 to 10 μm), and some being relatively short, such as the phasic (fast) abdominal muscles of crayfish and lobsters (s = 2 to 4 μm). Lobster claw-closer muscles have a constant long sarcomere length of 8 μm, whereas the cutter claw muscles have a mixed sarcomere population with lengths of 3 and 9 μm (Jahromi and Atwood 1971). Wray (1979a) has reported the possibility that some crustacean myosin filaments may not have four-stranded helical arrangements of myosin heads, but may be five-stranded. The striated adductor muscle of the scallop (*Pecten*) has a seven-stranded myosin cross-bridge array (Vibert and Craig 1983). All crustacean and insect visceral muscles have short fibres of small diameter. Thick filaments of 150 to 300 Å diameter are surrounded by 10 to 12 actin filaments (5:1 or 6:1 ratio) and the sarcomere lengths are 7 to 8 μm. Often the Z-bands of these muscles are composed of discrete dense bodies and there is little or no obvious M-band.

Very unusual invertebrate sarcomere behaviour is observed in the muscles of the horseshoe crab (*Limulus*), which is more closely related to spiders (Arachnids) than to the crustacean decapods (crabs and lobsters). Uniquely among the various striated muscles (vertebrate and invertebrate) that have been extensively studied, it appears that the myosin filaments in *Limulus* muscles can change their length from about 3.5 to 5 μm depending on the sarcomere length for s values up to about 7.5 μm (Dewey et al. 1973). Above this sarcomere length the thick filament length appears to remain constant, but systematic axial shearing between adjacent myosin filaments allows the A-band length to continue to increase up to above 6.5 μm. The reason for this unusual behaviour appears to be to allow the muscle to generate tension over a very broad range of sarcomere lengths and does not appear to be associated with a force-generating mechanism involving shortening thick filaments. It has been mentioned already that perforated Z-bands (Fig. 1h) in some invertebrate muscles provide an alternative mechanism for extending the physiological range of sarcomere lengths. Yet another mechanism for doing this is found in vertebrate smooth muscles, as described in a later section.

4 3-D Geometry of the Actin-Myosin Interaction in Striated Muscles

From the descriptions above of the known sarcomere ultrastructures of representative vertebrate and invertebrate muscles, it is very clear that the similar myosin and actin molecules in these different muscles are organised in a great variety of ways. It has been shown that some of the observed variations are related to muscle speed (2:1 or 3:1 filament lattices and short sarcomeres for high speeds), but even in cases where the filament ratios and sarcomere lengths are the same, as for example in the simple lattice and superlattice vertebrate structures and in the asynchronous flight muscles of bugs, bees and flies, there are different ways of placing the actin and myosin filaments into these lattices. Why should this be? One intriguing possibility put forward by Desherevskii (1971) and by Wray (1979b) concerns the lattice geometry in insect asynchronous flight muscles. One of the unusual features of these muscles, apart from the precise 3:1 lattice arrangement, is the fact that the axial repeats of the myosin

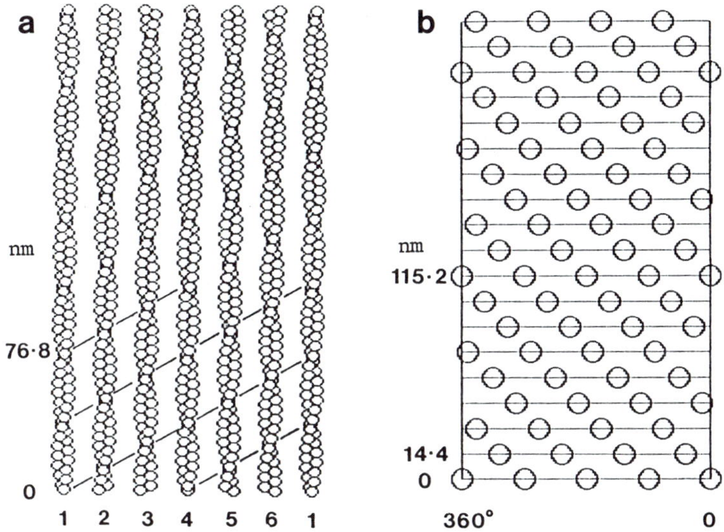

Fig. 11. **a** The lattice of actin filaments and **b** myosin cross-bridge origins seen by an imaginary observer on the axis of the myosin filament labelled O in Fig. 8c for *Lethocerus* flight muscle. (Squire 1992)

and actin filament helices are closely matched. The crossover repeat of the actin filament (385 Å) is exactly one-third of the repeat of the myosin filaments. In synchronous flight muscles and in vertebrate striated muscles the filament repeats are quite different. Desherevskii (1971) and Wray (1979b) attempted to correlate the matching of the filament spacings in *Lethocerus* flight muscle to the unusual oscillatory behaviour of these muscles. The muscles generate a low tension level when activated at a given muscle length, but when stretched and held they produce, after a short delay, a very much higher sustained tension level and a maintained increase in the rate of ATP consumption. This is the so-called stretch-activation phenomenon (Pringle 1967, 1978). Although similar stretch activation behaviour to that in insect muscles can be demonstrated in vertebrate skeletal and heart muscles under appropriate conditions (Armstrong et al. 1966; Ruegg and Schädler 1970; Steiger 1977), the effect is much less pronounced and any enhanced tension production or ATPase rate decays relatively rapidly after the stretch. The ability of many muscle types to display stretch-activation effects could be taken to imply that this phenomenon is an intrinsic property of the actin-myosin system, but Wray has argued that the highly developed stretch-activation property of asychronous flight muscles is related to the matching of the myosin and actin filament helices. However, it has now been shown (Squire 1992) that, although aspects of this general theme may be correct, the specific mechanism proposed by Wray (1979a) is probably not valid.

Figure 11 illustrates the symmetry and repeats of the actin filament array and the myosin filaments found in *Lethocerus* flight muscles. The actin filament array is based on the $P6_4$ symmetry determined from X-ray diffraction work and illustrated in Fig. 8c. If one imagines oneself to be on the axis of the myosin filament labelled O in Fig. 8c looking out at the surrounding actin filament array, what would be seen would be the kind of arrangement shown in Fig. 11a; the actin filaments present a helix of actin filaments (a left-handed 6/2 helix of repeat 768 Å) to the myosin filament. When viewed from the inside, the myosin filament has the origins of its myosin heads arranged as in Fig. 11b. An indication of the way that myosin heads might interact with the actin filament array can be obtained by superimposing the images (a) and (b) in Fig. 11 and estimating how far the myosin heads need to move to reach an actin monomer which is oriented appropriately for attachment to occur. By doing this, Wray was able to show that as the two arrays were moved axially relative to each other, at some positions the myosin heads lay directly over putative attachment sites on actin (match position), whereas with an axial shift of one-half of the common repeat of 385 Å the majority of heads would be well away from these sites (position of mismatch). The match position would return after a further axial shift of $1/2 \times 385$ Å.

This kind of deduction was put onto a more quantitative basis first by Haselgrove and Reedy (1984) and in more detail by Squire (1992). One can set up a computer model including the known filament and lattice geometries of a given muscle, and then use this to define the probability of myosin head attachment to actin under different conditions. In order to do this it is necessary to define both the freedom of movement of a myosin head (defined by $\Delta\theta$ and ΔZ in Fig. 12) and the range of azimuthal orientations that an actin monomer can have relative to the myosin head origin for attachment to occur (defined by ϕ and $\Delta\phi$ in Fig. 12). The total fraction of myosin heads that can attach to actin can then be determined as a function of the relative axial positions of the myosin filament and the actin array. The results from this computation for various conditions are summarised in Fig. 13b for *Lethocerus* muscle. Just as Wray suggested, there is a marked variation in attachment probability as the filament overlap changes; it oscillates with a period of 385 Å. The lower curve in Fig. 13b shows that under appropriate conditions it might be possible to go from a situation with no myosin head interaction at all to one with 25% attachment and substantial force production. This looks like a good stretch-activating system. Unfortunately, when further details of the A-band structure in *Lethocerus* are considered, it turns out that the Wray mechanism does not appear to work. As discussed earlier, the *Lethocerus* M-band, and by inference the whole A-band, contains myosin filaments with three different orientations 60° apart (Fig. 10a,b). However, the actin filament array appears to be regular right across the myofibril. Rotation of the myosin filament at O in Fig. 8c by 60° has a marked effect on the myosin head labelling pattern. A myosin head on the bottom left-hand origin in the lattice of Fig. 11b when superimposed on the actin array in (a) would lie opposite the crossover point at the bottom end of filament 1. Rotation of the myosin filament by 60° would bring this same head origin to just opposite the bottom end of actin filament 2. In order to make the same interaction with actin, the myosin filament would have to move axially by 768/6 Å to come opposite the first crossover on filament 2. Extending this to all of the myosin heads shows that exactly the same labelling patterns would occur as for the

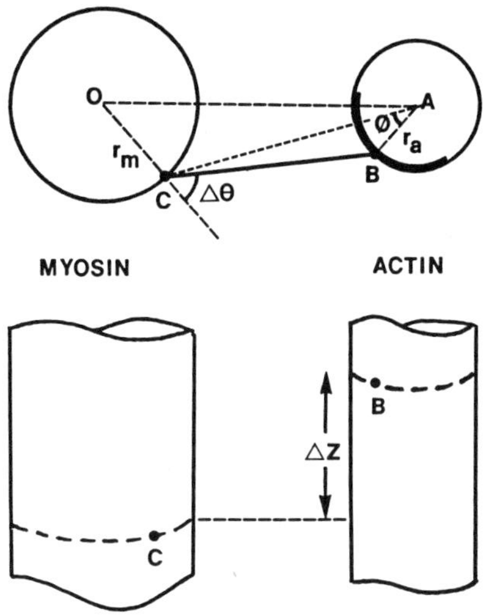

Fig. 12. Parameters describing the range of movement of a myosin head ($\Delta\theta$ and ΔZ) and the azimuthal location (ϕ) of the centre (B) of an actin target area. The *dark arc* around B shows the aximuthal extent (+/− $\Delta\phi$) of the chosen target area. Other chosen parameters: myosin backbone radius (r_m), 7 nm for vertebrate muscles, 9 nm for insect muscles; actin outer radius (r_a), 5 nm; filament separation OA, 27 nm for vertebrate muscle, 30 nm for insect muscles. A myosin head from C is assumed to bind at B, or to any actin site within the azimuthal range +/− $\Delta\phi$ of B, provided the $\Delta\theta$ and ΔZ values needed to reach this position are less than or equal to the prescribed maximum values. (Squire 1992)

unrotated filament except that the whole match/mismatch pattern would be shifted axially by 768/6 Å. By similar reasoning the filament rotated by 2 × 60° would make the same interactions when shifted axially by 768/3 Å. The net effect of all three filament orientations is illustrated in Fig. 13c; for each filament orientation there is an oscillatory labelling probability, but the three curves are shifted axially by multiples of 768/3 Å. Adding the three responses together gives a total labelling probability that is flat.

In Fig. 8d it was shown that the filament arrangement in the 3:1 lattice of bee asychronous flight muscle was different from that in *Lethocerus*. This time, the observer on the axis of the filament labelled O in Fig. 8d, on looking radially outwards, would see the same kind of myosin head array on a single filament as in *Lethocerus* but a different actin filament array; the actin filaments all present the same face to the observer (Fig. 14b). When the interaction probability is assessed in the same way as for *Lethocerus*, the result in Fig. 13d is obtained; there is a very slight variation of labelling with overlap, but this is so slight that it is probably unimportant; it appears insignificant compared with the oscillation in Fig. 13b. Since the bee myofibril appears to contain a single myosin filament orientation over a large region, and since the actin filament array appears to be regular, Fig. 13d represents the total sarcomere response for bee muscle; like the total response for *Lethocerus*, it is essentially flat.

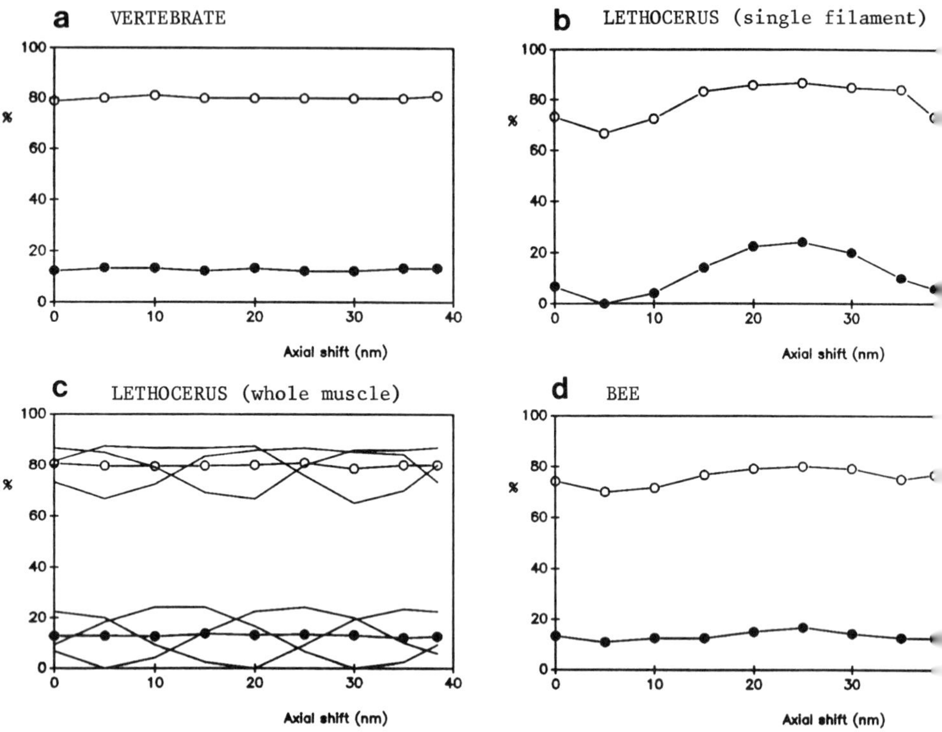

Fig. 13. Maximum labelling of actin target areas by myosin heads as a function of axial displacement between actin and myosin filaments for: **a** vertebrate muscles, **b** single *Lethocerus* myosin filaments, **c** three *Lethocerus* myosin filaments with 60° rotations (*solid lines*, individual filaments; *solid lines with circles*, total A-band interaction), **d** single *Apis* myosin filaments. Throughout: Upper traces (*open circles*) for $\phi = 270°$, $\Delta\Theta = \Delta\phi = 60°$, $\Delta Z = 6\,\text{nm}$; lower traces (*filled circles*) for $\phi = 270°$, $\Delta\theta = \Delta\phi = 30°$, $\Delta Z = 3\,\text{nm}$. The labelling is relatively insensitive to filament separation (about a 1% change in labelling for 1 nm change in spacing). Changing azimuth (ϕ) of the target area shifts the whole distribution axially; changing $\Delta\theta$, $\Delta\phi$ and ΔZ by the same proportion shifts the labelling curves vertically, with little change in shape, between the illustrated extremes. (Squire 1992)

Turning now to vertebrate muscles, with lattices such as those in Fig. 8a,b, this time the observer on the axis of the myosin filaments labelled O in either case would see the actin array shown in Fig. 14d. As in the case of *Lethocerus*, the actin array is helical around the myosin filament, but this time it is a right-handed 6/2 helix of repeat about 720 Å. The head arrays on the myosin filaments are three-stranded with a repeat of 429 Å and subunit repeat of 143 Å (Fig. 14d). The computed labelling pattern, using the same assumptions as for *Lethocerus* or *Apis*, is shown in Fig. 13a; it applies to both simple lattice and superlattice vertebrate muscles. As for the insect muscles, there is little variation in labelling probability with changing filament overlap. Of course, if the total actin and myosin filament overlap changes as the sarcomere length increases, then the total active tension that can be produced will gradually decrease. But the analysis in Fig. 13 shows that for each muscle type there are no fluctuations in

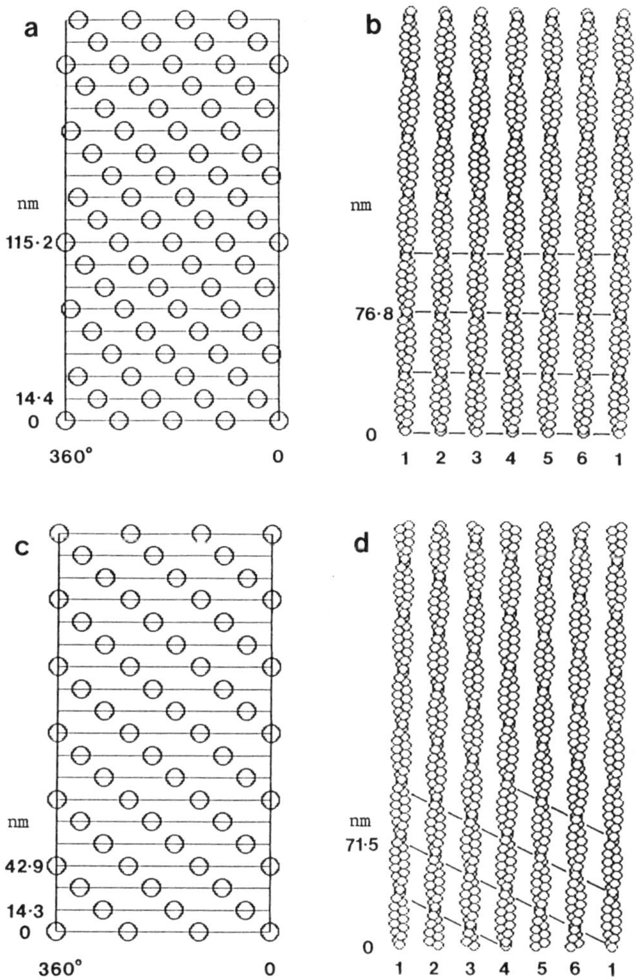

Fig. 14. Myosin head lattices **a** and **c** and actin filament lattices **b** and **d** similar to those in Fig. 11 but this time for *Apis* flight muscle (**a,b**) and vertebrate striated muscle (**c,d**). All are imagined to be viewed by an observer on the axis of the appropriate myosin filaments labelled O in Fig. 8. (Squire 1992)

labelling probability with changing filament position if the total filament overlap is constant, and even on the descending part of the length tension curve (e.g. between sarcomere lengths of 2.2 and 3.6 μm in vertebrate muscles; Gordon et al. 1966) there will be a steady reduction of tension without any superimposed fluctuations due to matching and mismatching of the myosin and actin filament arrays. Since the muscles mentioned all seem to have their different filament arrays organised in such a way as to maintain this smoothly varying labelling probability, one might well conclude that muscle likes to work this way; a conclusion reinforced by the fact that the axial staggers between myosin filaments in fly flight muscles would also produce a smooth labelling probability, as would the 4:1, 5:1 or 6:1 filament lattices in slow invertebrate muscles, where there are so many actin filaments available that there would be many appropriately oriented actin monomers available to each myosin head at any relative filament overlap.

The conclusion from this section with regard to the published "Wray model" of stretch activation must be that it does not work. However, stretch activation does occur and it is most highly developed in the asynchronous insect flight muscles. A marked oscillation in the mechanical properties of these muscles as a function of sarcomere length, and repeating about every 385 Å, has been observed (Abbott and Cage 1984). It could well be that Desherevskii and Wray were right to draw attention to the fact that it is only in this muscle type that the axial periods of the actin and myosin filaments are matched. If it is not match and mismatch of the helix of myosin heads with the actin array that causes stretch activation, perhaps it is related to the fact that with an actin filament pitch of 768 Å, which is identical to the troponin repeat along the actin filaments (Ebashi et al. 1969), the troponin molecules all along the actin filament will all have exactly the same azimuthal positions around the actin filament axis; their positions relative to an adjacent myosin filament will be identical for each repeat of the actin filament. Add to this the fact that there is the unusually large regulatory protein troponin-H in *Lethocerus*, being roughly equivalent to conventional troponin-I with added tropomyosin, and the fact that these muscles contain the unusual actin-like protein arthrin (actin with ubiquitin; Bullard et al. 1985), then a specialised thin filament regulatory system tailored to the matched repeats of the actin and myosin filaments seems to be very plausible.

5 Structure and Function of Smooth Muscles

5.1 Molluscan Smooth Muscles

A very common form of muscle in the lower invertebrates (e.g. jellyfish, cestodes, nematodes, sea cucumber, molluscs etc.) is smooth muscle, the most thoroughly studied examples being the molluscan smooth muscles. Extensive studies of smooth muscle ultrastructure have been carried out particularly on the adductor muscles of certain bivalves, such as mussels (*Mytilus*), oysters (*Ostrea*) and scallops (*Pecten*). The adductor muscles hold the two shells closed when necessary and can do so for many hours with little expenditure of energy. This remarkable example of muscle tenacity and endurance with great economy of effort is a unique property of some invertebrate smooth muscles and is known as the "catch" phenomenon (see review by Rüegg 1971). Smooth muscles in invertebrates apart from molluscs, for example in tapeworms and flukes, also display catch. Note that molluscan muscles are not all smooth; the main part of the adductor muscle in scallops is striated and is used for rapid opening and closing of the two shells to provide a jet-propelled "swimming" action. Some other molluscan muscles are obliquely-striated, with a similar structure to the body wall muscles in annelids and nematodes (Rosenbluth 1965, 1968).

All of these invertebrate muscles, whether smooth, striated or obliquely striated, have paramyosin in their myosin filaments. However, there is a large preponderance of paramyosin over myosin in the invertebrate smooth muscle paramyosin filaments and these filaments can be extremely large; for example they are up to 2000 Å in diameter and 25 μm in length in the anterior byssus retractor muscle (ABRM) of *Mytilus*. A detailed ultrastructural study of the ABRM by Sobiescek (1973) revealed what is very probably a general structural theme throughout this muscle type. Within the muscle cells, which are about 1.6 mm long and about 4 to 4.5 μm in diameter, the very large paramyosin filaments interact with pools of actin filaments, which are also very long (about 11 μm), and which form reasonably well-ordered quasi-hexagonal arrays known as "actin lattices". These typically have an inter-actin filament spacing of about 120 to 140 Å (Elliott and Lowy 1961; Lowy and Vibert 1972; Millman and Elliott 1973; Svendsen 1982). A bundle of parallel actin filaments is anchored at one end by an amorphous-looking dense body (Fig. 15a) which links at its other end to a second array of actin filaments of opposite polarity. One or two paramyosin filaments, together with two anti-parallel actin arrays, form rudimentary sarcomeres in the muscle, and such sarcomeres in series along the muscle produce rudimentary myofibrils or "contractile units". The muscles do not appear cross-striated because the sarcomeres have varying lengths and because sarcomeres in adjacent contractile units are not axially aligned.

There is no reason to doubt that, with the notable exception of the catch phenomenon, the contractile behaviour of these muscles is like that in the striated muscles, with a similar cross-bridge cycle on actin. Although the catch mechanism is not yet thoroughly understood, there is evidence that it is paramyosin-mediated (Szent-Györgyi et al. 1971) and it has been suggested that it may simply correspond to myosin cross-bridges becoming trapped in an extremely slow step in the cross-bridge cycle (Lowy and Millman 1963; Parry and Squire 1973; Achazi 1982; Cohen 1982). Unlike rigor, catch is a state that exists in the presence of ATP and that can be released by the action of 5-HT (5-hydroxytryptamine) on the muscle. Within the muscle cells, cyclic AMP leads to relaxation of the catch state. Early suggestions that this state may correspond to relatively long-lived interactions directly between adjacent paramyosin filaments (Rüegg 1958; Johnson 1962) have been largely discounted by the recent work of Bennett and Elliott (1989) who studied electron micrographs of rapidly frozen, freeze-substituted ABRM in the resting and phasically contracting states, as well as in rigor and catch. No evidence for fusing of the paramyosin filaments was seen. These authors also saw a certain amount of rearrangement of the actin filaments between states, from predominantly actin lattices in resting muscle, to predominantly rosettes plus a small amount of actin lattice in rigor muscle. Active and catch muscles showed intermediate distributions, with partial rosettes but still many actin filaments well away from paramyosin filaments.

Recently, it has been suggested that catch results from the altered binding between the head end of the myosin rod (myosin subfragment 2) and the underlying paramyosin molecules, depending on the state of phosphorylation of both molecules (Achazi 1979, 1982; Cohen 1982). This suggestion is supported by recent evidence that catch may involve a dephosphorylated state of myosin (Castellani and Cohen 1987a,b; Castellani et al. 1988).

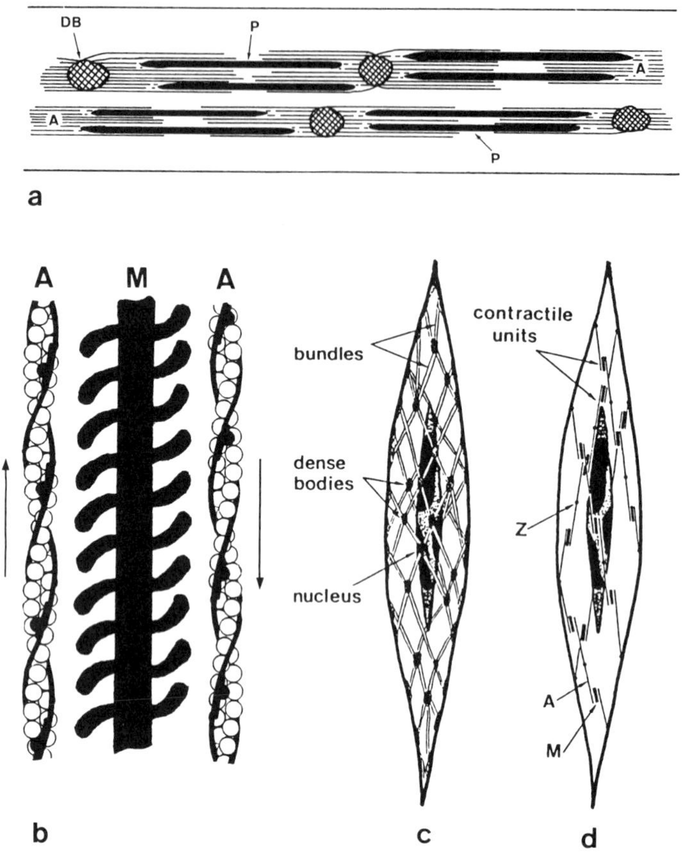

Fig. 15. **a** Illustration of the rudimentary contractile units described by Sobiesick (1973) for a molluscan smooth muscle (ABRM) (Squire 1986b). *A* Actin filament bundle; *P* paramyosin filament; *DB* dense body. **b** to **d** Illustrations of various features of vertebrate smooth muscles. In **b** the interaction between side polar or face polar myosin filaments (*M*) with oppositely directed actin filaments is indicated (after Small and Squire 1972). **c** Schematic illustration of the probable arrangement of the 100 Å filaments (intermediate filaments) and dense bodies in vertebrate smooth muscle cells to form an internal cytoskeleton. Within this network the contractile units in **a** and **b** would be located in a kind of spiral array around the nucleus, as in **d**. Part **c** is after Small (1977); **d** is after Small and Squire (1972). *A* Actin; *M* myosin; *Z* the location of the polarity change in the thin filament assembly, which may possibly coincide with the dense bodies in **c**

5.2 Vertebrate Smooth Muscles

The muscles in the walls of the blood vessels and various hollow organs (apart from the heart) in vertebrates all tend to be smooth muscles, with spindle-shaped cells as illustrated in Fig. 2c. Such cells have parallels with a number of other muscle types (Small and Squire 1972; Small 1977, 1987 Somlyo 1980;). Like cardiac muscles, there are electrical connections between cells, but conduction is less efficient than in cardiac muscles, so that waves of contraction pass along the relevant organ. Like molluscan smooth muscles, vertebrate smooth muscles contain actin lattices, dense bodies and myosin filaments organised into rudimentary contractile units. As in all muscles, they contain an extensive cytoskeleton comprising intermediate filaments. Finally, as in *Limulus* and some other arthropods, these muscles are designed to operate over a very large range of muscle lengths. The mechanism devised for this is totally different either from the filament shortening found in *Limulus* or from the Z-band perforation that occurs in other muscles. Small and Squire (1972) demonstrated an unusual form of myosin aggregate in the guinea pig taenia coli muscle. Here, instead of the myosin filaments being bipolar (Fig. 4d), with oppositely directed myosin molecules at opposite ends of the myosin filament, vertebrate smooth muscle myosin filaments can have what was originally called "face polarity", with oppositely directed myosin molecules on opposite sides of a myosin filament (Fig. 15b). The filaments seen by Small and Squire (1972), which clearly showed the 143 Å myosin spacing, were flat, about 100 to 150 Å thick but a few hundred Å wide. These may well have been artifactually wide, possibly being aggregates of smaller myosin filaments, and for this reason the idea of face polarity lost favour. Indeed, it was later reported that some vertebrate smooth muscles contain bipolar filaments (Somlyo et al. 1981). However, more recently, various published studies of vertebrate smooth muscle myosin aggregates and isolated myosin filaments have generally supported the unusual face-polar organisation of the myosin molecules (Craig and Megerman 1977; Trybus and Lowey 1987; Cooke et al. 1989). Although the term side-polar has been used instead of face-polar for these filaments, the meaning is the same. It is this unusual filament structure that leads to the ability of these muscles to produce significant tensions over a wide range of sarcomere lengths (Small and Squire 1972).

Studies of the locations of intermediate filaments (see Small 1987), dense bodies (Bond and Somlyo 1982), α-actinin (Fay et al. 1983) and the myosin and actin filaments (Small and Squire 1972; Small 1977), in various vertebrate smooth muscles, have shown that the intermediate filaments interconnect the dense bodies, which are dispersed through the cell and which are also located as "plaques" at the cell surface. The dense bodies, in turn, may act as actin-anchorage sites and sites of polarity reversal for the actin arrays (Bond and Somlyo 1982), although there has been some disagreement on this. Just as in molluscan smooth muscles, the actin and myosin filaments form rudimentary contractile units with actin filaments of opposite polarity interacting with opposite faces (or sides) of the myosin filaments. The resulting organisation is illustrated in Fig. 15b. It is easy to show (Small and Squire 1972) that long actin filaments and side (face) polar myosin filaments, permit a range of working lengths (at which 90% of maximum tension is achieved) of at least the observed value of 0.5 to 0.65 L_0, where L_0 is the muscle length at which maximum tension is developed (Lowy and Mulvany 1973). This compares with about 0.28 L_0 for vertebrate striated muscle (e.g. frog semitendinosus muscle). The total ratio of maximum to minimum muscle lengths of about 5:1 is also well modelled by this scheme (see Squire 1986b).

Note finally that results from other muscle types (e.g. vertebrate striated muscles) ought to provide a caution about pooling results from different vertebrate smooth muscles. The conclusions sketched briefly above are based on studies of a variety of different smooth muscles, but they are discussed as though all such muscles are alike. It could well be that myosin filaments in some smooth muscles are face (side) polar, whereas in others they are bipolar, and that some dense bodies link intermediate filaments while others serve as sites of polarity reversal for actin filaments. Like other muscles, each particular muscle should be studied in depth and results between muscles related with care.

6 Molecular Movements Involved in Contraction

Previous sections have shown that there is a great diversity of muscle structure across the animal kingdom, but in each case the molecular organisation is designed to take the same basic muscle motor, the myosin cross-bridge cycle on actin, and to harness this mechanism in an appropriate way according to the needs of the animal. However, the cross-bridge mechanism itself is only poorly understood and a fundamentally important task of muscle researchers is to define the molecular events which actually give rise to force production and directed movement. In recent years great advances have been made in the understanding both of the biochemistry of the actomyosin ATPase (Trentham et al. 1976; Brenner 1990) and in the transient mechanical properties of muscle (Huxley 1973; Huxley and Simmons 1973; Ford et al. 1977, 1981, 1985, 1986; Irving 1987; Brenner 1990). Unfortunately, until recently, knowledge of the structural events involved in contraction has advanced more slowly. This is not because of the lack of suitable techniques; one of the most effective structural methods, that of time-resolved X-ray diffraction from contracting muscle, was used successfully about 10 years ago to show molecular movements occurring in muscle on a millisecond time scale (Huxley et al. 1980, 1981, 1983; Kress et al. 1986). Although this technique has gradually developed over recent years, all of the groundwork for these experiments was carried out in the 1970s by H.E. Huxley, J. Haselgrove, K.C. Holmes, G. Rosenbaum, J. Bordas and their collaborators (Rosenbaum et al. 1971; Haselgrove et al. 1977; Bordas et al. 1980; Rosenbaum and Holmes 1980). Nevertheless, the enormous technical advances that this work represented still failed to yield definitive results about the cross-bridge mechanism. In particular, it could not be demonstrated conclusively whether or not myosin heads "swing" on actin to produce movement (Fig. 4e), as proposed in the historic paper by H.E. Huxley in 1969. Although the 20 years since then have provided many circumstantial indications that the mechanism may be correct, and nothing to date has disproved it, until recently nothing has confirmed it. One of the reasons for this shows up a clear application of the kind of comparative study of muscle structure that this chapter represents. All of the early time-resolved diffraction studies of vertebrate muscle made use of frog sartorius and semitendinosus muscles. At the time it was not known that these muscles, in common with most other vertebrates, have a disordered A-band superlattice structure (Fig. 8a). Such a structure makes X-ray diffraction studies problematical; X-ray diffraction patterns are incompletely "sampled" and thus are difficult to interpret (Squire 1981). When it was dis-

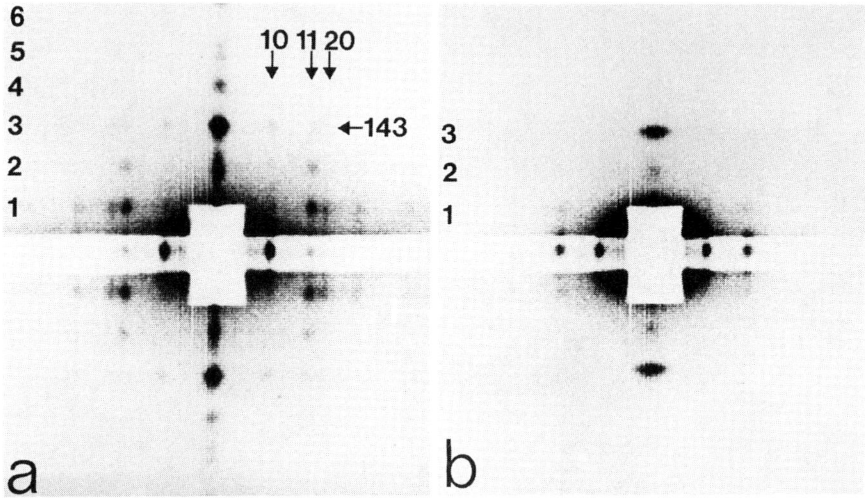

Fig. 16. Low angle X-ray diffraction patterns from turbot fin muscle (axis vertical) in **a** the resting state and **b** the fully contracting state (tetanus plateau). Layer lines due to the helical array of myosin heads around the myosin filaments (Fig. 4b) are labelled ($l =$) 1 to 6 in **a** and 1 to 3 in **b**; they are orders of a repeat of 429 Å, with the 3rd order ($l = 3$) at 143 Å^{-1} being arrowed. These layer lines are sampled at row lines due to the hexagonal lattice of filaments in the A-band. The 10, 11 and 20 row lines are indicated; they have strong features on many layer lines, and especially the equator where the 100 and 110 reflections are clear. They change considerably in intensity between the two patterns. Patterns recorded on a 3-m camera on line 2.1 at the Daresbury synchrotron. (Harford and Squire 1990)

covered by Luther and Squire (1980) and Harford and Squire (1986) that bony fish muscles have a simple lattice structure, it was realised that much better use of the time-resolved diffraction technique could be made. The diffraction patterns from these muscles are beautifully sampled (Fig. 16a); their interpretation is intrinsically easier (Harford and Squire 1986), and time-resolved diffraction studies are more informative (Harford et al. 1988; Harford and Squire 1990a,b,c). This is a clear example of the need to select an experimental preparation that is matched to the particular experiments involved. It is not appropriate to go into details of these results here, especially since they have been reviewed elsewhere (Harford and Squire 1990a), but the main conclusions are summarised below.

Figure 16a shows evidence of well-sampled (horizontal) layer-lines ($l = 1, 2, 3, 4$, etc.) which provide information about the organised helix of myosin heads on the myosin filament surface in resting muscle (Figs. 4a and 14c). It can be concluded from the intensity distribution on these layer lines that the myosin heads, which are relatively long, narrow structures, about 190Å long × 50Å across, are tilted axially so that they make an angle of about 40°–60° to a perpendicular from the filament surface. They may also be slightly slewed, so that they do not project radially outwards when viewed down the filament axis (Fig. 18d; Harford and Squire 1986). Diffraction patterns from rigor muscles,

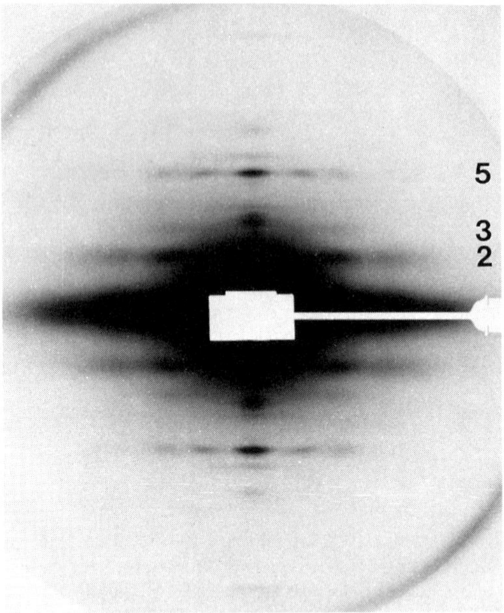

Fig. 17. X-ray diffraction patterns similar to those in Fig. 16 but for turbot fin muscle in the rigor state. Actin layer line orders of 2 × 360 Å are indicated as *2* (360 Å), *3* (240 Å) and *5* (144 Å). (Harford, Chew and Squire unpubli. data)

where most if not all of the myosin heads are attached to actin filaments, are dominated by the layer lines (orders of about 360 to 370 Å) arising from the actin filaments (Fig. 17). These layer lines in patterns from rigor muscle are much more intense than their counterparts in patterns from resting muscle, showing that extra material, presumably the myosin heads, is labelling the actin filaments in a regular way. In diffraction patterns from contracting muscles (Fig. 16b), a weaker enhancement of the actin layer lines occurs as well as a considerable intensity reduction of the myosin layer lines.

Movements of myosin heads from the helix on myosin in resting muscle, to their attachment and annealing into their preferred configuration on actin in rigor or during contraction, can be monitored using the equator (zero layer line; $l = 0$) of the diffraction pattern. This provides information about the whole A-band unit cell when viewed in projection down the filament axis. The equatorial patterns in Fig. 16a,b are very different. Using phasing of the 10, 11, 20, 21 and 30 equatorial reflections of 0, 0, π, 0, 0 (as justified elsewhere; Harford and Squire 1990), it is possible to reconstruct "images" of the muscle based on the different observed equatorial intensity distributions. Such images are shown in Fig. 18a–c for resting, fully active and rigor fish muscles. A gradual shift of mass from an annulus around the myosin filament backbone, to the general vicinity of the actin filaments, can be seen as more and more myosin heads become actin attached in a well-defined configuration. All of these various

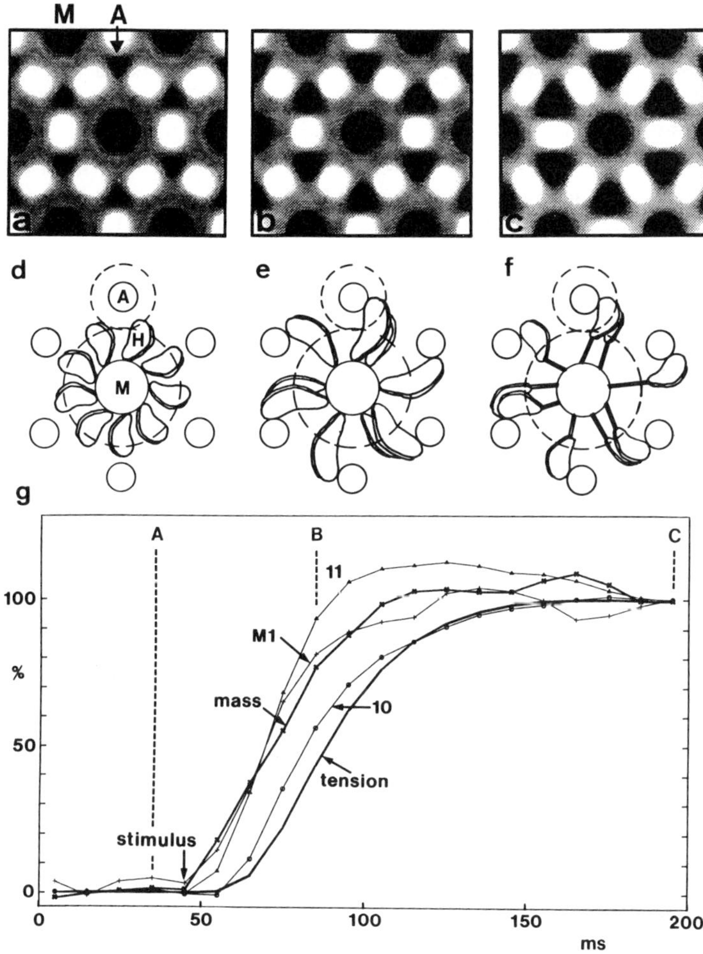

Fig. 18. Electron density distribution in the A-band of turbot fin muscle when viewed down the mascle long axis for **a** resting muscle, **b** fully active muscle and **c** rigor muscle. The maps were obtained by Fourier synthesis from the 100, 110, 200, 210 and 300 equatorial reflections in patterns such as those in Figs. 16 and 17. Observed amplitudes were combined with phases 0°, 0°, 180°, 0° and 0° to generate the maps, and an artifical temperature factor was applied to reduce series termination artifacts (see Harford and Squire 1990a,b). **d** to **f** Model interpretations (cf. similar diagram in Yu and Brenner 1989) of the myosin head (*H*) distributions in **a** resting muscle, **b** the initially (weakly) attached state during contraction (see Fig. 4e state 1), and **f** the end of the cross-bridge stroke (strongly attached, possibly rigor-like). In **a** the myosin heads are centred on the myosin filament (*M*) and form the myosin helix shown in Fig. 4b. In **e** the myosin heads are transiently attached to actin (*A*), but in a non-stereospecific manner, and their neck ends (close to the myosin rod) are still on the original myosin helix. They would give rise to a weak myosin layer line pattern with an altered intensity distribution along it. In **f**, the heads are stereospecifically bound to actin and are tilted such that their neck ends have moved away from the myosin filament surface. The *dashed circles* around the actin filaments show the region in which "actin mass" was counted for the mass plot in **g** below. **g** Time courses of various X-ray reflections, the muscle tension and the "actin mass' (see above) during the rising phase of a tetanus in turbot fin muscle. Time courses for the *10* and *11* equatorial reflections and the 11/20 combination on the first myosin layer line (*M1*; 429 Å, $l = 1$ in Fig. 16) are shown. The 10 and 11 reflections have quite different time courses, and the mass time course leads the tension time course at the level of 50% change, by about 15 to 20 ms. Position *A* is resting muscle (cf. **a** above), position *C* is fully active (cf. **b** above) and position *B* is a situation where the mass labelling actin is about the same as at *C* but the 10 and 11 reflections have quite different relative intensities. (Harford and Squire 1990)

results are consistent with the idea that myosin heads start in their resting helix on myosin and that when the muscle is activated they move off this helix (loss of myosin layer line intensity) and attach to actin (equatorial changes, actin layer line increases). However, the important feature to demonstrate in order to confirm the swinging cross-bridge model is that there are two *different*, *attached* states of myosin on actin during contraction. Early indications that this is so are being obtained in the current time-resolved diffraction studies of fish muscle (Harford et al. 1988; Harford and Squire 1990; 1992a,b). These indications are as follows:

1. Even in patterns from active muscle a remnant of the myosin layer line pattern is left with an altered intensity distribution along it, indicative of a new structure.
2. Stereospecific actin labelling is indicated by small increases of the 370 Å, 59 Å and 51 Å actin layer lines, apart from a large increase on the second actin layer line at 185 Å thought to be associated with tropomyosin movement and actin filament regulation (Haselgrove 1973; Huxley 1973; Parry and Squire 1973).
3. Intensity changes of the 11 equatorial reflections occur well ahead of tension changes during the rising phase of tetanic contractions (Fig. 18g), and the 10 and 11 reflections change relative intensity with different time courses.

Taken together, these results imply that two structurally different states of myosin heads exist during the contractile cycle (Harford and Squire 1990; 1992a,b). Images produced from the observed intensities suggest that one state (predominant at the early phase of the tetanus) has heads which have not moved a great deal radially from the myosin filament, but which are probably actin attached, whereas the second state could be much more like rigor, with the myosin head mass much more closely located around the actin filaments. Since the "new" myosin layer lines that occur in patterns from contracting muscle are consistent with the persistence of a population of the earlier (non-specific binding) state during steady contraction (Fig. 18e) and the intensification of the actin layer lines is consistent with the presence of a more stereospecific attached state of heads on actin (e.g. such as in rigor; Fig. 18f), the different radial positions of the myosin heads relative to the actin filament axis in these two states suggest that an axial component of head movement (swinging) really is involved. This intriguing possibility requires further analysis of this and other related data (Lowy and Poulsen 1987; Harford and Squire 1990; 1992a,b), but the swinging cross-bridge model has much to commend it.

Acknowledgments. The author is greatly indebted to his many colleagues in the Muscle Group at Imperial College (Drs. Michael Chew, Jeff Harford, Pradeep Luther, Ed Morris and Helen Pask, Mrs. Gwen Nneji, Mrs. Beatriz Marquez and Mr. John Barry), not only for providing some of the figures included here, but also for their helpful comments on the text and for the pleasure of daily interaction with them. In addition, the help of Mrs. Sarah Webb in preparing the typescript is greatly appreciated. Various aspects of the work of the Muscle Group have been supported in recent years by grants from the Medical Research Council, the Leverhulme Trust, the Wellcome Trust, the British Heart Foundation, the Muscular Dystrophy Association and the Agriculture and Food Research Council.

References

Abbott RH, Cage PE (1984) A possible mechanism of length activation in insect fibrillar flight Muscle. J Muscle Res Cell Motil 5:387–397

Achazi RK (1979) Phosphorylation of molluscan paramyosin. Pfluegers Arch Gesamte Physiol 379:197–201

Achazi RK (1982) Catch muscle. In: Twarog BM, Levine RJC, Dewey MM (eds) Basic biology of muscles. Raven, New York, pp 291–308

Armstrong CF, Huxley AF, Julian FJ (1966) Oscillatory responses in frog skeletal muscle fibres. J Physiol (Lond) 186:26P–27P

Ashhurst DE (1967) The fibrillar flight muscles of giant water-bugs: an electron-microscope study. J Cell Sci 2:435–444

Atwood HL (1972) Crustacean muscle. In: Bourne SH (ed) The structure and function of muscle, 2nd edn. Academic Press, New York, pp 421–489

Auber J (1965) 1. Sur le mécanisme de la formation des filaments primaires au cours de l'a ccroissement des myo fibrilles chez *Calliphora erythrocephala*. CR Acad Sci Ser D 260:668–670

Auber J (1967) Distribution of the two kinds of myofilaments in insect muscle. Am Zool 7:451–456

Auber J, Couteaux R (1963) Ultrastructure de la strie Z dans des muscles de diptères. J Microsc (Paris) 2:309–324

Beckett EB, Bourne GH (1972) Histochemistry of skeletal muscle and changes in some muscle diseases. In: Bourne GH (ed) The structure and function of muscle, 2nd edn. Academic Press, New York, pp 289–358

Beinbrech G, Ashton FT, Pepe FA (1988) Invertebrate myosin filament: subfilament arrangement in the wall of tubular filaments of insect flight muscles. J Mol Biol 201:557–565

Bennett PM, Elliott A (1989) The "catch" mechanism in molluscan muscle: an electron microscopy study of freeze-substituted anterior byssus retractor muscle of *Mytilus edulis*. J Muscle Res Cell Motil 10:297–311

Black RM (1970) The elements of palaeontology. Cambridge University Press, Cambridge

Blanchard A, Ohanian V, Critchley D (1989) The structure and function of α-actinin. J Muscle Res Cell Motil 10:280–289

Bond M, Somlyo AV (1982) Dense bodies and actin polarity in vertebrate smooth muscle. J Cell Biol 95:403–413

Bordas J, Koch MHJ, Clout PN, Dorrington E, Boulin C, Gabriel A (1980) A synchrotron radiation camera and data acquisition system for time resolved X-ray scattering studies. J Phys E:Sci Instrum 13:938–944

Brenner B (1990) Muscle mechanics and biochemical kinetics. In: Squire JM (ed) Molecular mechanisms in muscular contraction. Topics in molecular and structural biology 13, MacMillan, London, pp 77–149

Brenner B , Squire JM (1987) Rapid stiffness of single relaxed skinned fish muscle fibres: no detectable crossbridge attachment at low ionic stength. J Muscle Res Cell Motil 8:66–67

Brooke MH, Kaiser KK (1970) Muscle fiber types: how many and what kind? Arch Neurol 23:369–379

Bullard B, Bell J, Craig R, Leonard K (1985) Arthrin: a new actin-like protein in insect flight muscle. J Mol Biol 182:443–454

Bullard B, Leonard K, Larkins A, Butcher G, Karlik C, Fyrberg E (1988) Troponin of asynchronous flight muscle. J Mol Biol 204:621–637

Castellani L, Cohen C (1987a) Myosin rod phosphorylation and the catch state of molluscan muscles. Science 235:334–337

Castellani L, Cohen C (1987b) Rod phosphorylation favours folding in a catch muscle myosin. Proc Natl Acad Sci USA 84:4058–4062

Castellani L, Elliott BW Jr, Cohen C (1988) Phosphorylatable serine residues are located in a non-helical tailpiece of a catch muscle myosin. J Muscle Res Cell Motil 9:533–540

Chapman RF (1982) The insects, 3rd edn. Hodder and Stoughton, London

Cheng N, Deatherage JF (1989) Three-dimensional reconstruction of the Z-disk of sectioned bee flight muscle. J Cell Biol 108:1761–1774

Cohen C (1982) Matching molecules in the catch mechanism. Proc Natl Acad Sci USA 79:3176–3178
Cohen C, Vibert PJ (1987) Actin filaments: images and models. In: Squire JM, Vibert PJ (eds) Fibrous protein structure. Academic Press, London, pp 283–306
Cooke PH, Fay FS, Craig R (1989) Myosin filaments isolated from skinned amphibian smooth muscle cells are side-polar. J Muscle Res Cell Motil 10:206–220
Cooke R (1986) The mechanism of muscle contraction. CRC Crit Rev Biochem 21(1):53–118
Craig, R, Megerman J (1977) Assembly of smooth muscle myosin into side-polar filaments. J Cell Biol 75:990–996
Crowther RA, Luther PK (1984) Three-dimensional reconstruction from a single oblique section of fish muscle M-band. Nature 307:569–570
Deatherage JF, Cheng N, Bullard B (1989) Arrangement of filaments and cross-links in the bee flight muscle Z disk by image analysis of oblique sections. J Cell Biol 108:1775–1782
Desherevskii VI (1971) A kinetic theory of striated muscle contraction. Biorheology 7: 147–170
Dewey MM, Levine RJC, Colflesh DE (1973) Structure of *limulus* striated muscle. J Cell Biol 58:574–593
Dubowitz V (1981) Histochemistry of muscle disease. In: Walton J (ed) Disorders of voluntary muscle, 4th edn. Churchill/Livingstone, Edinburgh, pp 261–295
Dubowitz V, Brooke M (1973) Muscle biopsy: a modern approach. Saunders, Philadelphia
Ebashi S, Endo M, Ohtsuki I (1969) Control of muscle contraction. Q Rev Biophys 2:351–384
Edman A-C, Squire JM, Sjostrom M (1988) Fine structure of the A-band in cryo-sections. Diversity of M-band structure in chicken breast muscle. J Ultrastruct Mol Struct Res 100:1–12
Elliott A, Bennett PM (1984) Molecular organization of paramyosin in the core of molluscan thick filaments. J Mol Biol 176:477–493
Elliott GF (1965) Low-angle X-ray diffraction patterns from insect flight muscle. J Mol Biol 13:956–958
Elliott GF, Lowy J (1961) Low-angle X-ray reflections from living molluscan muscles. J Mol Biol 3:41–46
England PJ (1986) The phosphorylation of cardiac contractile proteins. In: Rupp H (ed) The regulation of heart function. Thieme, New York, pp 223–233
Fay FS, Fujiwara K, Rees DD, Fogarty KE (1983) Distribution of α-Actinin in single isolated smooth muscle cells. J Cell Biol 96:783–795
Ford LE, Huxley AF, Simmons RM (1977) Tension responses to sudden length change in stimulated frog muscle fibres near slack length. J Physiol 269:441–515
Ford LE, Huxley AF, Simmons RM (1981) The relation between stiffness and filament overlap in stimulated frog muscle fibres. J Physiol 311:219–249
Ford LE, Huxley AF, Simmons RM (1985) Tension transients during steady shortening of frog muscle fibres. J Physiol 361:131–150
Ford LE, Huxley AF, Simmons RM (1986) Tension transients during the rise of tetanic tension in frog muscle fibres. J Physiol 372:595–609
Franzini-Armstrong C (1970) Natural variability in the length of thin and thick filaments in single fibres from a crab, *Portunus depurator*. J Cell Sci 6:559–592
Franzini-Armstrong C (1973) The structure of a simple Z-line. J Cell Biol 58:630–642
Fraser RDB, MacRae TP (1973) Conformation in fibrous proteins. Academic Press, New York
Freundlich A, Squire JM (1983) 3-Dimensional structure of the insect (*Lethocerus*) flight muscle M-band. J Mol Biol 169:439–453
Fuller G, Squire JM (1987) 4-Stranded helical arrangement of myosin heads on insect (*Lethocerus*) flight muscle thick filaments. J Muscle Res Cell Motil 8:67
Furst DO, Osborn M, Nave R, Weber K (1988) The organization of titin filaments in the half-sarcomere revealed by monoclonal antibodies in immunoelectron microscopy: a map of ten nonrepetitive epitopes starting at the Z-line extends close to the M-line. J Cell Biol 106:1563–1572
Gordon AM, Huxley AF, Julian FJ (1966) The variation in isometric tension with sarcomere length in vertebrate muscle fibres. J Physiol 184:170–192

Grove BK, Kurer V, Lehere C, Doetschman TC, Perriard J-C, Eppenberger HM (1984) A new 185 000-dalton skeletal muscle protein detected by monoclonal antibodies. J Cell Biol 98:518–524

Harford JJ, Squire JM (1986) The "crystalline" myosin cross-bridge array in relaxed bony fish muscle: low-angle X-ray diffraction from plaice fin muscle and its interpretation. Biophys J Soc 50:145–155

Harford JJ, Squire JM (1990) Static and time-resolved X-ray diffraction studies of fish muscle. In: Squire JM (ed) Molecular mechanisms in muscular contraction. Topics in molecular and structural biology 13. Macmillan, London, pp 287–320

Harford JJ, Squire JM (1992a) X-ray diffraction from bony fish muscle in equilibrium states. (in preparation)

Harford JJ, Squire JM (1992b) Evidence for structurally different attached states of myosin crossbridges on actin during contraction of fish muscle. (submitted)

Harford JJ, Squire JM, Maeda Y, Towns-Andrews L (1988) Synchrotron radiation studies of the molecular basis of force production in fish muscle. Proc. 2nd Int. Conf. Biophysics and synchrotron radiation, Chester, pp 148–153

Haselgrove JC (1973) X-ray evidence for a conformational change in the actin-containing filaments of vertebrate striated muscle. Cold Spring Harbor Symp Quant Biol 37:341–352

Haselgrove JC, Reedy MK (1984) Geometrical constraints affecting crossbridge formation in insect flight muscle. J Muscle Res Cell Motil 5:3–24

Haselgrove JC, Faruqi AR, Huxley HE, Arndt VW (1977) The design and use of a camera for low-angle X-ray diffraction experiments with synchrotron radiation. J Phys E:Sci Instrum 10:1035–1044

Hirose K, Wakabayashi T (1988) Thin filaments of rabbit skeletal muscle are in helical register. J Mol Biol 204:797–801

Hodge AJ (1955) Studies on the structure of muscle III phase contrast and electron microscopy of dipteran flight muscle. J Biophys Biochem Cytol 1:361–380

Holmes KC, Tregear RT, Barrington-Leigh J (1980) Interpretation of the low angle X-ray diffraction from insect flight muscle in rigor. Proc R Soc Lond B 207:13–33

Horowits R, Podolsky RJ (1987) The positional stability of thick filaments in activated skeletal muscle depends on sarcomere length: evidence for the role of titin filaments. J Cell Biol 105:2217–2223

Hoyle G, McNeill PA (1968) Correlated physiological and ultrastructural studies on specialized muscle. J Exp Zool 167:495

Hoyle G, McAlear JH, Selverston A (1965) Mechanism of supercontraction in striated muscle. J Cell Biol 26:621–640

Huxley AF (1957) Muscle structure and theories of contraction. Prog Biophys Biophys Chem 7:255–318

Huxley AF (1973) A note suggesting that the cross-bridge attachment during muscle contraction may take place in two stages. Proc R Soc Lond Ser B 183:83–86

Huxley AF, Niedergerke R (1954) Structural changes in muscle during contraction: interference microscopy of living muscle fibers. Nature 173:971–973

Huxley AF, Simmons RM (1973) Mechanical transients and the origin of muscular force. Cold Spring Harbor Symp Quant Biol 37:669–680

Huxley HE (1957) The double array of filaments in cross-striated muscle. J Biophys Biochem Cytol 3:631–648

Huxley HE (1963) Electron microscope studies on the structure of natural and synthetic protein filaments from striated muscle. J Mol Biol 7:281–308

Huxley HE (1969) The mechanism of muscular contraction. Science 164:1356–1366

Huxley HE, Brown N (1967) The low-angle X-ray diagram of vertebrate striated muscle and its behaviour during contraction and rigor. J Mol Biol 30:383–434

Huxley HE, Hanson J (1954) Changes in the cross-striations of muscle during contraction and stretch and their structural interpretation. Nature 173:973–976

Huxley HE, Faruqi AR, Bordas J, Koch M, Milch JR (1980) The use of synchrotron radiation in time-resolved X-ray diffraction studies of myosin layer-line reflections during muscle contraction. Nature 284:140–143

Huxley HE, Simmons RM, Faruqi AR, Kress M, Bordas J, Koch MHJ (1981) Millisecond time-resolved changes in X-ray reflections from contracting muscle during rapid mechanical transients, recorded using synchrotron radiation. Proc Natl Acad Sci USA 78(4):2297–2301

Huxley HE, Simmons RM, Faruqi AR, Kress M, Bordas J, Koch M (1983) Changes in the X-ray reflections from contracting muscle during rapid mechanical transients and their structural implications. J Mol Biol 169:469–506

Irving M (1987) Muscle mechanics and probes of the crossbridge cycle: In: Squire JM, Vibert PJ (eds) Fibrous protein structure. Academic Press, London, pp 495–528

Itoh Y, Suzuki T, Kimura S, Ohashi K, Higuchi H, Sawada H, Shimizu T, Shibata M, Maruyama K (1988) Extensible and less extensible domains of connectin filaments in stretched vertebrate skeletal muscle sarcomeres as detected by immunofluorescence and immunoelectron microscopy using monoclonal antibodies. J Biochem (Tokyo) 104:504–508

Jahromi SS, Atwood HL (1971) Structural and contractile properties of lobster leg-muscle fibers. J Exp Zool 176:475–486

Johnson WH (1962) Tonic mechanisms in smooth muscle. Physiol Rev (Suppl) 5:113–143

Knappeis GG, Carlsen F (1962) The ultrastructure of the Z-disc in skeletal muscle. J Cell Biol 13:323–335

Kress M, Huxley HE, Faruqi AR, Hendrix J (1986) Structural changes during activation of frog muscle studied by time-resolved X-ray diffraction. J Mol Biol 188:325–342

Lazarides E (1980) Intermediate filaments as mechanical integrators of cellular space. Nature 283:249–256

Lee JM, Luther PK, Squire JM (1987) Three-dimensional image reconstruction of the vertebrate Z-line. J Muscle Res Cell Motil 8:90

Lee JM, Luther PK, Squire JM (1992) 3-Dimensional structure of the Z-band in frog sartorius muscle. (unpublished results)

Lehman W, Kendrick-Jones J, Szent-Gyorgyi AG (1973) Myosin-linked regulatory systems: comparative studies. Cold Spring Harbor Symp Quant Biol 37:319–330

Levine RJC, Elfvin M, Dewey MM, Walcott B (1976) Paramyosin in invertebrate muscles II. Content in relation to structure and function. J Cell Biol 71:273–279

Levine RJC, Chantler PD, Kensler RW (1988) Arrangement of myosin heads on *limulus* thick filaments. J Cell Biol 107:1739–1747

Locker RH, Leet NG (1975) Histology of highly stretched beef muscle I: the fine structure of grossly stretched single fibres. J Ultrastruct Res 52:64–75

Locker RH, Leet NG (1976) Histology of highly stretched beef muscle II: further evidence on the location and nature of gap filaments. J Ultrastruct Res 55:157–172

Lowy J, Millman BM (1963) The contractile mechanism of the anterior byssus retractor muscle of *Mytilus edulis*. Philos Trans R Soc Lond Ser B 246:105–148

Lowy J, Mulvany MJ (1973) Mechanical properties of guinea pig taenia coli muscles. Acta Physiol Scand 88:123–136

Lowy J, Poulsen FR (1987) Fast X-ray diffraction studies of muscle. In: Squire JM, Vibert PJ (eds) Fibrous protein structure. Academic Press, London, pp 451–494

Lowy J, Vibert PJ (1972) Studies of the low-angle X-ray pattern of a molluscan smooth muscle during tonic contraction and rigor. Cold Harbor Spring Symp Quant Biol 37:353–359

Luther PK (1990) Three-dimensional reconstruction of a simple (fish muscle) Z-band. J Cell Biol *113*, 1043–1055

Luther PK, Crowther R (1984) Three-dimensional reconstruction from tilted sections of fish muscle M-band. Nature 307:566–568

Luther PK, Squire JM (1980) 3-Dimensional structure of the vertebrate muscle A-band. II: The myosin filament superlattice. J Mol Biol 141:409–439

Luther PK, Squire JM (1992) Evolution of the vertebrate muscle A-band. (in preparation)

Luther PK, Munro PMG, Squire JM (1981) 3-Dimensional structure of the vertebrate muscle A-band. III: M-region structure and myosin filament symmetry. J Mol Biol 151:703–730

Maruyama K, Matsubara S, Natori R, Nonomura Y, Kimura S, Ohashi K, Murakami F, Handa S, Eguchi G (1977) Connectin, an elastic protein in muscle. Characterisation and function. J Biochem 82:317–337

Millman BM, Elliott GF (1973) An X-ray diffraction study of contracting molluscan smooth muscle. Biophys J 12:1405–1414
Morris EP, Nnegi G, Squire JM (1990) Three-dimensional structure of nemaline rod Z-crystals. J Cell Biol 111:2961–2978
Morris EP, Squire JM, Fuller G (1991) The 4-stranded helical arrangement of myosin heads on insect (*Lethocerus*) flight muscle thick filaments. J Struct Biol 107:234–249
Munro PMG (1986) Comparative ultrastructures of red and white muscle fibres of roach. PhD Thesis, Imperial College, London
Munro PMG, Luther PK, Squire JM (1990) Electron microscopy of fish muscle. Electron microscopy reviews (in preparation)
Offer G (1987) Myosin filaments: In: Squire JM, Vibert PJ (eds) Fibrous protein structure. Academic Press, London, pp 307–356
Osborne MP (1967) Supercontraction in the muscles of the blowfly larva: an ultrastructural study. J Insect Physiol 13:1471–1582
Parry DAD, Squire JM (1973) Structural role of tropomyosin in muscle regulation: analysis of the X-ray diffraction patterns from relaxed and contracting muscles. J Mol Biol 75: 33–55
Patterson C (1978) Evolution. British Museum (Natural History), London
Pierobon-Bormioli S, Betto R, Salviati G (1989) The organisation of titin (connectin) and nebulin in the sarcomeres: an immunocytolocalization study. J Muscle Res Cell Motil 10:446–456
Pollard TD (1975) Functional implications of the biochemical and structural properties of cytoplasmic contractile proteins. In: Inoue S, Stephens RE (eds) Molecules and cell movement. Raven, New York, pp 259–286
Price MG (1987) Skelemins: cytoskeletal proteins located at the periphery of M-discs in mammalian striated muscle. J Cell Biol 104:1325–1336
Pringle JWS (1967) The contractile mechanism of insect fibrillar muscle. Prog Biophys Mol Biol 17:1–60
Pringle JWS (1968) Mechano-chemical transformation in striated muscle. Aspects of cell motility. Cambridge University Press, Cambridge, pp 67–86
Pringle JWS (1972) Arthropod muscle. In: Bourne SH (ed) The structure and function of muscle, 2nd edn. Academic Press, London, pp 491–541
Pringle JWS (1978) The Croonian lecture 1977, stretch activation of muscle: function and mechanism. Proc R Soc Lond (Biol) 201:107–130
Reedy MK (1964) In discussion on "The physical and chemical basis of muscular contraction," by J. Hanson and J. Lowy. Proc R Soc Ser B 160:458–460
Reedy MK (1968) Ultrastructure of insect flight muscle. I: Screw sense and structural grouping in the rigor crossbridge lattice. J Mol Biol 31:155–176
Reedy MK, Holmes KC, Tregear RT (1965) Induced changes in orientation of the crossbridges of glycerinated insect flight muscle. Nature 207:1276–1280
Reedy MK, Bahr GF, Fischman DA (1973) How many myosins per crossbridge? I Flight muscle myofibrils from the blowfly, *Sarcophaga bullata*. Cold Spring Harbor Symp Quant Biol 37:397–421
Reedy MK, Leonard KR, Freeman R, Arad T (1981) Thick filament mass determination by electron scattering measurements with the scanning transmission electron microscope. J Muscle Res Cell Motil 2:45–64
Reger JF, Cooper DP (1967) A comparative study on the fine structure of the basalar muscle of the wing and the tibial extensor muscle of the leg of the Lepidopteran *Achalarus lyciades*. J Cell Biol 33:531–542
Rosenbaum G, Holmes KC (1980) Small-angle diffraction of X-rays and the study of biological structures. In: Winick H, Doniach S (eds) Synchrotron radiation research. Plenum, New York, pp 533–564
Rosenbaum G, Holmes KC, Witz J (1971) Synchrotron radiation as a source for X-ray diffraction. Nature 230:434–437
Rosenbluth J (1965) Ultrastructural organisation of obliquely striated muscle fibres in *Ascaris lumbricoides*. J Cell Biol 25:496–515

Rosenbluth J (1968) Obliquely striated muscle. IV. Sarcoplasmic reticulum, contractile apparatus and endomysium of the body muscle of a polychaete, *Glycera*, in relation to its speed. J Cell Biol 36:245–259
Rowe RWD (1973) The ultrastructure of Z-disks from white, intermediate and red fibres of mammalian striated muscles. J Cell Biol 37:261–277
Rüegg JC (1958) The possible function of invertebrate tropomyosin. Biochem J 69:46P
Rüegg JC (1971) Smooth muscle tone. Physiol Rev 51:201–248
Rüegg JC, Schädler M (1970) Mechanical activation of the contractile system in skeletal muscle. Pfluegers Arch 319:139–145
Sjöström M, Squire JM (1977a) Fine structure of the A-band in cryo-sections: the structure of the A-band of human skeletal muscle fibres from ultra-thin cryo-sections negatively stained. J Mol Biol 109:49–68
Sjöström M, Squire JM (1977b) Cryo-ultramicrotomy and myofibrillar fine structure: a review. J Microsc 111:239–278
Small JV (1977) Studies on isolated smooth muscle cells: the contractile apparatus. J Cell Sci 24:327–349
Small JV (1987) The cytoskeleton: selected topics. In: Squire JM, Vibert PJ (eds) Fibrous protein structure. Academic Press, London, pp 357–388
Small JV, Squire JM (1972) Structural basis of contraction in vertebrate smooth muscle. J Mol Biol 67:117–149
Smith DS (1961) The structure of insect fibrillar flight muscle. A study made with special reference to the membrane system of the fiber. J Biophys Biochem Cytol 10 (Suppl):123
Smith DS (1965) The organization of flight muscle in an aphid, *Megoura viciae* (Homoptera). With a discussion on the structure of synchronous and asynchronous striated muscle fibers. J Cell Biol 27:379–393
Smith DS (1966) The organization and function of the sarcoplasmic reticulum and T-system of muscle cells. Prog Biophys Mol Biol 16:107–142
Sobieszek A (1973) The fine structure of the contractile apparatus of the anterior byssus retractor muscle of *Mytilus edulis*. J Ultrastruct Res 43:313–343
Somlyo AV (1980) Ultrastructure of vascular smooth muscle. In: Bohr D, Somlyo A, Sparks H (eds) Handbook of physiology, vol 2, sect 2. American Physiological Society, Bethesda, pp 33–67
Somlyo AV, Butler TM, Bond M, Somlyo AP (1981) Myosin filaments have non-phosphorylated light chains in relaxed smooth muscle. Nature 294:567–569
Squire JM (1972) General model of myosin filament structure. II. Myosin filaments and cross-bridge interactions in vertebrate striated and insect flight muscles. J Mol Biol 72:125–138
Squire JM (1974) Symmetry and three-dimensional arrangement of filaments in vertebrate striated muscle. J Mol Biol 90:153–160
Squire JM (1981) The structural basis of muscular contraction. Plenum, New York
Squire JM (1986a) Muscle myosin filaments: internal structure and crossbridge organisation. Comments Mol Cell Biophys 3(3):155–177
Squire JM (1986b) Muscle: design, diversity and disease. Benjamin/Cummings, Menlo Park, California
Squire JM (1987) Muscle contraction, mechanics and control. Encyclopedia of Neuroscience, Birkhauser, Boston, pp 712–716
Squire JM (1992) Muscle filament lattices and stretch-activation: the match/mismatch model reassessed. J Muscle Res Cell Motil (in press)
Squire JM, Luther PK, Trinick J (1987) Muscle myofibril architecture. In: Squire JM, Vibert PJ (eds) Fibrous protein structure. Academic Press, London, pp 423–448
Squire JM, Freundlich A, Peckham M, Luther PK (1992a) Lattice organisation in *Lethocerus* and *Apis* muscles (unpublished results)
Squire JM, Lee JM, Luther PK, Morris EP (1992b) Symmetry and bridge interactions in vertebrate muscle Z-bands. (in preparation)
Steiger GJ (1977) Stretch activation and tension transients in cardiac, skeletal and insect flight muscle. In: Tregear RT (ed) Insect flight muscle. Proc Oxford Symposium, North Holland, Amsterdam, pp 221–268

Storey TI, Usinger RL (1965) General zoology. McGraw-Hill, New York
Strehler EE, Carlson E, Eppenberger HM, Thornell, L-E (1983) Ultrastructural location of M-band proteins in chicken breast muscle as revealed by combined immunocytochemistry and ultramicrotomy. J Mol Biol 166:141–158
Stromer MH, Tabatabai LB, Robson RM, Goll DE, Zeece MG (1976) Nemaline myopathy, an integrated study: selective extraction. Exp Neurol 50:402–421
Svendsen KH (1982) Ultrastructure of a molluscan smooth muscle during phasic contractions and in the relaxed state. Int J Biol Macromol 4:43–49
Szent-Györgyi AG, Cohen C, Kendrick-Jones J (1971) Paramyosin and the filaments of molluscan "catch" muscles. II. Native filaments: isolation and characterisation. J Mol Biol 56:239–258
Thorson JW, White DCS (1969) Distributed representations for actin-myosin interaction in the oscillatory contraction of muscle. Biophys J 9:360–390
Toselli PA, Pepe FA (1968) Fine structure of the ventral intersegmental abdominal muscles of the insect *Rhodnius prolixus* during the molting cycle. I. Muscle structure at molting. J Cell Biol 37:445–461
Tregear RT, Squire JM (1973) Myosin content and filament structure in smooth and striated muscle. J Mol Biol 77:279–290
Trentham DR, Eccleston JF, Bagshaw CR (1976) Kinetic analysis of ATPase mechanisms. Q Rev Biophys 9:217–281
Trinick J, Offer G (1979) Cross-linking of actin filaments by heavy meromyosin. J Mol Biol 133:549–556
Trinick J, Knight P, Whiting A (1984) Purification and properties of titin. J Mol Biol 180:331–356
Trombitas C, Tigyi-Sebes A (1977) Fine structure and mechanical properties of insect muscle. In: Tregear RT (ed) Insect flight muscle. Proc Oxford Symposium. North Holland, Amsterdam, pp 79–90
Trombitas K, Tigyi-Sebes A (1974) Direct evidence for connecting C filaments in flight muscle of honey bee. Acta Biochem Biophys Acad Sci Hung 9:243–253
Trybus KM, Lowey S (1987) Assembly of smooth muscle myosin minifilaments: effects of phosphorylation and nucleotide binding. J Cell Biol 105:3007–3019
Varriano-Marston E, Franzini-Armstrong C, Haselgrove JC (1987) Structure of the M-band. J Electron Microsc Tech 6:131–141
Vibert P, Craig R (1983) Electron microscopy and image analysis of myosin filaments from scallop striated muscle. J Mol Biol 165:303–320
Walliman T, Schlosser T, Eppenberger H (1982) ATP regeneration potential of M-band creatine kinase. J Muscle Res Cell Motil 3:503–504
Wang K, McClure J, Tu A (1979) Titin: major myofibrillar component of striated muscle. Proc Natl Acad Sci USA 76:3698–3707
White DCS (1967) Structural and mechanical properties of insect fibrillar flight muscle in the relaxed and rigor states. PhD Thesis, University of Oxford
White DCS, Thorson J (1973) The kinetics of muscle contraction. Prog Biophys 27:175–255
Wray JS (1979a) Structure of the backbone in myosin filaments of muscle. Nature 277:37–40
Wray JS (1979b) Filament geometry and the activation of insect flight muscle. Nature 280:325–326
Yamaguchi M, Robson RM, Stromer MH, Dahl DS, Oka T (1978) Actin filaments form the backbone of nemaline myopathy rods. Nature 271:265–267
Yu LC, Brenner B (1989) Structures of actomyosin crossbridges in relaxed and rigor muscle fibres. Biophys J 55:441–453

Chapter 5

Molecular Mechanism of Actin-Myosin Interaction in Muscle Contraction

H. Sugi

Contents

1 Introduction

Muscle contraction results from the relative sliding between the thick and thin myofilaments in all kinds of muscle. Since myofilament sliding is coupled with ATP hydrolysis, the physiological function of muscle is to convert energy derived from chemical reactions into mechanical work and heat. The thick filament consists mainly of myosin, which is a large molecule (MW 450 000) with

Department of Physiology, School of Medicine, Teikyo University, Itabashi-ku, Tokyo 173, Japan

Advances in Comparative and Environmental Physiology
Vol. 12. Edited by H. Sugi

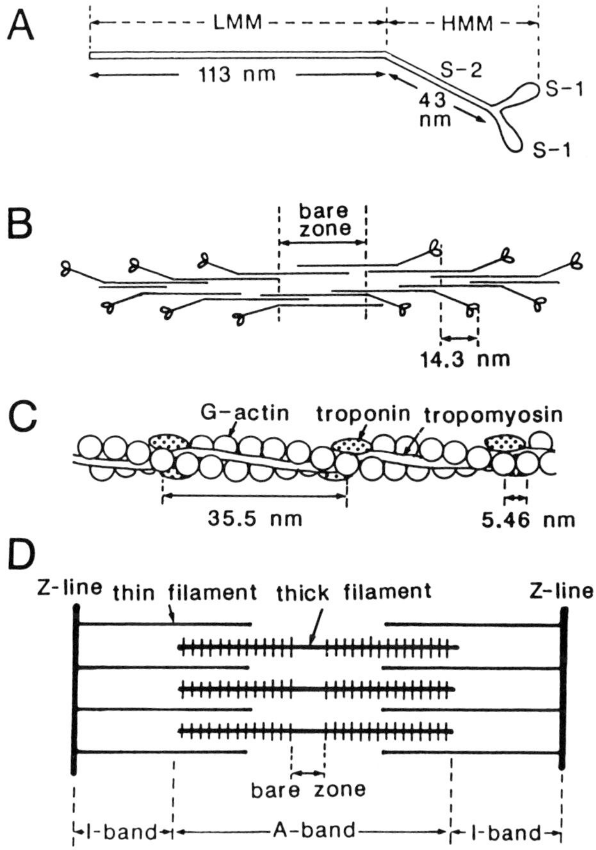

Fig. 1A–D. Ultrastructure of the thick and thin filaments, and their arrangement within a sarcomere. **A** Diagram of myosin molecule. **B** Longitudinal arrangement of myosin molecules in the thick filament. **C** Structure of the thin filament. **D** Longitudinal arrangement of the thick and thin filaments within a sarcomere. Note that the half-sarcomere is the structural and functional unit of muscle

two pear-shaped "heads"and a "rod" of 156 nm long. As shown in Fig. 1A, a myosin molecule can be split by light tryptic digestion into two parts: (1) the rod of 113 nm long, known as light meromyosin (LMM), and (2) the rest of the myosin molecule, containing the two heads and the rod of 43 nm long, known as heavy meromyosin (HMM). HMM can be further digested into two separate heads, known as subfragment-1 (S-1), and the rod, known as subfragment-2 (S-2). In the thick filament, LMM aggregates to form the filament backbone which is polarized in opposite directions on either side of the central point, while the S-1 heads extend laterally from the filament backbone with an axial interval of 14.3 nm to form so-called cross-bridges, except for a region without S-1 heads, i.e. the bare zone, at the center of the thick filament (Fig. 1B). The S-2 rod is believed to serve as a "hinge" between the S-1 head and the filament, thus enabling the head to swing away from the filament.

On the other hand, the thin filament consists primarily of two helical strands of globular actin monomers (G-actin, MW 41700), which are wound around each other with a pitch of 35.5 nm. The axial separation of the G-actin in the filament is 5.46 nm (Fig. 1C). In vertebrate skeletal muscle, the thin filament also contains tropomyosin and troponin. The tropomyosin molecule is rod-shaped, and lies in the grooves between the G-actin strands, while troponin is located in every pitch of the strands. In relaxed muscle, the interaction between the S-1 heads (cross-bridges) and the thin filament is inhibited by tropomyosin. When calcium ions bind to troponin, it removes the inhibitory effect of tropomyosin to cause contraction (see Chap. 1).

As shown in Fig. 1D, the thin filaments extend in either direction from the Z-line to penetrate in between the thick filaments, located centrally in each sarcomere (see Chap. 4). The region containing only the thin filaments is called the I-band, whereas the region containing the thick filaments as well as part of the thin filaments is called the A-band. It has been firmly established that the filament lengths remain constant during active shortening and passive stretch of muscle fibers, based on (1) light microscopy of living fibers and isolated myofibrils (H.E. Huxley and Hanson 1954; A.F. Huxley and Niedergerke 1954); (2) electron microscopy of the thick and thin filaments (Page and Huxley 1963); and (3) X-ray diffraction of living fibers in which various periodicities of the filaments can be measured (for a review, see Wray and Holmes 1981). These findings underlie the "sliding filament theory", which denies the possibility that muscle contraction is caused by folding of long chain molecules extending over the entire muscle length.

Thus, the central problem in understanding the molecular mechanism of muscle contraction is: what makes the thick and thin filaments slide past each other? Since the ATPase activity and the actin binding sites are localized in the S-1 heads of the myosin molecule (Young 1967), it is generally agreed that the myosin S-1 heads, extending from the thick filaments towards the thin filaments, probably play a key role in the chemomechanical energy conversion processes taking place in contracting muscle.

2 Basic Characteristics of Contracting Muscle

In each sarcomere of contracting muscle, the relative sliding between the thick and thin filaments takes place symmetrically with respect to the center of the A-band. As muscle consists of thousands of sarcomeres connected in series, the half-sarcomere is regarded as the structural and functional unit of muscle. Muscle contraction is therefore ascribed to the myofilament sliding in each half-sarcomere as a result of the actin-myosin interaction coupled with ATP hydrolysis. Thus, the physiological characteristics of contracting muscle reflect the nature of the actin-myosin interaction, provided the contraction occurs uniformly along the entire muscle length. In this section, the most basic characteristics of contracting vertebrate skeletal muscle, shown diagrammatically in Fig. 2, will be briefly described. All the experiments have been done on frog skeletal muscle.

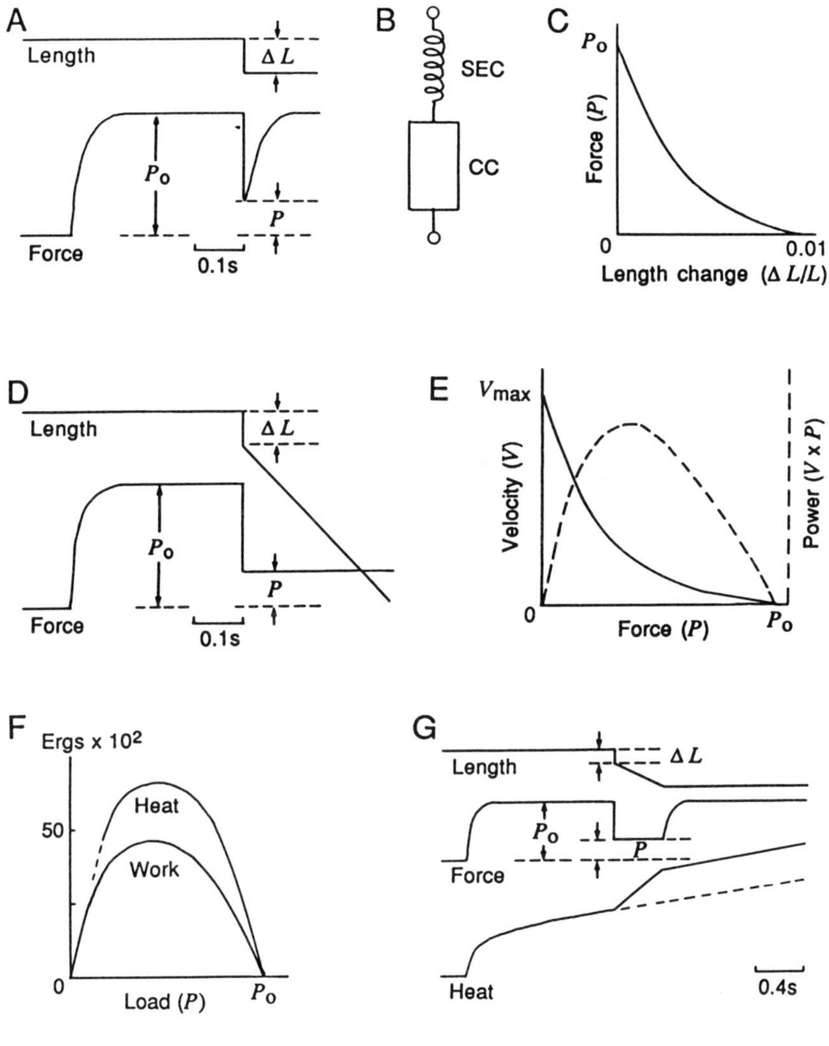

Fig. 2A–G. Basic characteristics of contracting muscle. **A** Effect of a quick release (ΔL) on the isometric force (P_0). The force decreases quickly from P_0 to P during the release, and then redevelops to the initial level. **B** Two-component model consisting of *SEC* and *CC* connected in series with each other. **C** Force-extension curve of the SEC determined by quick release applied to an isometrically contracting muscle. The force immediately after a quick release (P) is plotted against the amount of quick release (ΔL relative to the initial muscle length L). **D** Effect of a quick decrease in the force from P_0 to P on the muscle length. The muscle length initially shortens quickly for a distance (ΔL), and then shortens with a constant velocity. **E** Force-velocity relation. The velocity of shortening (V) is maximum (V_{max}) at $P = 0$, and zero at $P = P_0$. Power ($V \times P$) is also shown by a *broken line*. **F** Dependence of the amount of work done and the amount of heat produced on the amount of load (P) in a short tetanus or a twitch (Fenn 1924). *Heat* represents the heat produced above that in the isometric condition. **G** Heat production during a tetanus including a period of isotonic shortening. Note an increase in the rate of heat production during the period of isotonic shortening

2.1 The Series Elastic Component

When the muscle length is quickly decreased by ΔL (a quick release) during the plateau of isometric tetanus, the tetanic force (P_0) decreases quickly to a lower value (P) during the release, and after the completion of the release, starts to return to its original level with a time course analogous to the rise of tetanic force in response to stimulation (Gasser and Hill 1924; Fig. 2A). This fact has been explained by assuming a spring, called the series elastic component (SEC) connected in series with the contractile component (CC; Fig. 2B). During a sufficiently quick release, the CC does not have time to change its length, so that the force in muscle decreases along the force-extension curve of the SEC. After the completion of release, the CC starts to shorten by stretching the SEC to result in force redevelopment towards the original level. The force extension curve of the SEC is shown in Fig. 2C, where the force immediately after the quick release is plotted against the amount of quick release. The SEC includes the compliance of tendons and the recording system, but when these external components are subtracted or removed experimentally, the SEC is shown to distribute uniformly along the entire muscle length (Jewell and Wilkie 1958). The extension of the above internal SEC when the CC exerts its maximum isometric force (P_0) is about 1% of muscle length, as indicated by the intersection of the force-extension curve with the abscissa (Fig. 2C).

2.2 The Force-Velocity Relation

If a muscle, generating P_0 in the isometric condition, is suddenly allowed to shorten against an external load ($P < P_0$), it first shortens quickly for a short distance (ΔL) due to elastic recoil of the SEC, and then shortens with a constant velocity over a considerable distance (10% of muscle length or more) (Fig. 2D). Since the length of the SEC remains unchanged under a constant load, the above constant-velocity shortening (without acceleration) reflects the characteristic of the CC, i.e. the characteristic of the myofilament sliding. This indicates that, during the course of isotonic shortening, the force generated by the muscle is equal to the load. The relation between the velocity of shortening (V) and the amount of load (P) (= force) is hyperbolic in shape, and can be expressed by the Hill equation, $(P + a)(V + b) = (P_0 + a)b$, where a and b are constants (Fig. 2E; Hill 1938). The velocity is zero when $P = P_0$, and shows the maximum value (V_{max}) when $P = 0$ (zero external load).

Since the power is the product of velocity and force, the power versus load relation during isotonic shortening can be directly obtained from the force-velocity relation. The relation is bell-shaped (dotted line, Fig. 2E), as the power is zero when the load is zero or P_0. This indicates that a muscle can vary its power output depending on the amount of external load.

2.3 Fenn Effect

Using the technique of measuring muscle heat production developed in Hill's laboratory, Fenn (1923, 1924) measured both the amount of work done and the amount of heat produced (above the heat produced in the isometric condition) during a short tetanus under various external loads. As shown in Fig. 2F, he found that both the amount of work done and the amount of heat produced increased parallel to each other when the load was increased from zero to a certain value, thus giving bell-shaped work and heat versus load relations. This means that a muscle can change its energy output (work + heat) depending on the amount of external load.

2.4 Heat Production During Contraction

Figure 2G is a diagram showing the rate of heat production during a tetanus including a period or isotonic shortening. Though the isometric force reaches the steady level of P_0 shortly after the beginning of tetanic stimulation, heat is continuously produced. The heat continuously produced during isometric contraction is called the maintenance heat (Hill 1949). When an isometrically contracting muscle is allowed to shorten isotonically, the rate of heat production increases above the maintenance heat rate. The shortening heat rate (h) increases linearly with the shortening velocity (V), though the h versus V relation tends to be less steep at high velocities (Hill 1964).

It should be noted that the above-mentioned characteristics of contracting muscle are obtained from frog skeletal muscle initially held at its slack length, so that the experiments are performed within the range of muscle lengths where the resting force is negligible. The length range roughly corresponds to the plateau region of the sarcomere length-force diagram of Gordon et al. (1966b) described later. In the next section, the basic characteristics of contracting muscle will be explained in terms of the actin-myosin interaction coupled with ATP hydrolysis, mainly on the basis of A.F. Huxley's 1957 contraction model.

3 Muscle Mechanics and Theories of Contraction

The basic behavior of contracting muscle, especially its energetic aspects, suggests strongly that, in contracting muscle, some chemical processes responsible for contraction are going on continuously. As early as 1923, Fenn already pointed out that a muscle behaves like a motor, which continuously converts energy into mechanical work and heat at a rate determined by the external load. In the light of the sliding filament mechanism, the motor corresponds to each myosin S-1 head (cross-bridge) interacting with actin in the thin filament, utilizing the free energy derived from ATP hydrolysis. Among a number of theories built up on this basis, A.F. Huxley's contraction model (A.F. Huxley 1957) has been central in the field of muscle research. In this chapter, I shall first give a brief account of the Huxley (1957) contraction model, and then describe the phenomena observed in the field of muscle mechanics, which are accounted for, or are difficult to account for, by this model, together with the additional assumptions made by the authors concerned.

3.1 A.F. Huxley's 1957 Contraction Model

The main assumptions of Huxley's contraction model are as follows:

1. Muscle contraction results from cyclic attachment and detachment between the myosin heads and the sites on the thin filament, so that the cross-bridges have only two states, i.e. attached and detached. A myosin head splits one ATP molecule in each attachment-detachment cycle. The free energy derived from ATP hydrolysis is partly converted into mechanical work, while the rest is dissipated as heat and entropy.

2. As shown in Fig 3A, a myosin head (M), connected to the thick filament with springs (S_1 and S_2), is fluctuating around its equilibrium position (0) due to thermal agitation. When it attaches to an active site (A) on the thin filament, it exerts either positive or negative force (F) proportional to its distance (x) from 0 by pulling S_1 or S_2 ($F = 0$ when $x = 0$, and $F = \mathrm{k}x$ when $x > 0$ or $x < 0$, k is the stiffness of the spring).

3. The rate constants for attachment of M to, and its detachment from, the thin filament (f and g, respectively) are functions of x as shown in Fig. 3B. M attaches to A if M and A are within the range of x where f has finite values ($0 \leqslant x \leqslant h$) to exert positive force for the filament sliding by pulling S_1. If the A-M link moves across 0 to the negative x region, it breaks fairly rapidly because f is zero and g has a large value.

In Fig. 3C, the steady-state distribution of A-M links (expressed by fractional numbers $0 \leqslant n \leqslant 1$) is shown against x at various steady shortening velocities V (expressed relative to V_{max}). In the isometric condition ($P = P_0$, no sliding between the filaments), A-M links are distributed only in the positive x region where f has finite values ($0 \leqslant x \leqslant h$). As the steady shortening velocity increases, the number of A-M links exerting positive forces decreases as a result of a moderate value of f; the proportion of M that slides past A without formation of an A-M link increases as the velocity of sliding between the filaments increases. On the other hand, A-M links brought into the negative x region tend to exist over a larger distance as the sliding velocity increases. Finally, the sliding velocity reaches a value, under which the positive and negative forces exerted by A-M links are equal. This velocity corresponds to the maximum shortening velocity (V_{max}) under zero external load.

The Huxley's contraction model can explain most of the basic characteristics of contracting muscle summarized in Fig. 2. In this model, the SEC originates from S_1 and S_2 between M and the thick filament (Fig. 3A). During the plateau of isometric tetanus, A-M links generate positive forces by pulling S_1. The quick decrease in force in response to a quick release is due to elastic recoil of stretched S_1, producing a rapid shift of the A-M link distribution into the negative x region. If the shift amounts to $h/2$, the positive and negative forces on

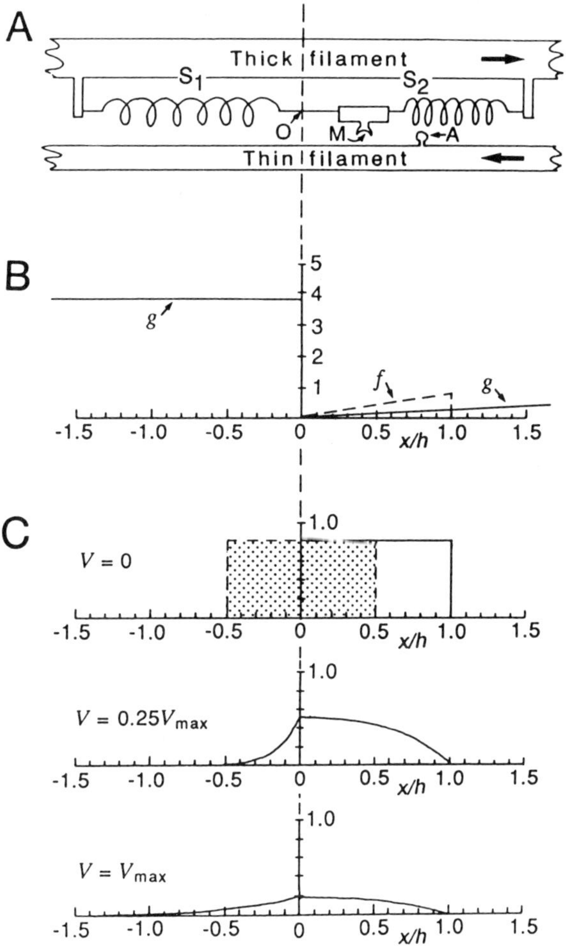

Fig. 3A–C. A.F. Huxley's contraction model (A.F. Huxley 1957). **A** Diagram showing the myosin head (*M*) connected to the thick filament with springs (S_1 and S_2), and the active site (*A*) on the thin filament. *Arrows* indicate the direction of relative sliding between the filaments. **B** Dependence of *f* and *g* on the distance (*x*) of M from its equilibrium position (0). M can only attach to A at $0 \leqslant x \leqslant h$ where the rate constant of attachment (*f*) has finite values ($f = f_1$ x/h, *broken line*). The rate constant of detachment (*g*) is $g = g_1$ x/h in the positive x region, and $g = g_2$ in the negative x region. **C** Diagrams showing distribution of A-M links with respect to x at three different steady shortening velocities. The *upper diagram* shows the A-M link distribution during the plateau of isometric tetanus; *broken line* and *shaded area* show the rapid shift of the A-M link distribution in response to a quick release to just drop the force from P_0 to zero. The *lower diagrams* show the A-M link distribution changes with increasing steady shortening velocity

S_1 and S_2 become equal immediately after the release, and the force just drops from P_0 to zero. Since the amount of quick release required to drop P_0 to zero is about 1% of muscle length (i.e. about 10 nm in each 1-μm-long half-sarcomere), the value of h is about 20 nm (15.6 nm, according to Huxley's estimate).

If the values of f and g are appropriately chosen, the model can account for the steady-state force-velocity relation and heat production during contraction at least qualitatively. As the values of g are assumed to be small in the positive x region, the rate of the A-M attachment-detachment cycle in the isometric condition is relatively slow, being consistent with the maintenance heat rate in the isometric condition. When the filaments slide past each other, A-M links formed in the positive x region are brought into the negative x region to break rapidly due to a large value of g. Thus, the overall rate of the attachment-detachment cycle coupled with ATP hydrolysis increases, compared to the isometric condition, to result in an increased heat rate during shortening.

Since the monumental publication of Huxley's 1957 contraction model, the contraction characteristics have been studied by a number of investigators more precisely using isolated single muscle fibers. As described in the following sections, some results can be accounted for within the framework of the Huxley model, but some results cannot readily be explained by it. Unless otherwise stated, the experiments are made on frog skeletal muscle fibers.

3.2 The Sarcomere Length-Force Relation

It has long been known that the isometric force in contracting muscle is maximum at its slack length, and decreases at longer or shorter lengths (Ramsey and Street 1940). The length-force relation was examined by Gordon et al. (1966a,b) on single frog skeletal muscle fibers, using a novel method in which the length of a fiber segment with reasonably uniform sarcomere lengths could be kept constant during the course of an isometric tetanus, thus avoiding complications arising from nonuniform sarcomere length changes along the fiber length. Their results are shown in Fig. 4A. When the sarcomere length is increased above 2.2 μm, the isometric force decreases linearly to reach zero at a sarcomere length of 3.65 μm. This part of the sarcomere length-force relation is called the "descending limb". The descending limb has been confirmed by many authors with more precise methods of clamping short fiber segments (Edman and Reggiani 1987; Bagni et al. 1988; Granzier and Pollack 1990). Since the number of the myosin heads, which can interact with the thin filament, gradually decreases to zero as the amount of overlap between the thick and thin filament is reduced from 2.2 to 3.65 μm, the linear descending limb indicates that the amount of isometric force in the fiber is proportional to the number of the myosin heads interacting with the thin filaments within a half-sarcomere. This implies that the myosin heads behave as independent force generators. Due to the presence of the bare zone at the center of the thick filament, there is a plateau region in the sarcomere length-force relation over a sarcomere length range of 2–2.2 μm.

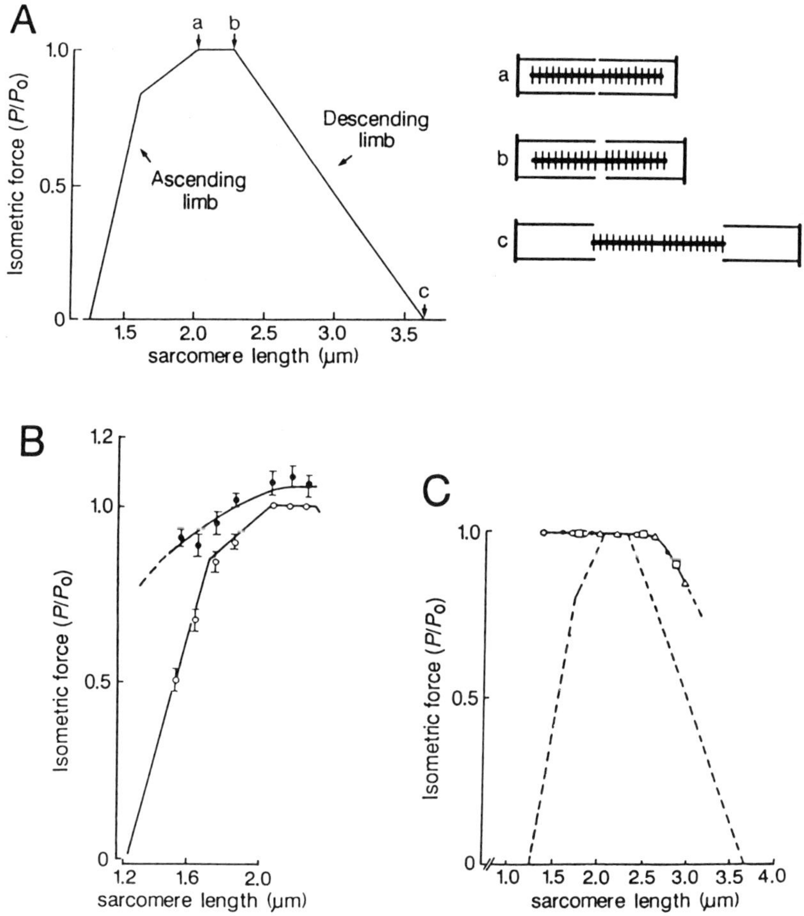

Fig. 4A–C. Sarcomere length-force relation. **A** Diagram of the sarcomere length-force relation in single muscle fibers (Gordon et al. 1966b) together with the arrangement of the thick and thin filaments in sarcomeres corresponding to the points *a*, *b*, and *c* along the curve. **B** Zn^{2+}-induced changes in the ascending limb of the sarcomere length-force relation in single muscle fibers. *Open circles*, normal Ringer; *filled circles*, Ringer containing Zn^{2+} (after Lopez et al. 1981). **C** Absence of the ascending limb in the sarcomere length-force relation of glycerinated myofibril bundles fully activated with Ca^{2+} (Sugi et al. 1983). *Broken line* indicates the sarcomere length-force relation of Gordon et al. (1966)

If, on the other hand, the fiber is made to contract isometrically at sarcomere lengths below 2 μm, the isometric force decreases steeply with decreasing sarcomere length, approaching zero around a sarcomere length of 1.3 μm. This part of the sarcomere length-force relation is called the "ascending limb". The ascending limb was explained as being due to double overlap of the thin filaments and collision of the thick filament against the Z-line (Gordon et al. 1966b). Evidence has been accumulating, however, that the ascending limb results from incomplete activation of the myofibrils within an intact muscle fiber at short sarcomere lengths. The ascending limb is reported to be less conspicuous in the presence of twitch potentiators such as Zn^{2+} (Fig. 4B; Lopez et al. 1981). The ascending limb virtually disappears up to a sarcomere length of 1.4 μm when glycerinated rabbit psoas myofibrils are fully activated with Ca^{2+} (Fig. 4C; Sugi et al. 1983), indicating that the myosin heads can generate the maximum force even at extremely short sarcomere lengths.

3.3 The Isotonic Velocity Transients

Civan and Podolsky (1966) found that, when the force in tetanized muscle fibers was suddenly reduced from P_0 to P (isotonic release), it first exhibited oscillatory changes in shortening velocity before starting to shorten isotonically. Similar phenoma have also been observed by Armstrong et al. (1966) and Sugi and Tsuchiya (1981a), as shown in Fig. 5A. The initial nonsteady phase of isotonic shortening has been called the "isotonic velocity transient"; the velocity is large immediately after the release, but it then exhibits alternate decrease and increase until reaching a constant value. As the oscillatory velocity transients cannot be explained by the Huxley model which only deals with the steady isotonic shortening, Podolsky et al. (1969) and Podolsky and Nolan (1973) modified the Huxley model by altering the shape of f and g as functions of x (Podolsky-Nolan model; Fig. 5C) to simulate the transients. The main feature of these functions are: (1) f has large values so that there is no M that slides past A without forming an A-M link; (2) in the negative x region, there is a gap between f and the major part of g, which is essential in producing the oscillatory transients; and (3) to account for the heat production during contraction, ATP hydrolysis is assumed to be coupled with the breaking of A-M links in the "gap" region (with a small g), but not with the breaking of A-M links in the major part of the g region.

Sugi and Tsuchiya (1981a) made a converse type of experiment, in which the force in muscle fibers was suddenly increased from P_0 to $P > P_0$. They found that with relatively small increases in force, the fibers showed remarkable length oscillations, consisting of alternate lengthening and shortening (Fig. 5B). To calculate these oscillations, they modified the Podolsky-Nolan model by adding a gap between f and the major part of g in the positive x region, with an additional assumption that a sudden increase in force produces an instantaneous uniform increase of g followed by its exponential return to the initial value at every value of x (Fig. 5D).

Such modified contraction models are only useful if they can suggest some experiments to test the validity of the assumptions made. In this connection, the Podolsky-Nolan model suggests the presence of an A-M attachment-detachment cycle without ATP hydrolysis during isotonic shortening, whereas the Sugi-Tsuchiya model implies that a sudden small-amplitude force increase might induce a synchronized alternate formation and breaking of A-M links that give rise to the alternate fiber lengthening and shortening. These possibilities should be examined in future research work.

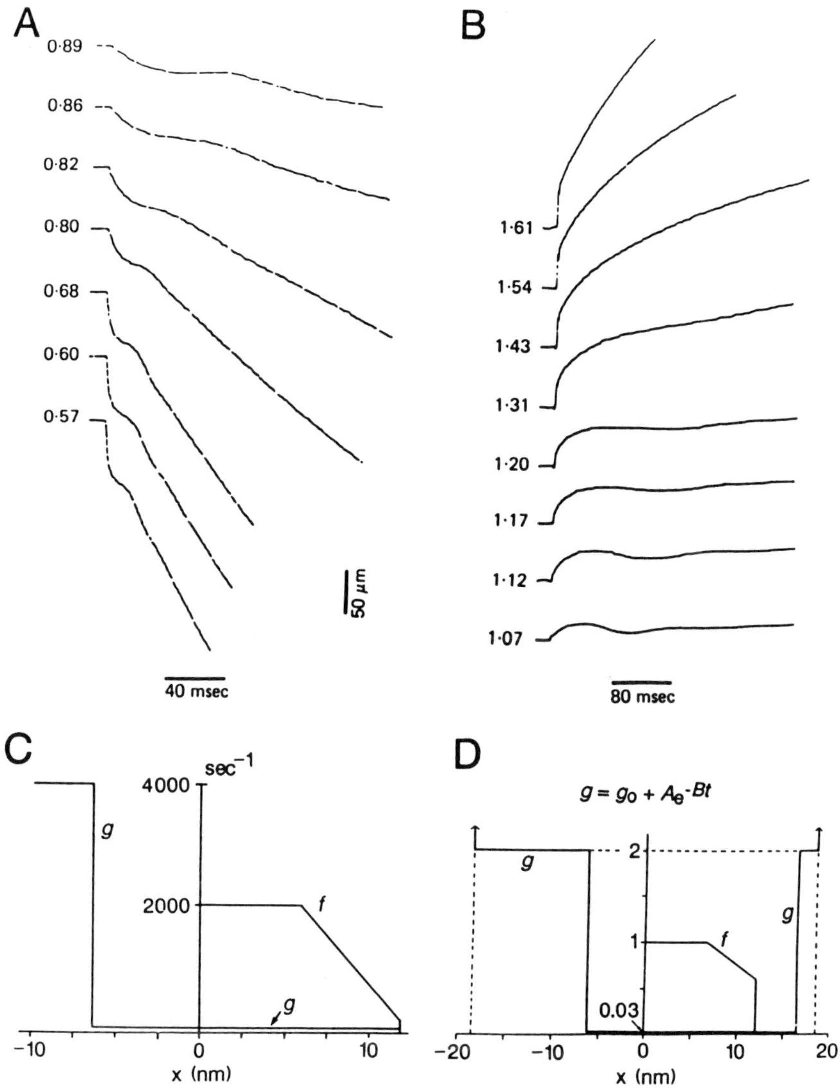

Fig. 5A–D. Isotonic velocity transients and the contraction models to simulate the transients. **A** Isotonic velocity transients following quick decreases in force (= load) from P_0 to $P < P_0$. Note oscillatory changes in shortening velocity before steady velocities are reached. **B** Isotonic velocity transients following quick increases in force from P_0 to $P > P_0$. Note oscillatory changes in fiber length in response to small increases in force. In both A and B, the values of P relative to P_0 are shown alongside each record (Sugi and Tsuchiya 1981a). **C** Podolsky-Nolan model for calculating the velocity transients following quick decreases in force. Note large values of f and a small value of g between f and the major part of g in the negative x region (Podolsky and Nolan 1973). **D** Sugi-Tsuchiya model for calculating the isotonic velocity transients following both quick decreases and quick increases in force. Note an additional gap between f and the major part of g in the positive x region together with the assumption that a quick increase in force produces an instantaneous increase in g followed by exponential recovery at every value of x. (Sugi and Tsuchiya 1981a)

3.4 Muscle Fiber Stiffness and the Location of Elasticity in the Sarcomere Structures

In Huxley's 1957 model, the only elastic structure in each sarcomere is assumed to be the springs connecting M to the thick filament. On this basis, small length changes applied to a contracting muscle fiber would be taken up by each spring via an A-M link, provided the compliance of tendons and the experimental apparatus is reasonably reduced or removed technically. Since A-M links may not break with small applied displacements, the muscle fiber stiffness, i.e. the amplitude of the force changes in response to small sinusoidal vibrations in kHz region (about 0.1% of fiber length) is generally taken as a measure of the number of A-M links (corresponding to the area showing the A-M link distribution in Fig. 3C).

The muscle fiber stiffness measured in this way actually provides information about the behavior of A-M links in various states. During the development of isometric force, the stiffness rises ahead of the force (Cecchi et al. 1984; Hatta et al. 1988), indicating that when an A-M link is formed, a finite time is required for it to exert force on the thin filament (Fig. 6A). This is not consistent with Huxley's 1957 model, in which M is continuously moving by thermal agitation so that it exerts force on the thin filament as soon as it attaches to A. The stiffness during steady isotonic shortening decreases as the amount of steady force (= load) is decreased from P_0 to zero, approaching a minimum value about 20% of that during an isometric tetanus (Julian and Sollins 1975). This result is in fairly good agreement with the number of A-M links predicted by the Huxley model (Fig. 3C).

Though the thick and thin filaments are assumed to be rigid in the Huxley model, it is possible that the compliance exists in some sarcomere structures other than the elastic links between the myosin heads and the thick filaments, and the applied length changes are also taken up by such structures. This possibility was considered by Ford et al. (1981) with the conclusion that the degree of relative contribution to the stiffness of the sarcomere structures cannot be readily determined. Suzuki and Sugi (1983) examined the extensibility of the sarcomere structures by electron microscopic observation of glycerinated rabbit muscle fibers stretched in rigor state. They found that the thin filaments in the nonoverlap region and the bare zone of the thick filament were extensible. Tentative calculation based on the electron microscopic data and the force-extension curve of the SEC suggests that about 40% of the compliance originates from the thin filaments in the I-band while the remaining 60% originates from the elastic links between the myosin heads and the thick filaments. Recently, Bagni et al. (1990) measured the muscle fiber stiffness at the plateau region of the sarcomere length-force curve, and concluded that about 19% of the sarcomere compliance may originate from the thin filaments.

Meanwhile, Sugi and coworkers have developed a novel method for measuring the stiffness with ultrasonic waves of MHz region (Tamura et al. 1982; Hatta et al. 1988). An unexpected finding by this method is that the muscle transverse stiffness, as determined from the propagation velocity of ultrasonic waves across the muscle, i.e. the transverse stiffness, decreases during

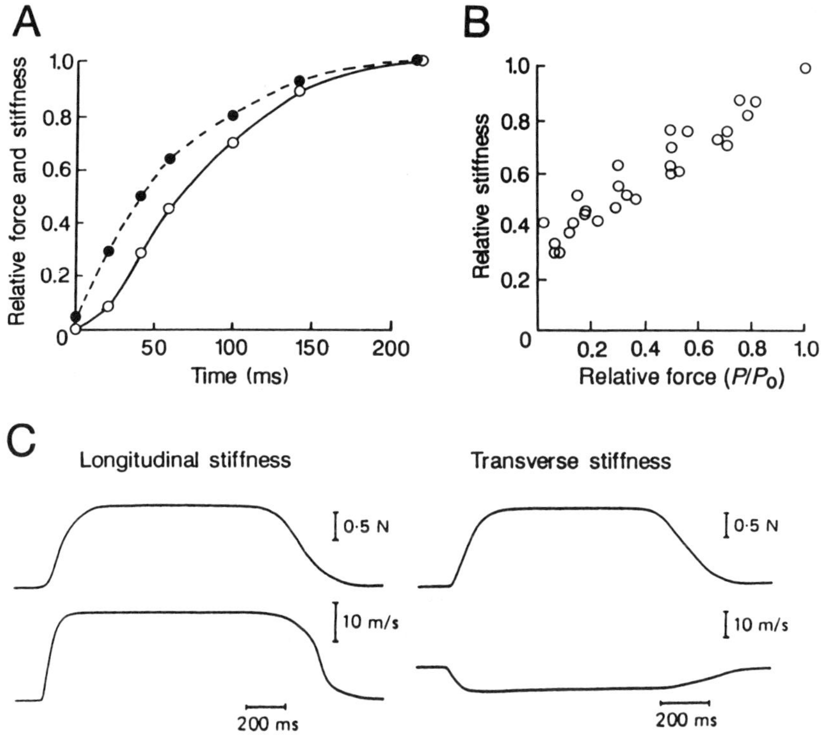

Fig. 6A–C. Muscle stiffness changes during contraction. **A** Changes in the force (*open circles*) and the stiffness (*filled circles*) of a single muscle fiber during the rise of isometric tetanic force (Cecchi et al. 1984). **B** Dependence of the stiffness of a muscle fiber during steady isotonic shortening on the amount of force (= load) relative to P_0 (Julian and Sollins 1975). The stiffness was measured with small sinusoidal vibrations of 4 kHz in **A** and 0.5–1 kHz in **B**. **C** Changes in the force (*upper traces*) and the stiffness (*lower traces*) of a muscle during isometric tetanus. The stiffness is measured as the propagation velocity of ultrasonic waves (3–7 MHz) between the two piezo-electric transducers. The upward and downward deflections of the stiffness record represent the increase and decrease of the stiffness, respectively. Note that the transverse stiffness decreases during contraction. (Hatta et al. 1988)

isometric contraction, while the longitudinal stiffness increases (Fig. 6C). The amount of decrease in the transverse muscle stiffness was proportional to the amount of overlap between the filaments. This striking phenomenon would reflect either the nature of A-M links or the change in state of water around the A-M links.

3.5 Stretch-Induced Enhancement of Mechanical Performance

It has been well known that, when a contracting muscle is slowly stretched, the force rises to a peak, which is reached at the completion of stretch, and then decays to a level higher than the initial isometric level (Abbott and Aubert 1952;

Sugi 1972, see Fig. 7A). The most striking feature of the effect of slow stretch is that, during stretch, the rate of heat production increases above the isometric maintenance heat rate, but the heat increment was less than the amount of work done on the stretched muscle by the experimental apparatus (Abbot and Aubert 1951; Hill and Howarth 1959). This phenomenon may result partly from the reduced turnover rate of A-M links during stretch, as evidenced by a reduced rate of ATP splitting (Curtin and Davies 1975), and partly from the storage of work in some sarcomere structures (Cavagna and Citterio 1974; Edman et al. 1978).

As the amplitude of slow stretches applied is very large (5–10% or more) compared to the interval between adjacent A sites in each half-sarcomere, each M would repeat attachment to, and detachment from, A sites many times without ATP hydrolysis (A.F. Huxley 1957; H.E. Huxley 1960). The possibility of storage of work in stretched muscle is supported by the fact that (1) the steady force attained after the completion of stretch is higher than the initial isometric force level in spite of a decrease in the amount of overlap between the filaments (which should decrease the isometric force after stretch along the descending limb of the sarcomere length-force relation, and (2) the high steady force attained after stretch is associated with a shift of the force-velocity relation towards higher force values (Edman et al. 1978). A similar shift of the force-velocity relation is also observed in tetanized muscle fibers during isotonic lengthening under loads $> P_0$ (Sugi and Tsuchiya 1981b). These results indicate that the mechanical performance of a tetanized muscle fiber increases as a result of slow stretches.

Sugi and Tsuchiya (1988) simultaneously measured the force and stiffness changes in single muscle fibers in response to slow stretches applied at the plateua of isometric tetanus. Figure 7A summarizes the results. The fiber was stretched uniformly along its entire length, from a sarcomere length of 2.6 to 2.75 μm with three different velocities. The stiffness first showed an abrupt increase at the beginning of stretch, and then decreased linearly though the stretch went on and the force continued to rise. On the basis of the Huxley model, the initial stiffness increase might result from a synchronous distortion of A-M links, and the subsequent linear stiffness decrease during stretch might reflect a decrease in the number of A-M links as the amount of overlap between the filaments decreases. After the completion of stretch, the stiffness decreased to a steady value equal to the steady value during ordinary isometric tetanus at a sarcomere length of 2.75 μm, i.e. at sarcomere length after stretch, though the force attained after stretch was higher than steady isometric force at a sarcomere length of 2.6 or 2.75 μm. These results suggest that the structures, in which the work done on the fiber during stretch is stored, may not be A-M links.

Sugi and coworkers also examined the equatorial X-ray diffraction pattern from tetanized whole muscles during slow stretches, and found that the intensity of 1,1 reflection decreases during stretch, while the intensity of 1,0 reflection remained almost unchanged (Amemiya et al. 1988; for the X-ray diffraction technique, see Chap. 4). Based on these findings, it has been suggested that a slow stretch would cause a disorder of the hexagonal lattice of the thick and thin filaments by displacing the thin filament from its trigonal position in each unit cell (Fig. 7B). This kind of disorder would result in an increase in the

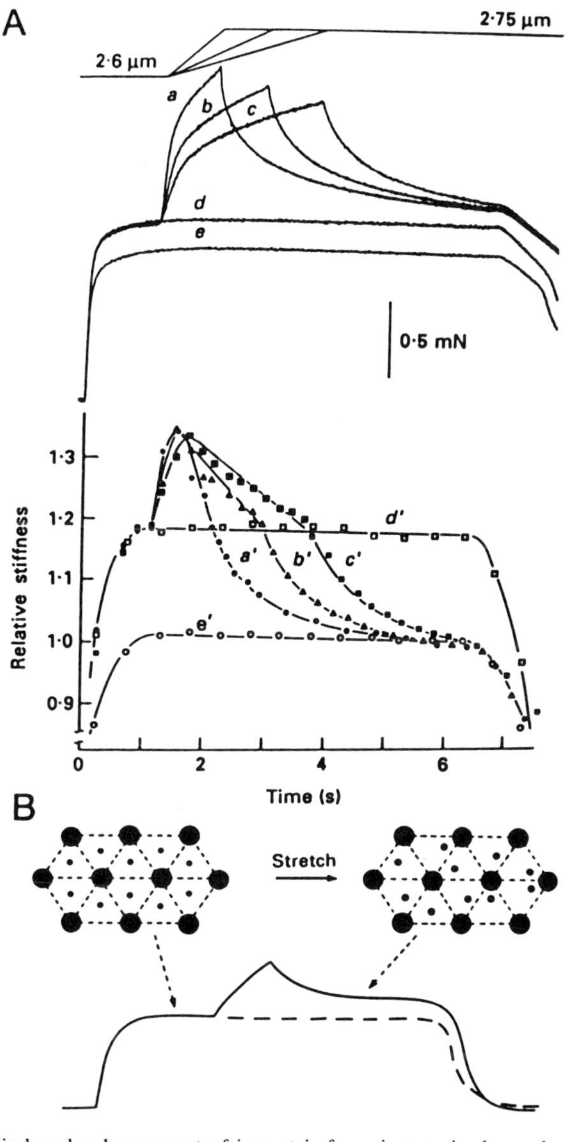

Fig. 7A,B. Mechanism of stretch-induced enhancement of isometric force in tetanized muscle fibers. **A** Force and stiffness changes of a tetanized muscle fiber in response to slow stretches. The fiber was first tetanized isometrically at sarcomere length 2.6 μm, and then stretched to sarcomere length 2.75 μm with three different velocities. Traces *a, b* and *c* in the force records correspond to traces a', b' and c' in the stiffness records respectively. Traces *d* and *e* are records of ordinary isometric tetanus at sarcomere length 2.6 μm and sarcomere length 2.75 μm, respectively, and correspond to traces d' and e' in the stiffness records. Note that the stiffness first increases abruptly at the beginning of stretch, then decreases linearly during stretch, and after the completion of stretch decreases to a steady level equal to that of ordinary isometric tetanus at sarcomere length 2.75 μm, while the isometric force attained after stretch is appreciably higher than the level of ordinary isometric force at sarcomere length 2.6 μm (Sugi and Tsuchiya 1988). **B** Stretch-induced disordering of the myofilament lattice as a possible cause of stretch-induced enhancement of isometric force. (Amemiya et al. 1988)

overall electrostatic repulsion forces between the filaments. As the distance between the filaments is shortened when the fiber is stretched due to the constant-volume behavior of the filament lattice (Elliott et al. 1963), the increased electrostatic repulsion forces in the lattice might show up as the force increment above the ordinary isometric force in the isotonic condition (Fig. 7A) or as the shift of the force-velocity curve towards higher force values during isotonic shortening. In fact, the stretch-induced disorder of the myofilament lattice is observed electron microscopically (Suzuki et al. 1989).

3.6 Shortening-Induced Deficit of Isometric Force

In contrast with the stretch-induced mechanical performance, slow releases (shortening) applied to a contracting muscle are known to reduce the subsequent isometric force development. When slow releases are applied at the plateau of isometric tetanus, the force decreases during release, and after the completion of release redevelops to a steady level lower than the level of ordinary isometric tetanus at the same muscle length (Abbott and Aubert 1952; Maréchal and Plaghki 1979); the slower the rate of release, the smaller the isometric force attained after release (see Fig. 8B).

Using experimental techniques to monitor the length of a central segment of single muscle fibers, Julian and Morgan (1979a,b) observed that, during the applied slow release, the shortening was taken up only by the fiber end segments, while the length of the central segment remained unchanged, suggesting that the deficit of isometric force after the slow release is largely due to nonuniformity of sarcomere length along the fiber length. Their observation has been confirmed by Sugi and Tsuchiya (1988), who recorded the length changes of the fiber segments along the entire fiber length in response to a slow release with a high-speed video system. As can be seen in Fig. 8A, the shortening applied to an isometrically contracting fiber is largely taken up by the two end segments (*a* and *e*), while the central segments (*b* and *c*) do not show appreciable shortening during release, and the above nonuniform segmental lengths persisted after the completion of release as long as tetanic stimulation continued. In contrast with the stiffness changes during slow stretches in which the force and the stiffness exhibited different time courses (Fig. 7A), the stiffness was observed to change always in parallel with the force during and after slow releases (Fig. 8B), suggesting that the (average) number of A-M links changes in parallel with the force. This may imply that the deficit of isometric force after slow releases results from a decrease in the force exerted by each A-M link (deactivation). As the force in a muscle fiber is the same in every fiber cross-section, the results shown in Fig. 8A may indicate that the deactivation of A-M links takes place not only in the slowly shortened segments, but also in the segments whose length stayed unchanged during slow releases.

The deficit of isometric force development also occurs if a tetanized muscle fiber is allowed to shorten at or near the maximum shortening velocity V_{max}. The shortening-induced deactivation of isometric force development is very

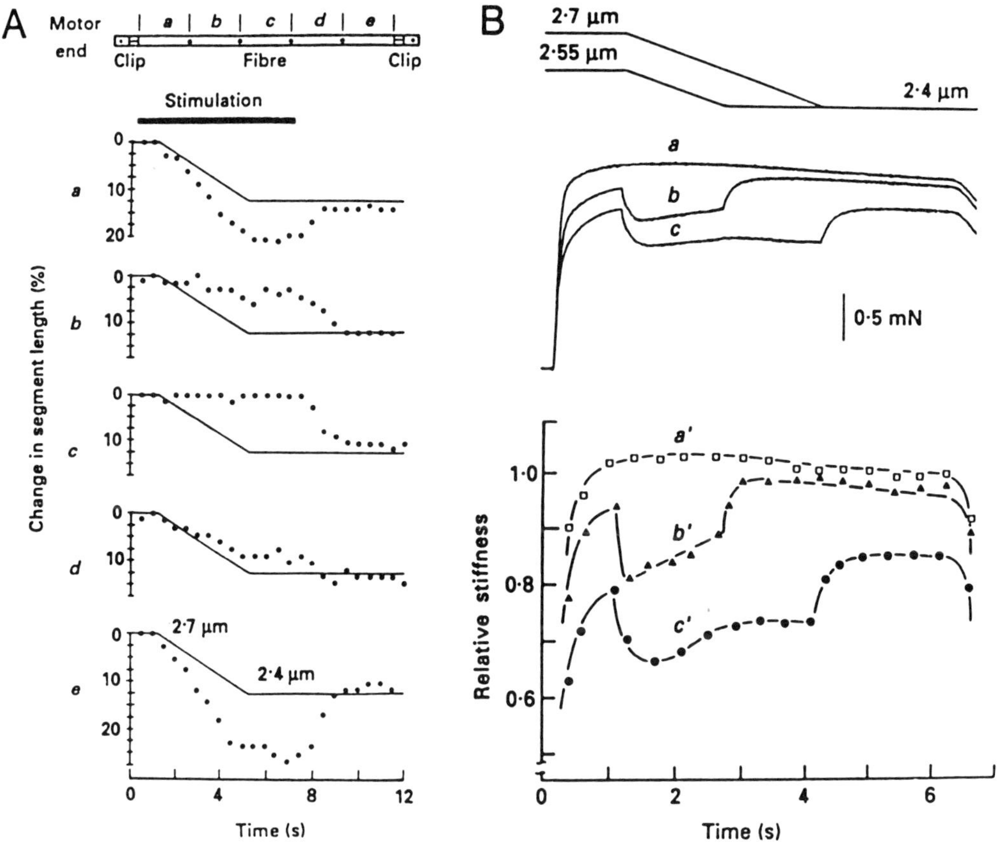

Fig. 8A,B. Shortening-induced deficit of isometric force in tetanized muscle fibers. **A** Segmental length changes of a tetanized muscle fiber in response to a slow release from sarcomere length 2.7 μm to sarcomere length 2.4 μm. The fiber was divided into five segments (*a*–*e*) with carbon surface markers. Note that the applied shortening is mostly taken up by the two end segments (*b* and *e*) while the length of the two central segments (*b* and *c*) does not change appreciably (Sugi and Tsuchiya 1988). **B** Force and stiffness changes of a tetanized muscle fiber in response to slow releases. The fiber was first tetanized isometrically at sarcomere length 2.55 μm (traces *b* and *b'*) or at sarcomere length 2.7 μm (traces *c* and *c'*), and then released to sarcomere length 2.4 μm. Traces *a* and *a'* are obtained from ordinary isometric tetanus at sarcomere length 2.4 μm. Note that force and stiffness change parallel to each other. (Sugi and Tsuchiya 1988)

marked in isometric twitches and unfused tetani (Edman 1975), but is much less pronounced in fused tetani (Edman 1980). Using glycerol-extracted muscle fibers, in which ions in external solution diffuse freely into the myofibrillar space, Ekelund and Edman (1982) found that the shortening-induced deactivation of Ca^{2+}-induced isometric force development could be reduced or totally eliminated by increasing the pCa of the surrounding medium, suggesting that the deactivation is caused by a decrease in troponin-Ca^{2+} binding in the thin filaments as a result of the preceding filament sliding.

3.7 Detailed Features of the Force-Velocity Relation

The value of the maximum shortening velocity V_{max} is normally determined by fitting the Hill equation to the force-velocity data points, and then extrapolating this equation to zero load (Hill 1938; Edman et al. 1976). Another method to determine the shortening velocity at zero load is to apply quick releases of various amplitudes to an isometrically tetanized fiber to make it slack and measure the time required for the fiber to take the slack up (Edman 1979). The shortening velocity V_0, determined in this way has been reported to be larger than V_{max} (Edman 1979; Julian and Moss 1981). According to Julian et al. (1986a), however, this discrepancy disappears when experiments are performed precisely. They also showed that shortening velocities under loads below 0.2 P_0 were definitely larger than the values obtained by simply fitting the data points with higher loads to the Hill equation.

An important argument concerning V_{max} in Ca^{2+}-activated skinned fibers is whether it remains constant (Podolsky and Teichholz 1970; Thames et al. 1974; Gulati and Podolsky 1978; Brenner 1980) or changes with pCa (Julian 1971; Julian and Moss 1981; Julian et al. 1986b). If V_{max} is independent of pCa, this implies that the kinetic properties of A-M links do not change with pCa, being consistent with the Huxley contraction model in which V_{max} is independent of the number of A-M links (Fig. 3C). If, however, V_{max} changes with pCa, it follows that the kinetic properties of A-M links are influenced by pCa.

Edman et al. (1976) noted that the force-velocity relation of single fibers deviated from the hyperbola at loads above 0.8 P_0. This deviation in the large load region has been carefully examined by Edman (1988) by recording the shortening velocity of the whole fiber and that of a fiber segment simultaneously under various isotonic loads, as shown in Fig. 9A. The force-velocity relation consisted of two regions, each being fitted to a hyperbola. The transition between the two regions of the force-velocity curve occurred at about 0.8 P_0, suggesting that the kinetics or performances of A-M links may change as the load exceeds 0.8 P_0.

Hill (1938) thought that the force-velocity relation in contracting muscle was instantaneous, so that the relation was valid not only in isotonic or isovelocity shortening but also under conditions in which the load was continuously changing with time. Iwamoto et al. (1990) determined the force-velocity relations of single muscle fibers in the auxotonic condition, i.e. the condition in which the fiber shortens against a Hookean spring. As shown in Fig. 9B, the force-velocity curves obtained were convex upwards in shape, and differed from one another depending on the compliance of the auxotonic load, indicating that under conditions in which the load is changing with time, the shortening velocity for a given amount of load changes depending on the past history of auxotonic shortening.

3.8 The Huxley-Simmons Contraction Model

A.F. Huxley and Simmons (1971) applied small quick length changes to an isometrically tetanized muscle fiber, and analyzed the resulting force changes;

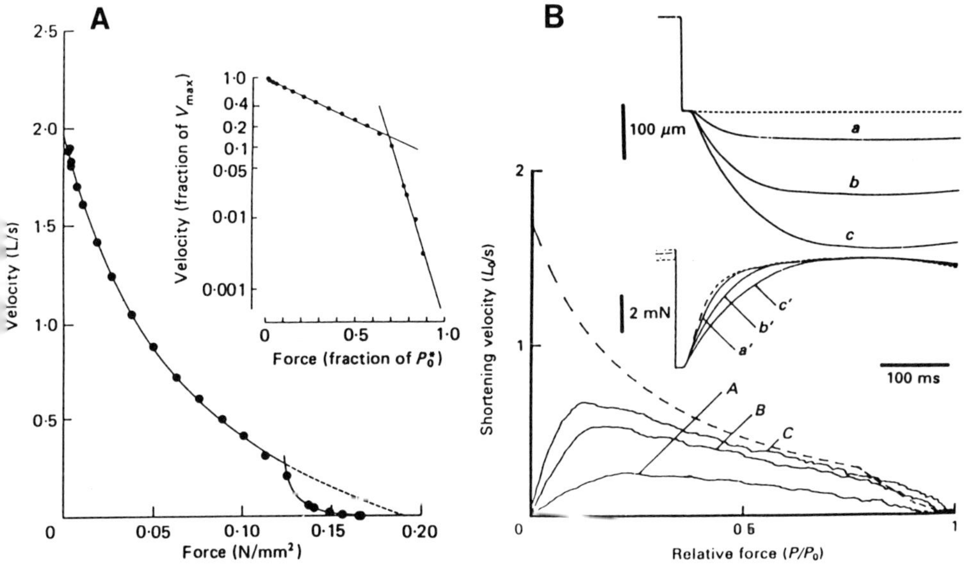

Fig. 9A,B. Force-velocity relations in single muscle fibers. **A** Force-velocity relation obtained in the isotonic condition. Note the deviation from hyperbola of the data points in the high force region. In the *inset*, shortening velocity (logarithmic scale) is plotted against isotonic load, showing the double-hyperbolic nature of the force-velocity curve (Edman 1988). **B** Force-velocity relation obtained in the auxotonic condition. Curves A, B and C were obtained from the length (traces a, b and c) and force (traces a', b' and c') records shown in the *inset*, respectively. The fiber was first tetanized isometrically, and then slackened to let it shorten against three different auxotonic loads [compliance, 1.2 (records a-a'), 3.5 (records b-b') and 5.9% fiber length/P_0 (records c-c'), respectively]. Note different curves for different auxotonic loads. *Broken line* represents isotonic force-velocity curve. (Iwamoto et al. 1990)

the length changes of a middle fiber segment between two surface markers were continuously recorded, thus eliminating the compliance of tendons and experimental apparatus (spot follower method). The force changes following a quick release (isometric force transients) consisted of three phases, i.e. (1) the initial quick force decrease from T_0 (= P_0) to T_1; (2) a quick force recovery from T_1 to T_2; and (3) the subsequent, much slower, force recovery towards the initial T_0 level (Fig. 10A). The phase 1 is coincident with the applied length change, and is considered the elastic recoil of the spring. Based on the analysis of phase 2 in connection with the amplitude of length change, Huxley and simmons proposed a model diagrammatically shown in Fig. 10B,C. Here, the myosin S-1 head (M) and the spring connecting it to the myosin filament are drawn more realistically than Huxley's 1957 model (Fig. 3A). Both M and the thin filament are assumed to have more than one combining site (M_1 to M_4 and A_1 to A_4). M first attaches to A by forming links M_1A_1 and M_2A_2, while the spring is at its slack length. Then, according to the assumption of the potential energy diagram for each attachment site A_1 to A_4, M changes its attachment sites from M_1A_1-M_2A_2 to M_2A_2-M_3A_3 and then to M_3A_3-M_4A_4. As a result, M rotates along the thin filament to pull the spring, and the resulting force in the spring shows up as isometric force if the thick and thin filaments are not allowed

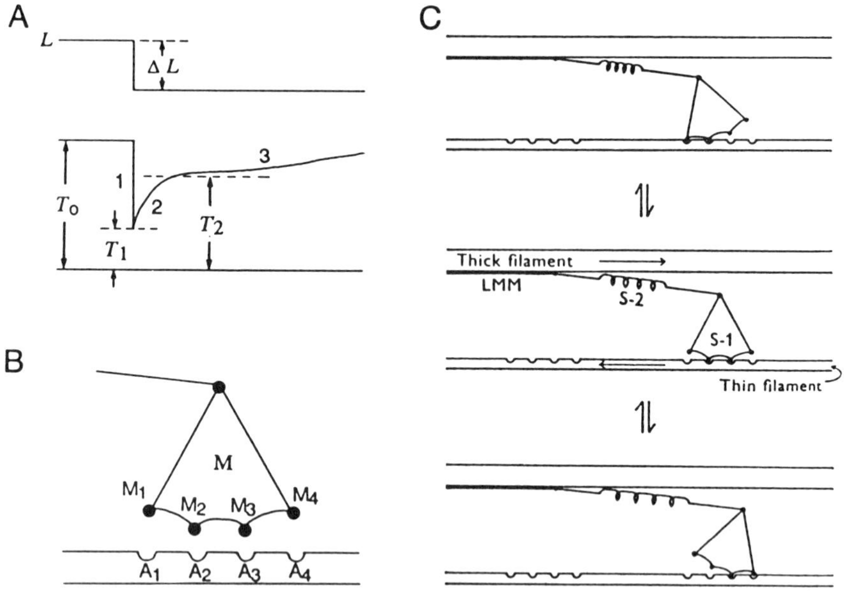

Fig. 10A–C. Huxley-Simmons contraction model (A.F. Huxley and Simmons 1971). **A** Diagram showing the components of the isometric force transients following a quick release. **B** Diagram showing the myosin S-1 head (M) with four binding sites (M_1-M_4) and part of the thin filament with four binding sites (A_1-A_4). **C** Mechanism of force generation by rotation of the myosin head on the thin filament by pulling the spring (S-2) between the myosin head and the thick filament. (A.F. Huxley 1974)

to slide past each other, or it causes the relative filament sliding under external loads $< P_0$. The ATP splitting is assumed to be coupled with the detachment of M from A. A quick release applied to an isometrically tetanized fiber causes a decrease in force in the spring, resulting in a quick force decrease (phase 1, Fig. 10A). This is followed by the increased probability that the attached M will rotate on the thin filament to pull back the spring, resulting in the quick force recovery (phase 2). The subsequent slow force recovery (phase 3) is due to the turnover of A-M links.

The main difference between the Huxley-Simmons model and Huxley's 1957 model is that in the former the A-M link has more than one state, including the initial nonforce-generating state, while in the latter the A-M link has only one state and generates force at the moment of its formation. The presence of the nonforce-generating A-M links is in accord with the results that both the stiffness increase and the transfer of mass from the thick to the thin filament (as evidenced by the change in the equatorial X-ray diffraction pattern, for a review, see Huxley and Faruqi 1983) occur ahead of the force development. As will be discussed later, the Huxley-Simmons model is useful in relating the hypothetical states of the A-M link to the enzyme kinetics of actomyosin ATPase in solution. The myosin head rotation as a possible mechanism of force generation has also attracted the attention of many investigators, and a number of experiments have been made to test the validity of this hypothesis. The amount of extension of the spring, when the force in a myosin head increases from zero to the maximum value, has been carefully estimated by A.F. Huxley and his coworkers (Ford et al. 1977, 1981) to be less than 4 nm.

In the studies of the isometric force transients, however, the length of a fiber segment is well controlled, but it is impossible to record the "segmental" force at the fiber end. Thus, if the length changes applied to one fiber end are so rapid that the time required for the transmission of a mechanical impulse along the entire fiber length is not negligible, the resulting rapid force change does not necessarily reflect the events taking place within the length-controlled segment. Sugi and Kobayashi (1983) examined the sarcomere length changes in response to a quick release along the length of a tetanized muscle fiber with a high-sensitivity laser diffraction technique. The propagation velocity of a mechanical wave along the tetanized fiber was about 180 m/s (Schoenberg et al. 1974) and was found to be faster than the propagation velocity of sarcomere length change along the fiber. Thus, the quick force recovery following a quick release was observed to start at the fixed fiber end while the sarcomere shortening was still going on near the fixed fiber end. These results indicate that the isometric tension transients are accompanied by nonuniform sarcomere length changes along the fiber, and may not give direct information about the behavior of A-M links in each sarcomere, though this does not reduce the value of the model per se.

4 Biochemical Aspects of Actin-Myosin Interaction

The molecular mechanism of contraction can be approached biochemically by studying the ATPase activity of myosin and the actin-myosin complex (actomyosin) in solution. To relate the enzyme kinetics of actomyosin in solution to the characteristics of muscle contraction described in the preceding sections, however, the following limitations should be kept in mind.

1. The function of actomyosin to convert energy from chemical reactions into mechanical work is lost in solution, so that all the free energy derived ATP hydrolysis is dissipated as heat and entropy.
2. No theory has hitherto been developed to deal with the probability of a reaction between actin and myosin in the myofilament lattice in muscle.
3. Both the ionic strength and the effective protein concentrations are much higher in muscle than in solutions in which the enzyme kinetics can be studied.

In spite of the above limitations, it has now become possible to discuss the biochemical results in connection with the mechanism of force generation in muscle to a certain extent.

4.1 Enzyme Kinetics of Myosin ATPase

As already mentioned, myosin is an ATPase. Its normal substrate for ATP hydrolysis in solution is Mg-ATP (Lymn and Taylor 1970). The kinetics of myosin ATPase is studied by examining the time course of ATP hydrolysis by myosin, HMM, or S-1 extracted from rabbit skeletal muscle. For convenience, the letter M will hereafter be used to denote a myosin S-1 head, in which the ATPase activity and the actin-binding site are localized, irrespective of whether myosin, HMM, or S-1 is used. Following mixing of ATP with M, there is an initial, very rapid liberation of inorganic phosphate (P_i) (phosphate burst), which is followed by a much slower P_i liberation (Tonomura et al. 1962; Lymn and Taylor 1970). This phenomenon is accounted for by the reaction

$$M + ATP \rightleftharpoons M \cdot ATP \rightleftharpoons M \cdot ADP \cdot P_i \xrightarrow{\text{slow}} M + ADP + P_i. \quad (1)$$

When ATP binds to M, it is rapidly hydrolyzed to form a ternary complex $M \cdot ADP \cdot P_i$. In the quenched-flow method used in this kind of experiment, the measured P_i consists of free P_i in solution and P_i bound to M. Therefore, the rapid formation of $M \cdot ADP \cdot P_i$ is the cause of the initial P_i burst. This is followed by a much slower rate-limiting release of ADP and P_i from M (0.02 mol of Pi per mol of M per s).

Later studies with the stopped-flow method, in which the liberation of P_i or ADP is measured spectroscopically, showed that Eq. (1) can be extended to the following equation (Trentham et al. 1972, 1976; Bagshow and Trentham 1973; Bagshaw et al. 1974):

$$M + ATP \rightleftharpoons M \cdot ATP \rightarrow M^* \cdot ATP \rightleftharpoons M^{**} \cdot ADP \cdot P_i \xrightarrow{\text{slow}} M^* \cdot ADP + P_i \rightleftharpoons M + ADP. \quad (2)$$

Asterisks indicate increases in the fluorescence of M arising from tryptophane residues in the S-1 head (the more asterisks, the stronger the fluorescence). The fluorescence changes reflect conformational changes of M as a result of interaction with ATP. The rate-limiting step in Eq. (2) is either due to slow protein isomerization or to slow liberation of P_i.

4.2 Enzyme Kinetics of Actomyosin ATPase

Actin monomers (G-actin) normally form two helical strands wound around each other (F-actin). In the absence of ATP, actin and myosin form actomyosin gel at low ionic strength due to tight rigor A-M links. When ATP is added to the actomyosin gel, it first clears due to dissociation of A-M links, and later precipitates rapidly (superprecipitation). As the whole myosin molecules tend to form heterogeneous clumps with actin in solution, the actomyosin ATPase activity is normally studied with homogeneous actin plus HMM (acto-HMM) or actin plus S-1 (acto-S-1) systems.

When acto-HMM is mixed with ATP, the amount of initial Pi burst is similar to that when only myosin is mixed with ATP (Lymn and Taylor 1971), while the subsequent steady-state ATPase activity increases more than 200 times (Eisenberg and Moos 1968, 1970). This indicates that the rate-limiting step, i.e. $M \cdot ADP \cdot Pi \rightarrow M \cdot ADP + P_i$, is accelerated by actin. Taking these and other results into consideration, the most probable sequence of reaction taking place in contracting muscle may be the following:

$$
\begin{array}{l}
\qquad\qquad\qquad\qquad\qquad\qquad\qquad M^{*}ATP \rightarrow M^{**} \cdot ADP\ P_i \\
\qquad\qquad\qquad\qquad\qquad A \ \leftarrow \uparrow \qquad\qquad\qquad\qquad \downarrow \ \leftarrow \ A \\
AM + ATP \rightarrow AM \cdot ATP \rightarrow AM^{*} \cdot ATP \quad AM^{*} \cdot ADP \cdot P_i \rightarrow AM \cdot ADP \cdot P_i \\
\uparrow \qquad\qquad\qquad\qquad \rightarrow AM \cdot ADP + P_i \rightarrow AM + ADP.
\end{array}
\qquad (3)
$$

The above reaction cycle includes attachment of actin (A) to, and its detachment from, the binding site on M, thus providing a simple correspondence to the contraction models described in the previous sections.

Since Eq. (3) is derived from experiments in which actin concentrations were much lower than that required to bring the steady-state actin-activated ATPase to its maximum value, Eisenberg and his coworkers carried out experiments to test the validity of Eq. (3) under steady-state actomyosin ATPase activity (by using very low ionic strength), and introduced an additional pathway in Eq. (4) in which ATP is hydrolyzed by the acto-S-1 complex (Eisenberg et al. 1972; Eisenberg and Keilly 1973; Stein et al. 1981):

$$
\begin{array}{l}
\qquad\qquad\qquad\qquad\qquad\qquad\qquad M^{*} \cdot ATP \rightarrow M^{**} \cdot ADP \cdot P_i \\
\qquad\qquad\qquad\qquad\qquad A \qquad\qquad\qquad\qquad\qquad \downarrow \ \leftarrow \ A \\
AM + ATP \rightarrow AM \cdot ATP \rightarrow AM^{*} \cdot ATP \rightarrow AM^{*} \cdot ADP \cdot P_i \ \rightarrow \\
\uparrow \qquad\qquad\qquad\qquad AM \cdot ADP \cdot P_i \rightarrow AM \cdot ADP + P_i \rightarrow AM + ADP.
\end{array}
\qquad (4)
$$

The above reaction schemes have stimulated muscle physiologists to perform experiments linking the kinetics of actomyosin ATPase and muscle mechanics data.

4.3 Experiments Relating the Kinetics of Actomyosin ATPase to Contraction

A question that arises regarding the reaction schemes of actomyosin ATPase is which step is coupled with force generation in contracting muscle? It is now generally agreed that $M \cdot ATP$ and $M \cdot ADP \cdot P_i$ bind to actin weakly, so that the steps $M \cdot ATP \rightarrow AM \cdot ATP$ and $M \cdot ADP \cdot P_i \rightleftharpoons AM \cdot ADP \cdot P_i$ are in rapid equilibrium. On the other hand, M or $M \cdot ADP$ can bind to actin very tightly (for a review, see Hibbard and Trentham 1986). This implies that force generation in muscle may be coupled with the transition of the myosin head from the weakly binding state to the strongly binding state. Apart from the above biochemical evidence for the two bound states of the myosin head, Brenner et al. (1982) found a new type of A-M link in skinned rabbit psoas muscle fibers at low ionic strengths and in the absence of Ca^{2+}. The fiber stiffness was extremely sensitive to the velocity of stretch, indicating that A-M links are rapidly equilibrating rather than static. X-ray diffraction studies demonstrated that A-M links at low ionic strengths are different in configuration from rigor links (Matsuda and Podolsky 1984). It remains, however, to be determined whether the configuration of the low ionic strength A-M link is the same as that of the weakly binding intermediates in the normal actomyosin ATPase cycle.

In the absence of Ca^{2+}, the addition of ATP to skinned muscle fibers in rigor state breaks rigor A-M links. The kinetics of this transition, AM + ATP $\rightarrow$ A + M · ATP, were studied by Goldman et al. (1982) using "caged ATP". This compound has no effect on the actin-myosin interaction in muscle, but it releases ATP by photolysis with a rate constant of about $100\,s^{-1}$. Thus, if caged ATP is made to diffuse uniformly into skinned fibers and then is photolysed with a brief intense laser pulse, the resulting changes in force and stiffness reflect the kinetics of the ATP-induced breaking of rigor A-M links. The rate of the ATP-induced detachment of A from M was found to be far more rapid than that expected from the Huxley contraction model, in which the turnover rate of A-M links in the isometric condition is limited by their rate of detachment (small value of g at $0 \leqslant x \leqslant h$ in Fig. 3B). More recently, Danzig et al. (1987) studied the effect of photolysis of "caged P_i" on isometrically contracting glycerinated fibers, and showed Pi-induced exponential decay of isometric force with rates increasing with Pi concentration. This result is consistent with the idea that the increased Pi concentration reverses the force-generating steps in Eq. (4), i.e. $AM^* \cdot ADP \cdot P_i \rightleftharpoons AM\ ADP \cdot P_i \rightleftharpoons AM \cdot ADP + P_i$. The use of such "caged" compounds seems to be very promising in connecting the kinetics of actomyosin ATPase to the characteristics of muscle contraction.

Though not directly relevant to the reaction schemes described in this section, Harrington and his coworkers (Ueno and Harrington 1981; Harrington et al. 1988) made enzyme-probe experiments which showed local melting of the α-helical structure in the S-2 region of rabbit psoas muscle fibers when S-2 was detached from the thick filament surface, suggesting the possibility that if S-2 swings away from the thick filament surface, the S-2 melting would shorten its length to result in force generation when the myosin head attaches to the thin filament. Pollack (1990) even considers the possibility that contraction may result from shortening of the thick filament per se, based on in vitro thick filament shortening observed in horseshoe crab muscle (Dewey et al. 1979). Sugi and Gomi (1981, 1984), however, examined the sarcomere shortening of this muscle and concluded that thick filament shortening is not associated with physiological contraction.

4.4 Possible Cooperativity of the Two Myosin Heads in Muscle Contraction

The fact that myosin has two S-1 heads suggests the possibility that the two heads work cooperatively in muscle contraction. Tonomura, a pioneer in studies of myosin and actomyosin ATPase, insisted that the two heads were not identical but differed from each other in both structure and function (Tonomura 1972; Inoue et al. 1979). According to Tonomura and his coworkers, only one head of myosin hydrolyzes ATP via $M \cdot ADP \cdot P_i$ to cause the initial P_i burst, while the other head hydrolyzes ATP without forming this complex. They put forward a highly speculative mechanism in which the two myosin heads produce

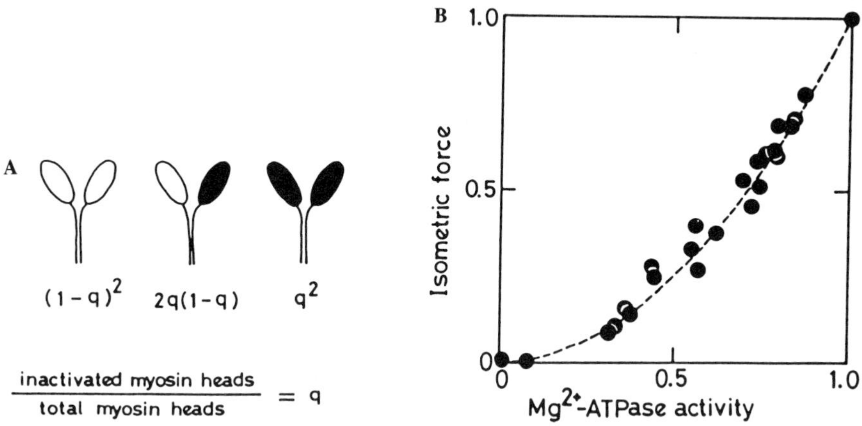

Fig. 11A,B. Evidence for cooperativity of two myosin heads in muscle contraction. **A** Three different species of myosin molecules and their respective fractions calculated on the assumption of an independent reaction of the two heads with PDM. *Filled head* represents PDM-inactivated one. **B** Relation between the Ca^{2+}-activated force development and the ATPase activity in partially PDM-inactivated glycerinated muscle fibers. Both the force and the ATPase activity are expressed relative to the control value. (Chaen et al. 1986)

sliding between the thick and thin filaments in a cooperative manner. Unfortunately, their result that the amount of P_i burst was 1 mol per mol of myosin was not confirmed by other investigators, and the concept of nonidentical myosin heads is not widely accepted.

Evidence has been presented, on the other hand, that the two myosin heads are identical in structure and function, and work cooperatively in muscle contraction. Tokiwa and Morales (1971) explored the possible cooperativity between the two myosin heads by using an ATP analog that attaches to myosin heads and causes loss of ATPase activity. If the two heads are identical, a myosin solution partially inactivated by this analog will be a mixture of three species of myosin molecules with zero, one and two heads inactivated, and their concentrations are $(1 - q)^2$, $2q(1 - q)$, and q^2, respectively (q is the total fraction of inactivated heads) (Fig. 11A). They found that the ATPase activity of a partially inactivated myosin solution was proportional to the total fraction of native heads $(1 - q)$, while the rate of superprecipitation was proportional to the fraction of myosin molecules with two native heads $(1 - q)^2$. This suggests that the two myosin heads work cooperatively in muscle contraction, since superprecipitation is generally regarded to be comparable with muscle contraction.

More recently, Chaen et al. (1986) made simultaneous measurements of the Ca^{2+}-activated isometric force and the ATPase activity of glycerinated rabbit psoas fibers, which were treated with various concentrations of p-phenylenedimaleimide (PDM) reacting with sulfhydryls on the myosin head to cause loss of ATPase activity and binding to actin (Reisler et al. 1974; Burke et al. 1976; Lamed et al. 1976). As shown in Fig. 11B, it was found that the magnitude of isometric force was proportional to the square of ATPase activity.

On the basis of the probabilistic calculation of the concentrations of three myosin head species (Fig. 11A), these results are in accord with Tokiwa and Morales's results, strongly suggesting that myosin molecules in the thick filaments can interact with the thin filaments to generate force only when their two heads are native, while there is no cooperativity between the two heads in catalyzing ATP hydrolysis.

5 Conformational Changes of the Myosin Head During Muscle Contraction

In spite of extensive physiological and biochemical studies, the question as to what makes the myofilaments slide past each other, is not yet answered. Since the most important process for converting chemical energy into mechanical work in muscle may be conformational changes of the S-1 head of myosin (cross-birdge) attached to the thin filament, a number of attempts have hitherto been made to detect such changes.

5.1 Electron Microscope Studies

Reedy et al. (1965) were the first to obtain clear electron micrographs of the myosin heads in ultrathin sections of insect flight muscle prepared by the conventional OsO_4 fixation technique. They found that the angle between the myosin head and the thick filament axis was predominantly 90° in relaxed muscle and predominantly 45° in rigor muscle (inclined towards the center of the thick filament). This finding has been widely taken to suggest that the myosin head first attaches to the thin filament at 90° and then rotates on the thin filament to generate force for myofilament sliding. This implies that the 45° rigor orientation of the myosin heads corresponds to their orientation at the end of their "powerstroke". Based on this idea, the orientation of the myosin head in insect flight muscle has been examined electron microscopically (with X-ray diffraction monitoring) under various "rigor-like" conditions including the use of nonhydrolyzable ATP analogs (Kuhn 1973; Reedy et al. 1983; Tregear et al. 1990); however, no clear conclusion has been reached. According to Heuser and Cooke (1983), the rigor myosin head orientation may be dependent on myofilament-lattice spacing which is sensitive to the conditions of fixation.

Recent development of quick-freezing techniques, coupled with freeze-fracture or freeze-substitution procedures, has made it possible to obtain clear myosin head images in vertebrate skeletal muscle fibers. Using these techniques, Tsukita and Yano (1985) studied the myosin head orientation in rabbit skeletal muscle fibers during isometric force generation and in rigor. They examined the optical diffraction patterns of the electron micrographs, and found that, during isometric force generation, almost all the myosin heads were regularly arrayed around the thin filaments, maintaining their adxial 14.3 nm spacings, while in rigor state the myosin heads no longer maintained their regular array. This finding seems to be inconsistent with the expectation that if isometric force is generated by "powerstroke" of the myosin heads working indepently of one another, their regular array would disappear during contraction. In this connection, it is of interest that during an isometric tetanus, muscle stiffness increases in the longitudinal direction and decreases in the transverse direction, while it increases in both directions in rigor state (Hatta et al. 1988). These results may be taken to suggest that there would be little or no myosin heads taking the rigor orientation during contraction.

When rigor muscle fibers are stretched, the average myosin head angle does not change appreciably in either insect and frog muscles (Suzuki et al. 1986; Trombitas et al. 1988), indicating that the cross-bridges can not be readily rotated by externally applied forces.

5.2 X-Ray Diffraction Studies

X-ray diffraction provides a powerful tool for studying the behavior of the myosin heads in living muscle in a noninvasive manner. Due to the recent development of the position-sensitive photon counter and the intense X-ray sources from synchrotron radiation, it is now possible to investigate the changes in X-ray diffraction patterns from contracting muscle with high time resolutions. The equatorial X-ray diffraction patterns of vertebrate skeletal muscle are produced by the hexagonal myofilament lattice, and is sensitive to the distribution of mass within the unit cell, reflecting the myosin head orientation.

Since the myosin heads bind tightly with the thin filaments in rigor state, the heads may change their orientation through the application of external forces by stretching rigor fibers. Naylor and Podolsky (1981), however, observed no appreciable effect of stretch on the intensity ratio of the 1,0 and 1,1 equatorial reflections from rigor rabbit muscle fibers. Tanaka et al. (1984) observed only a small stretch-induced decrease of the 1,1 reflection intensity, which could be due to a disorder of the 1,1 lattice, but not to the change in myosin head orientation. These results seem to be in good agreement with the electron microscope studies, indicating that, in rigor muscle fibers, the attachment angle of the myosin heads to the thin filament is so stable that it cannot be appreciably altered by external force. On the other hand, no appreciable changes in the intensity of the 1,0 and 1,1 equatorial reflections have been observed in contracting frog muscle in response to a quick release to drop the force from P_0 to zero (Amemiya et al. 1980; Sugi et al. 1980; Huxley et al. 1981, 1983). This may indicate that the force generation in muscle may not be associated with the rotation of the whole myosin head.

The intensity of the 14.3 nm meridional reflection, arising from the axial spacing of the myosin heads on the thick filament, falls at the onset of contraction, but then increases beyond the resting value during the plateau of isometric tetanus (Yagi et al. 1981; Huxley et al. 1982). This result seems to correspond with Tsukita and Yano's electron microscope observation that most of the myosin heads maintain their regular array during contraction. The intensity of the 14.3 nm meridional reflection decreases markedly in response to a quick release applied to a contracting myscle (Huxley et al. 1983). According to them, this reflection might arise from the axial spacings of the junction between the myosin S-1 head and the S-2, which is relatively fixed in an axial position during contraction due to relatively small S-2 length changes (4 nm, Ford et al. 1977) during the working cycle of the myosin head; a quick release would cause buckling in a considerable proportion of the S-2 links, thus disrupting the 14.3 nm axial repeat of the S-1-S-2 junction.

5.3 Other Studies

The behavior of the myosin heads during muscle contraction has also been studied intensively with a variety of physicochemical methods. Morales and his coworkers succeeded in measuring changes in the polarization of tryptophane fluorescence from the myosin heads in relaxed, contracting, and rigor muscle fibers, and showed different myosin head orientations corresponding to the three states (Aronson and Morales 1969; dos Remedios et al. 1972). They further attached a fluorescent probe to sulfhydryls on the myosin head, and detected fluctuations of polarized fluorescence arising from changes in orientation of the probes during muscle contraction. This indicates that the myosin heads rotate repetitively during contraction (Borejdo et al. 1979), though it remains to be determined whether the motion occurs in the myosin heads attached to the thin filament.

Yanagida (1984) also studied the motion of the myosin head by using fluorescent ATP and ADP analogs attached to the myosin head in muscle. During isometric contraction, the bound nucleotides were ordered with respect to the fiber axis, suggesting a very limited amplitude of the myosin head "powerstroke" during contraction. The orientation of the nucleotides bound to the myosin heads did not change by sudden length changes of contracting fibers (Yanagida 1985). Cooke et al. (1982, 1984) used electron paramagnetic resonance spectroscopy to examine the behavior of spin labels attached to sulfhydryls on the myosin heads in muscle. During isometric contraction, 80% of the probes showed a random angular distribution, while the remaining 20% were highly ordered. They believe that the ordered component arises from the heads attached to the thin filament, again implying that the amplitude of powerstroke may be very limited. The angle of spin labels attached to the myosin head in rigor fibers did not change by applying static stress (Cooke 1981).

Thus, the results of electron microscope, X-ray diffraction, and physicochemical studies described in this section all seem to indicate that the powerstroke of the myosin head may not be associated with the rotation of the whole myosin head. If the position of the reactive sulfhydryls on the head is taken into consideration, the conformational changes of the myosin head are likely to be localized in the region near the junction between S-1 and S-2. For a precise detection of S-1 structural changes, measurement of distances between amino acid residues in the S-1 head using the technique of fluorescent energy transfer (Morales and Botts 1979; Botts et al. 1989) seems to be promising.

6 In Vitro Assay Systems for Studying the Actin-Myosin Interaction

Recently, methods have been developed with which the force and movement due to the ATP-dependent actin-myosin interaction can be assayed under a light microscope. The in vitro assay systems have proved to be extremely useful in eliminating the gap between muscle physiology and muscle biochemistry, because these systems enable us to directly observe force generation and movement involving very small numbers of actin and myosin molecules.

6.1 In Vitro Movement Assay Systems

Oplatka and his coworkers (Oplatka et al. 1974; Borejdo and Oplatka 1976) were the first to demonstrate that force generation occurs in a system without continuous three-dimensional myofilament-lattice structures. They extracted the thick filaments from glycerinated rabbit muscle fibers, and showed that this system could develop force after irrigation with HMM, single-headed HMM, or S-1. This important finding has stimulated a number of investigators to construct various in vitro assay systems for studying the actin-myosin sliding. Two such systems have proven useful.

One system, developed independently by Sheetz and Spudich (1983) and Shimmen and Yano (1984), consists of well-oriented actin filament arrays (actin cables) in the giant internodal cells of green algae and small latex beads coated with rabbit skeletal muscle myosin. In the presence of ATP, the myosin-coated beads continue to move on the actin cables in the direction determined by the polarity of the actin cables (i.e. similar to that of protoplasmic streaming). The unloaded velocity of the bead movement is comparable to that of myofilament sliding at V_{max}.

In the other system, fluorescently labeled F-actin filaments are made to slide randomly on the samples of myosin (native or single-headed), HMM, or S-1 fixed on a glass surface in the presence of ATP (Kron and Spudich 1986; Harada et al. 1987; Toyoshima et al. 1987). An interesting issue which was brought forth by this assay system is that single-headed myosin or S-1 can cause the F-actin sliding, contrary to the view of Chaen et al. (1986) that one-head inactivated myosin molecules can no longer generate force in muscle. This apparent discrepancy may arise from the fact that the one-head inactivated myosin is different from the single-headed myosin both structurally and functionally. The one inactivated head may exert inhibitory influence on the other native head of the same molecule, while each of the two heads has the potential ability to generate force. In fact, if the one inactivated head is detached from a myosin molecule, the other native head is shown to contribute to force generation (Chaen et al. unpublished). In the single-headed myosin or S-1 samples, it seems possible that the two adjacent S-1 heads would work cooperatively to produce the F-actin sliding in a manner similar to that between the two heads of a myosin molecule, though Harada and Yanagida (1988) claim that this is unlikely. It should also be noted that with physiological ionic strength, the ATP-dependent

F-actin sliding occurs with native myosin, but not with single-headed myosin (Yanagida and Harada 1988).

Another interesting point in the fluorescent F-actin versus myosin system is concerned with the distance of the actin-myosin sliding per hydrolysis of one ATP molecule (for a review, see Huxley 1990). According to Yanagida and his coworkers, the estimated value (100–300 nm) is more than ten times larger than the amplitude of the "powerstroke" of the myosin head, suggesting that a myosin head would repeat powerstrokes many times, while one ATP molecule is hydrolyzed. Spudich's group is of the opinion that the value is 5–30 nm, being within the framework of the contraction models of Huxley (1957) and Huxley and Simmons (1971). Although the actin and myosin samples in this system are chemically well defined, it is not clear what factors determine the actin-myosin sliding velocity. In addition, this system seems to be unsuitable for systematically studying the relation between force and sliding velocity.

More recently, another useful in vitro movement assay system has been developed independently by Sellers and Kachar (1990) and Yamada et al. (1990). In this system, the fluorescently labeled F-actin filament is made to slide along an extremely long, thick filament isolated from a molluscan smooth muscle in the presence of ATP. An F-actin filament can slide along the entire length of the thick filament region, but its velocity differs markedly by a factor of about ten across the bare region. Since the polarity of the myosin head arrangement on the thick filament is reversed across the bare region, the above result indicates that the polarity of myosin heads affects the velocity of the ATP-dependent actin-myosin sliding, while the sliding direction is determined by the polarity of the thin filament.

6.2 In Vitro Force-Movement Assay Systems

Two in vitro assay systems have been developed in our laboratory, with which the relation between force and movement of the ATP-dependent actin-myosin interaction can be studied systematically. One system consists of a glass microneedle coated with rabbit skeletal muscle myosin and the actin cables of green algae (Chaen et al. 1989). In the presence of ATP, the myosin-coated needle moves along the actin cables until it stops due to a balance between forces generated by the actin-myosin head contacts and by the bent needle (Fig. 12A). Since the force on the needle is calculated from its displacement and stiffness, the force-velocity relation of the actin-myosin sliding can be obtained in the auxotonic condition, i.e. the condition in which the load on the myosin heads increases with the progress of the actin-myosin sliding. The force-velocity curve thus obtained was convex upwards in shape (Fig. 12B), being analogous to the force-velocity curve of muscle fibers obtained in the auxotonic condition (Fig. 9B). This indicates that the force-movement assay system retains the basic characteristics of muscle contraction.

Using this system, Oiwa et al. (1991) measured the amount of work done by the ATP-induced actin-myosin interaction. The myosin-coated needle first attached to the actin cables due to rigor A-M links, and upon application of a

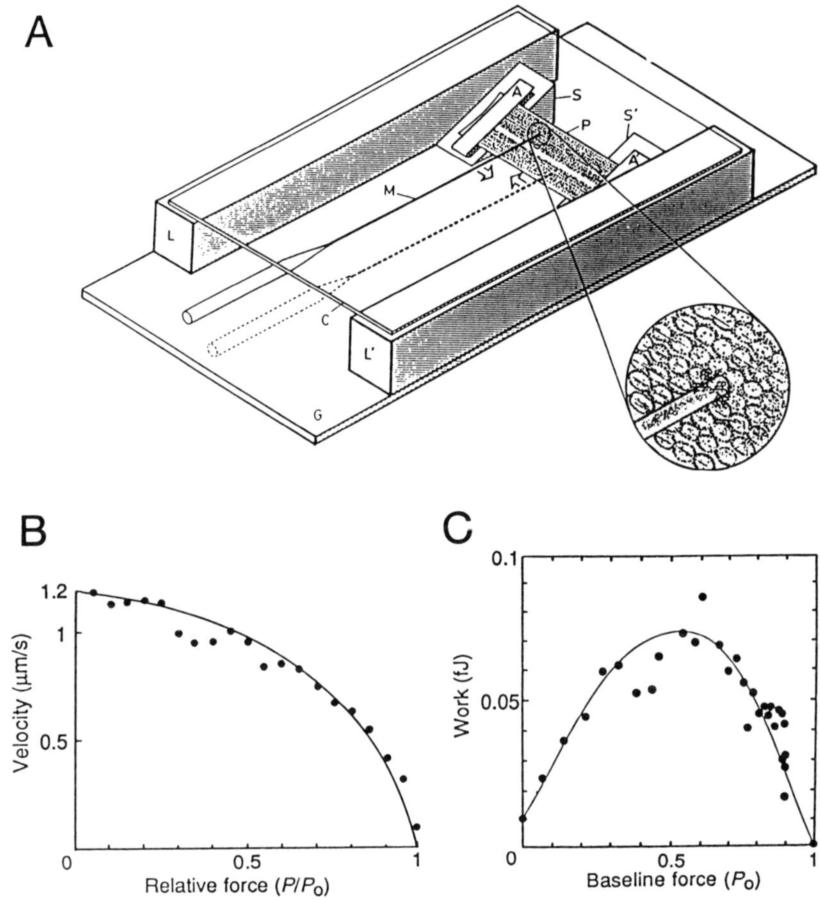

Fig. 12A–C. In vitro force-movement assay system consisting of a myosin-coated microneedle and actin cables. **A** Diagram of experimental arrangement. The tip of a myosin-coated needle (M) is placed in contact with the actin cables on an algal cell strip preparation (P). *Arrows* indicate the direction of needle movement determined by the actin cable polarity, which is reversed across the indifferent zone located at the center of the preparation (see Fig. 13A; Chaen et al. 1989). **B** Force-velocity curve of the needle sliding along the actin cables (Chaen et al. 1989). **C** Relation between the amount of work done by the actin-myosin interaction induced by a constant amount of ATP application and the initial baseline force from which the needle starts moving. (Oiwa et al. 1991)

current pulse to release ATP from the ATP electrode, it moved on the actin cables for a distance, and stopped at the new position due to reformation of rigor links when the applied ATP diffused away from the needle. Thus, it was possible to induce ATP-induced needle movement, starting at various initial baseline forces. As shown in Fig. 12C, the amount of work done by the needle with a constant amount of ATP application increased as the initial baseline force was increased from zero to a certain value, and then decreased towards zero as the baseline force approached the maximum "isometric" force P_0 generated by

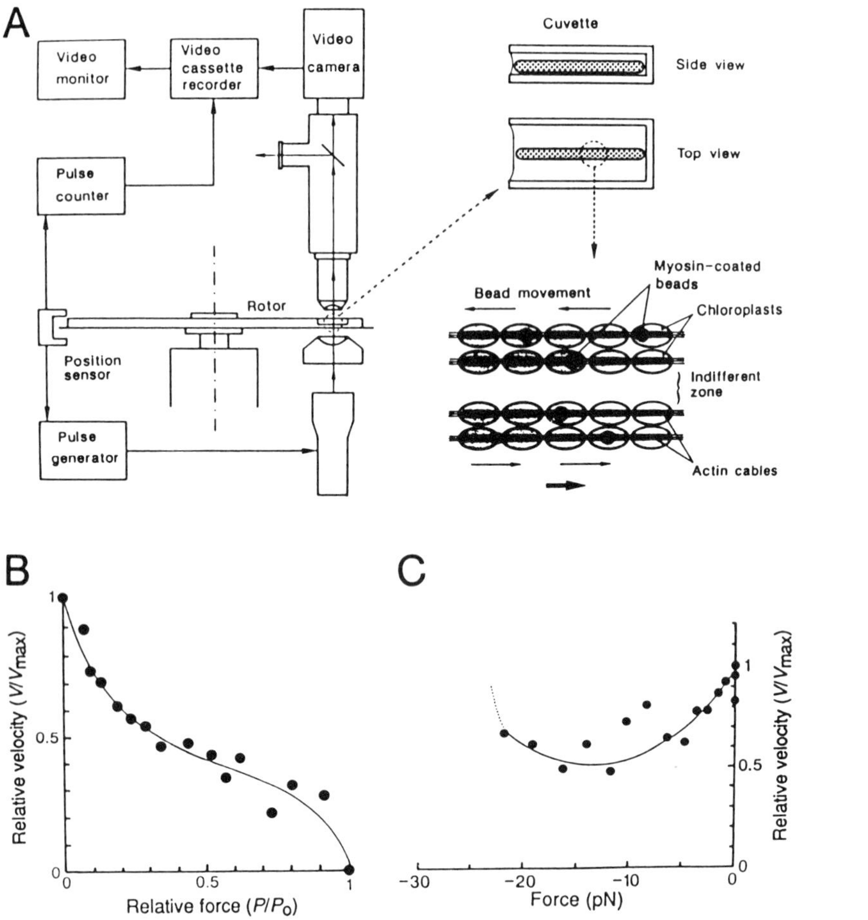

Fig. 13A–C. In vitro force-movement assay system combined with a centrifuge microscope. **A** Diagram of experimental arrangement. The cuvette containing an algal cell segment is mounted on the rotor of a centrifuge microscope equipped with a video system. The direction of bead movement is reversed across the indifferent zone, where no chloroplasts and no actin cables exist, because of the reversal of the actin cable polarity. **B** Steady-state force-velocity relation of a moving bead. Note that the curve is hyperbolic in the low force region, but deviates from the hyperbola at the high force region. **C** Steady-state force-velocity relation for "negative" loads. Note that the velocity of bead movement decreases markedly with increasing "negative" load. (Oiwa et al. 1990)

the actin-myosin interaction. The bell-shaped work versus baseline force curve resembles closely the work versus load relation in contracting muscle (Fig. 2F), indicating again that the interaction between the randomly oriented myosin heads and the actin cables retains the basic characteristics of contracting muscle.

Another force-movement assay system is a myosin-coated beads versus actin-cable system mounted on a centrifuge microscope (Oiwa et al. 1990; Fig. 13A). By using polystyrene beads, it was possible to effectively apply centrifugal forces to the beads moving on the actin cables. By applying constant centrifugal forces directed opposite to the bead movement, the steady-state force-velocity relation was obtained (Fig. 13B). The force-velocity curve was hyperbolic in shape in the low load region, but deviated from the hyperbola in the high load region, being analogous to the force-velocity curve of single muscle fibers (Edman 1988; Fig. 9A). An unexpected result was obtained with centrifugal forces in the same direction as that of the bead movement, serving as "negative" loads. As can be seen in Fig. 13C, the velocity of the bead movement decreased markedly with an increasing amount of negative load until the bead was eventually detached from the actin cable to flow away in the direction of centrifugal force. In the Huxley contraction model, V_{max} is determined by a balance between positive and negative forces exerted by the myosin heads (Fig. 3C). It is therefore difficult to account for the above unexpected result on the basis of the Huxley model. In this system, the amount of P_0 mostly ranged from 10 to 20 pN, suggesting that the number of myosin molecules responsible for the bead motion was only five to ten. If this is taken into consideration, the continuous force-velocity curve would reflect the basic properties of individual myosin heads rather than the change in the number of myosin heads involved in the bead movement.

Thus, the in vitro force-movement assay systems seem to be extremely useful, not only in connecting muscle physiology and muscle biochemistry, but also in solving new "mysteries" in the molecular mechanism of muscle contraction.

References

Abbott BC, Aubert XM (1951) Changes of energy in muscle during very slow stretches. Proc R Soc Lond Ser B 139:104–117

Abbott BC, Aubert XM (1952) The force exerted by active striated muscle during and after change of length. J Physiol 117:77–86

Amemiya Y, Tameyasu T, Tanaka H, Hashizume H, Sugi H (1980) Time-resolved X-ray diffraction from frog skeletal muscle during shortening against an inertial load and a quick release. Proc Jpn Acad 56(B):235–240

Amemiya Y, Iwamoto H, Kobayashi T, Sugi H, Tanaka H, Wakabayashi K (1988) Time-resolved X-ray diffraction studies on the effect of slow length changes on tetanized frog skeletal muscle. J Physiol 407:231–241

Armstrong CM, Huxley AF, Julian FJ (1966) Oscillatory responses in frog skeletal muscle fibres. J Physiol 186:26–27P

Aronson JF, Morales MF (1969) Polarization of tryptophane fluorescence from glycerol-extracted muscle fibers. Biochemistry 8:4517–4522

Bagni MA, Cecchi G, Colomo F, Tesi C (1988) Plateau and descending limb of the sarcomere length-tension relation in short length-clamped segments of frog muscle fibres. J Physiol 401:581–595

Bagni MA, Cecchi G, Colomo F, Poggesi C (1990) Tension and stiffness of frog muscle fibres at full filament overlap. J Muscle Res Cell Motil 11:371–377

Bagshaw CR, Trentham DR (1973) The reversibility of adenosin triphosphate cleavage by myosin. Biochem J 133:323–328

Bagshaw CR, Eccleston JF, Eckstein F, Goody RS, Gutfreund H, Trenthan DR (1974) The magnesium ion-dependent adenosine triphosphatase of myosin. Two-step processes of

adenosine triphosphate association and adenosine dephosphate dissociation. Biochem J 141:351–364
Borejdo J, Oplatka A (1976) Tension development in skinned glycerinated rabbit psoas fiber segments irrigated with soluble myosin fragments. Biochim Biophys Acta 440:241–258
Borejdo J, Putnam S, Morales MF (1979) Fluctuations in polarized fluorescence: evidence that muscle cross bridges rotate repetitively during contraction. Proc Natl Acad Sci USA 76:6346–6450
Botts J, Thomason JF, Morales MF (1989) On the origin and transmission of force in actomyosin subfragment-1. Proc Natl Acad Sci USA 86:2204–2208
Brenner B (1980) Effect of free sarcoplasmic Ca^{2+} concentration on maximum unloaded shortening velocity: measurements on single glycerinated rabbit psoas muscle fibres. J Muscle Res Cell Motil 1:409–428
Brenner B, Schoenberg M, Chalovich JM, Greene LE, Eisenberg M (1982) Evidence for cross-bridge attachment in relaxed muscle at low ionic strength. Proc Natl Acad Sci USA 79:7288–7291
Burke M, Reisler E, Harrington WF (1976) Effect of bridging the two essential thiols of myosin on its spatial and actin-binding properties. Biochemistry 15:1923–1927
Cavagna GA, Citterio G (1974) Effect of stretching on the elastic characteristics and the contractile component of frog striated muscle. J Physiol 239:1–14
Cecchi G, Griffiths PJ, Taylor SR (1984) The kinetics of cross-bridge attachment and detachment studied by high frequency stiffness measuurements. In: Pollack GH, Sugi H (eds) Contractile mechanisms in muscle. Plenum, New York, pp 641–655
Chaen S, Oiwa K, Shimmen T, Iwamoto H, Sugi H (1989) Simultaneous recordings of force and sliding movement between a myosin-coated glass microneedle and actin cables in vitro. Proc Natl Acad Sci USA 86:1510–1514
Chaen S, Shimada M, Sugi H (1986) Evidence for cooperative interactions of myosin heads with thin filament in the force generation of vertebrate skeletal muscle fibers. J Biol Chem 261:13632–13636
Chaen S, Oiwa K, Shimmen T, Iwamoto H, Sugi H (1989) Simultaneous recordings of force and sliding movement between a myosin-coated glass microneedle and actin cables in vitro. Proc Natl Acad Sci USA 86:1510–1514
Civan MM, Podolsky RJ (1966) Contraction kinetics of striated muscle fibres following quick changes in load. J Physiol 184:511–534
Cooke R (1981) Stress does not alter the conformation of a domain of the myosin cross-bridge in rigor muscle fibres. Nature 294:570–571
Cooke R, Crowder MS, Thomas DD (1982) Orientation of spin labels attached to cross-bridges in contracting muscle fibres. Nature 300:776–778
Cooke R, Crowder MS, Wendt CH, Barnett VA, Thomas D (1984) Muscle cross-bridges: do they rotate? In: Pollack GH, Sugi H (eds) Contractile mechanisms in muscle. Plenum, New York, pp 413–427
Curtin NA, Davies RE (1975) Very high tension with very little ATP breakdown by active skeletal muscle. J Mechanochem Cell Moti l3:147–154
Danzig JA, Lacktis JW, Homsher E, Goldman YE (1987) Mechanical transients initiated by photolysis of caged P_i during active skeletal muscle contractions. Biophys J 51:3a
Dewey MM, Levine RJC, Colfresh DE, Walcott B, Bann L, Baldwin A, Brink P (1979) Stractural changes in thick filament during sarcomere shortening in *Limulus* striated muscle. In: Sugi H, Pollack GH (eds) Cross-bridge mechanism in muscle contraction. University of Tokyo Press, Tokyo, pp 3–22
dos Remedios CG, Millikan RGC, Morales MF (1972) Polarization of tryptophan fluorescence from single striated muscle fibers. J Gen Physiol 59:103–120
Edman KAP (1975) Mechanical deactivation induced by active shortening in isolated muscle fibres of the frog. J Physiol 246:255–275
Edman KAP (1979) The velocity of unloaded shortening and its relation to sarcomere length and isometric force in vertebrate muscle fibres. J Physiol 291:143–159
Edman KAP (1980) Depression of mechanical performance by active shortening during twitch and tetanus of vertebrate muscle fibres. Acta Physiol Scand 109:143–159

Edman kAP (1988) Double-hyperbolic force-velocity relation in frog muscle fibres. J Physiol 494:301–321
Edman KAP, Reggiani C (1987) The sarcomere length-tension relation determined in short segments of intact muscle fibres of the frog. J Physiol 385:709–732
Edman KAP, Mulieri LA, Scubon-Mulieri B (1976) Non-hyperbolic force-velocity relationship in single muscle fibres. Acta Physiol Scand 98:143–156
Edman KAP, Elzinga G, Noble MIM (1978) Enhancement of mechanical performance by stretch during tetanic contractions of vertebrate skeletal fibres. J Physiol 281:139–155
Eisenberg E, Kielly WW (1973) Evidence for a refractory state of heavy meromyosin and subfragment-1 unable to bind to actin in the presence of ATP. Cold Spring Harbor Symp Quant Biol 37:145–152
Eisenberg E, Moos C (1968) The adenosine triphosphatase activity of acto-heavy meromyosin. A kinetic analysis of actin activation. Biochemistry 7:1486–1489
Eisenberg E, Moos C (1970) Actin activation of heavy meromyosin adenosine triphosphatase. Dependence on adenosine triphosphatate and actin concentrations. J Biol Chem 245:2451–2456
Eisenberg E, Dobkin L, Kielley WW (1972) Heavy meromyosin: evidence for a refractory state unable to bind to actin in the presence of ATP. Proc Natl Acad Sci USA 69:667–671
Ekelund MC, Edman KAP (1982) Shortening induced deactivation of skinned fibres of frog and mouse striated muscle. Acta Physiol Scand 116:189–199
Elliott GF, Lowy J, Millman BM (1963) An X-ray and light diffraction study of the filament lattice of striated muscle in the living state and in rigor. J Mol Biol 6:295–305
Fenn WO (1923) A quantitative comparison between the energy liberated and the work performed by the isolated sartorius muscle of the frog. J Physiol 58:175–203
Fenn WO (1924) The relation between the work performed and the energy liberated in muscular contraction. J Physiol 58:373–395
Ford LE, Huxley AF, Simmons RM (1977) Tension responses to sudden length change in simulated frog muscle fibres near slack length. J Physiol 269:441–515
Ford LE, Huxley AF, Simmons RM (1981) The relation between stiffness and filament overlap in stimulated frog muscle fibres. J Physiol 311:219–249
Gasser HS, Hill AV (1924) The dynamics of muscular contraction. Proc R Soc Lond Ser B 96:398–437
Goldman YE, Hibberd MG, McCray JA, Trentham DR (1982) Relaxation of muscle fibres by photolysis of caged ATP. Nature 300:701–705
Gordon AM, Huxley AF, Julian FJ (1966a) Tension development in highly stretched vertebrate muscle fibres. J Physiol 184:143–169
Gordon AM, Huxley AF, Julian FJ (1966b) The variation in isometric tension with sarcomere length in vertebrate muscle fibres. J Physiol 184:170–192
Granzier HLM, Pollack GH (1990) The descending limb of the force-sarcomere length relation revisited. J Physiol 421:595–615
Gulati J, Podolsky RJ (1978) Contraction transients of skinned muscle fibers: effects of calcium and ionic strength. J Gen Physiol 72:701–716
Harada Y, Yanagida T (1988) Direct observation of molecular motility by light microscopy. Cell Motil Cytoskel 10:71
Harada Y, Noguchi A, Kishino A, Yanagida T (1987) Sliding movement of single actin filaments on one-headed myosin filaments. Nature 326:605–608
Harrington WF, Ueno H, Davis JS (1988) Helix-coil melting in rigor and activated cross-bridges of skeletal muscle. In: Sugi H, Pollack GH (eds) Molecular mechanism of muscle contraction. Plenum, New York, pp 307–318
Hatta I, Sugi H, Tamura Y (1988) Stiffness changes in frog skeletal muscle during contraction recorded using ultrasonic waves. J Physiol 403:193–209
Heuser JE, Cooke R (1983) Actin-myosin interction visualized in freeze-etch, deep-etch replica technique. J Mol Biol 169:97–122
Hibbard MG, Trentham DR (1986) Relationships between chemical and mechanical events during musclular contraction. Annu Rev Biophys Chem 15:119–161
Hill AV (1938) The heat of shortening and the dynamic constants of muscle. Proc R Soc B 126:136–195

Hill AV (1949) The heat of activation and the heat of shortening in a muscle twitch. Proc R Soc B 136:195–211
Hill AV (1964) The effect of load on the heat of shortening of muscle. Proc R Soc B 159:297–318
Hill AV, Howarth JV (1959) The reversal of chemical reactions in contracting muscle during an applied stretch. Proc R Soc B 151:169–193
Huxley AF (1957) Muscle structure and theories of contraction. Prog Biophys Biophys Chem 7:255–318
Huxley AF (1974) Review lecture, muscular contraction. J Physiol 243:1–43
Huxley AF, Niedergerke R (1954) Interference microscopy of living muscle fibres. Nature 173:971–973
Huxley AF, Simmons RM (1971) Proposed mechanism of force generation in striated muscle. Nature 233:533–538
Huxley HE (1960) Muscle cells. In: Brachet J, Mirsky JE (eds) The cell. Academic Press, New York, pp 365–481
Huxley HE (1990) Sliding filaments and molecular motile systems. J Biol Chem 265:833347–8350
Huxley HE, Faruqi AR (1983) Time-resolved X-ray diffraction studies on vertebrate striated muscle. Annu Rev Biophys Bioeng 12:381–417
Huxley HE, Hanson J (1954) Changes in the cross-striations of muscle during contraction and stretch and their structural interpretation. Nature 173:973–976
Huxley HE, Simmons RM, Faruqi AR, Kress M, Bordas J, Koch MHJ (1981) Millisecond time-resolved changes in X-ray reflections from contracting muscle during rapid mechanical transients, recorded using synchrotron radiation. Proc Natl Acad Sci USA 78:2297–2301
Huxley HE, Farugi AR, Kress K, Bordas J, Koch MHJ (1982) Time-resolved X-ray diffraction studies of the myosin layer-line reflections during muscle contraction. J Mol Biol 158:637–684
Huxley HE, Simmons RM, Faruqi AR, Kress M, Bordas J, Koch MHJ (1983) Changes in the X-ray diffractions from contracting muscle during rapid mechanical transients and their structural implications. J Mol Biol 169:469–506
Inoue A, Takenaka H, Arata T, Tonomura Y (1979) Functional implications of the two-headed structure of myosin. Adv Biophys 13:1–194
Iwamoto H, Sugaya R, Sugi H (1990) Force-velocity relation of frog skeletal muscle fibres shortening under continuously changing load. J Physiol 422:185–502
Jewell BR, Wilkie DR (1958) An analysis of the mechanical components in frog's striated muscle. J Physiol 143:515–540
Julian FJ (1971) The effect of calcium on the force-velocity relation of briefly glycerinated frog muscle fibres. J Physiol 218:117–145
Julian FJ, Morgan DL (1979a) Intersarcomere dynamics during fixed-end tetanic contractions of frog muscle fibres. J Physiol 293:365–378
Julian FJ, Morgan DL (1979b) The effect on tension of non-uniform distribution of length changes applied to frog muscle fibres. J Physiol 293:379–392
Julian FJ, Moss RL (1981) Effects of calcium and ionic strength on shortening velocity and tension development in frog skinned muscle fibres. J Physiol 311:179–199
Julian FJ, Sollins MR (1975) Variation of muscle stiffness with force at increasing speeds of shortening. J Gen Physiol 66:287–302
Julian FJ, Rome LC, Stephenson DG, Striz S (1986a) The maximum speed of shortening in living and skinned frog muscle fibres. J Physiol 370:181–199
Julian FJ, Rome JC, Stephenson DG, Striz S (1986b) The influence of free calcium on the maximum speed of shortening in skinned frog muscle fibres. J Physiol 380:257–273
Kron SJ, Spudich JA (1986) Fluorescent actin filaments move on myosin fixed to a glass surface. Proc Natl Acad Sci USA 83:6272–6276
Kuhn HJ (1973) Transformation of chemical energy into mechanical energy by glycerol-extracted fibres of insect flight muscle in the absence of nucleosidetriphosphate hydrolysis. Experientia 29:1086–1088

Lamed R, Oplatka A, Reisler E (1976) Affinity chromatography of heavy meromyosin subfragment-1 reacted with thiol reagents. Biochim Biophys Acta 427:688–695

Lopez JR, Wank LA, Taylor SR (1981) Skeletal muscle: length-dependent effects of potentiating agents. Science 214:79–82

Lymn RW, Taylor EW (1970) Transient state phosphate production in the hydrolysis of nucleoside triphosphates by myosin. Biochemistry 9:2975–2983

Lymn RW, Taylor EW (1971) Mechanism of adenosine triphosphate hydrolysis by actomyosin. Biochemistry 10:4617–4624

Maréchal G, Plaghki L (1979) The deficit of the isometric tetanic tension redeveloped after a release of frog muscle at constant velocity. J Gen Physiol 73:453–467

Matsuda T, Podolsky RJ (1984) X-ray evidence for two structural states of the actomyosin cross-bridge in muscle fibers. Proc Natl Acad Sci USA 81:2364–2368

Morales MF, Botts J (1979) On the molecular basis for chemomechanical energy transduction in muscle. Proc Natl Acad Sci USA 76:3857–3859

Naylor GRS, Podolsky RJ (1981) X-ray diffraction of striated muscle fibers in rigor. Proc Natl Acad Sci USA 78:5559–5563

Oiwa K, Chaen S, Kamitsubo E, Shimmen T, Sugi H (1990) Steady-state force-velocity relation in the ATP-dependent sliding movement of myosin-coated beads on actin cables in vitro studied with a centrifuge microscope. Proc Natl Acad Sci USA 87:7893–7897

Oiwa K, Chacn S, Sugi H (1991) Measurement of work done by ATP-induced sliding between rabbit muscle myosin and algal cell actin cables in vitro. J Physiol 437:751–763

Oplatka A, Gadasi H, Borejdo J (1974) The contraction of "ghost" myofibrils and glycerinated muscle fibers irrigated with heavy meromyosin subfragment-1. Biochem Biophys Res Commun 58:905–912

Page SG, Huxley HE (1963) Filament lengths in striated muscle. J Cell Biol 19:369–390

Podolsky RJ, Nolan AC (1973) Muscle contraction transients, cross-bridge kinetics, and the Fenn effect. Cold Spring Harbor Symp Quant Biol 37:661–668

Podolsky RJ, Teichholz LE (1970) The relation between calcium and contraction kinetics in skinned muscle fibres. J Physiol 211:19–35

Podolsky RJ, Nolan AC, Zavaler SA (1969) Cross-bridge properties derived from muscle isotonic velocity transients. Proc Natl Acad Sci USA 64:504–511

Pollack GH (1990) Muscle and molecules: uncovering the principles of biological motion. Ebner & Sons, Seattle.

Ramsey RW, Street SF (1940) The isometric length-tension diagram of isolated skeletal muscle fibers of the frog. J Cell Comp Physiol 15:11–34

Reedy MK, Goody RS, Hofmann W, Rosenbaum G (1983) Co-ordinated electron microscopy and X-ray studies of glycerinated insect flight muscle. I. X-ray diffraction monitoring during preparation for electron microscopy of muscle fibres fixed in rigor, in ATP and in AMPPNP. J Muscle Res Cell Motil 4:25–53

Reedy MK, Holmes KC, Tregear RT (1965) Induced changes in orientation of the cross bridges of glycerinated insect flight muscle. Nature 207:1276–1280

Reisler E, Burk M, Himmelfarb S, Harrington WF (1974) Spatial proximity of the two essential sulfhydryl groups of myosin. Biochemistry 13:3837–3840

Schoenberg M, Wells JB, Podolsky RJ (1974) Muscle compliance and the logitudinal transmission of mechanical impulse. J Gen Physiol 64:623–642

Sellers JR, Kachar B (1990) Polarity and velocity of sliding filaments: control of direction by actin and of spead of myosin. Sience 249:406–408

Sheetz MP, Spudich JA (1983) Movement of myosin-coated fluorescent beads on actin cables in vitro. Nature 303:31–35

Shimmen T, Yano M (1984) Active sliding movement of latex beads coated with skeletal muscle myosin on Chara actin bundles. Protoplasma 121:132–137

Stein LA, Chock PB, Eisenberg E (1981) Mechanism of actomyosin ATPase: effect of actin on the ATP hydrolysis step. Proc Natl Acad Sci USA 78:1346–1350

Sugi H (1972) Tension changes during and after stretch in frog muscle fibres. J Physiol 225:237–253

Sugi H, Gomi S (1981) Changes in the A-band width during contraction in horseshoe crab striated muscle. Experientia 37:65–67
Sugi H, Gomi S (1984) Cinematographic studies on the A-band length changes during Ca-activated contraction in horseshoe crab muscle myofibrils. In: Pollack GH, Sugi H (eds) Contractile mechanisms in muscle. Plenum, New York, pp 107–118
Sugi H, Kobayashi T (1983) Sarcomere length and tension changes in tetanized frog muscle fibers after quick stretches and release. Proc Natl Acad Sci USA 80:6422–6425
Sugi H, Tsuchiya T (1981a) Isotonic velocity transients in frog muscle fibres following quick changes in load. J Physiol 319:239–252
Sugi H, Tsuchiya T (1981b) Enhancement of mechanical performance in frog muscle fibres after quick increases in load. J Physiol 319:239–252
Sugi H, Tsuchiya T (1988) Stiffness changes during enhancement and deficit of isometric force by slow length changes in frog skeletal muscle fibres. J Physiol 407:215–229
Sugi H, Amemiya Y, Hashizume H, Tameyasu T, Tanaka H (1980) Kinetic properties of the cross-bridges in frog skeletal muscle as studied by time-resolved X-ray diffraction. In: Ebashi S, Maruyama K, Endo M (eds) Muscle contraction. Its regulatory mechanisms. Jpn Sci Soc Press, Tokyo, Springer Berlin Heidelberg New York, pp 53–63
Sugi H, Ohta T, Tameyasu T (1983) Development of the maximum isometric force at short sarcomere lengths in calcium-activated muscle myofibrils. Experientia 39:147–148
Suzuki S, Sugi H (1983) Extensibility of the myofilaments in vertebrate skeletal muscle as revealed by stretching rigor muscle fibers. J Gen Physiol 81:531–546
Suzuki S, Oshimi Y, Sugi H (1986) Ultrastructural observation of the cross-bridge movement during muscle contraction. Proc XIth Int Congr Electron microsc, Kyoto pp 3111–3112
Suzuki S, Tsuchiya T, Oshimi Y, Takei T, Sugi H (1989) Electron microscopic studies on the stretch-induced disordering of the myofilament lattice in tetanized frog skeletal muscle fibers. J Electron Microsc 38:60–63
Tamura Y, Hatta I, Matsuda T, Sugi H, Tsuchiya T (1982) Changes in muscle stiffness during contraction recorded using ultrasonic waves. Nature 299:631–633
Tanaka H, Sugi H, Hashizume H (1984) Effect of stretch on the equatorial X-ray diffraction pattern from frog skeletal muscle in rigor. In: Pollack GH, Sugi H (eds) Contractile mechanisms in muscle. Plenum, New York, pp 203–205
Thames MD, Teichholz LE, Podolsky RJ (1974) Ionic strength and the contraction kinetics of skinned muscle fibres. J Gen Physiol 63:509–530
Tokiwa T, Morales MF (1971) Independent and comparative reactions of myosin heads with F-actin in the presence of adenosine triphosphate. Biochemistry 10:1722–1727
Tonomura Y (1972) Muscle protein, muscle contraction and cation transport. University of Tokyo Press, Tokyo
Tonomura Y, Kitagawa S, Yoshimura J (1962) The initial phase of myosin a adenosine triphosphatase and the possible phosphorylation of myosin A. J Biol Chem 237:3660–3660
Toyoshima YY, Kron SJ, McNally EM, Niebling KR, Toyoshima C, Spudich JA (1987) Myosin subfragment-1 is sufficient to move actin filaments in vitro. Nature 328:536–539
Tregear RT, Wakabayashi K, Tanaka H, Iwamoto H, Reedy MC, Reedy MK, Sugi H, Amemiya Y (1990) X-ray diffraction and electron microscopy from *Lethocerus* flight muscle partially relaxed by adenylylimidodiphosphate and ethylene glycol. J Mol Biol 214:129–141
Trentham DR, Bardsley RG, Eccleston JF, Weeds AG (1972) Elementary processes of the magnesium ion-dependent adenosine triphosphatase activity of heavy meromyosin. Biochem J 126:635–644
Trentham DR, Eccleston JF, Bagshaw CR (1976) Kinetic analysis of ATPase mechanisms. Q Rev Biophys 9:217–281
Trombitas K, Baasteu PHWW, Pollack GH (1988) Effect of tension on the rigor cross-bridge angle. In: Sugi H, Pollack GH (eds) Molecular mechanism of muscle contraction. Plenum, New York, pp 17–30
Tsukita S, Yano M (1985) Actomyosin structure in contracting muscle detected by rapid freezing. Nature 317:182–184
Ueno H, Harrington WF (1981) Cross-bridge movement and the conformational state of the myosin hinge in skeletal muscle. J Mol Biol 149:619–640

Wray JS, Holmes KC (1981) X-ray diffraction studies of muscle. Annu Rev Physiol 43:553–565

Yagi N, O'Brien J, Matsubara I (1981) Changes of thick filament structure during contraction of frog striated muscle. Biophys J 33:121–138

Yamada A, Ishii N, Takahashi K (1990) Direction and speed of actin filaments moving along thick filaments isolated from molluscan smooth muscle. J Biochem 108:341–343

Yanagida T (1984) Angles of fluorescently labelled myosin heads and actin monomers in contracting and rigor stained muscle fiber. In: Pollack GH, Sugi H (eds) Contractile mechanisms in muscle. Plenum, New York, pp 397–411

Yanagida T (1985) Angle of active site of myosin heads in contracting muscle during sudden length changes. J Muscle Res Cell Motil 6:43–52

Yanagida T, Harada Y (1988) Minimum structural unit required for energy transduction in muscle. In: Sugi H, Pollack GH (eds) Molecular mechanism of muscle contraction. Plenum, New York, pp 277–287

Young DM (1967) Studies on the structural basis of the interaction between myosin and actin. Proc Natl Acad Sci USA 58:2393–2400

Chapter 6

Mechanisms of Cytoplasmic Streaming and Amoeboid Movement

T. Shimmen

Contents

1 Introduction

The molecular mechanism of the generation of motive force by actomyosin has been studied most extensively in skeletal muscle. However, the skeletal muscle is a rather highly specialized example of such systems. It is becoming evident that actomyosin is commonly distributed in nonmuscle cells, including plant

Department of Life Science, Faculty of Science, Himeji Institute of Technology, 1479-1 Kanaji Kamigōrichō, Akōgun Hyogo 678-12, Japan

Advances in Comparative and Environmental Physiology
Vol. 12. Edited by H. Sugi

cells, although the extent of the intracellular organization of the actomyosin molecules is less significant in nonmuscle cells than in skeletal muscle. Cytoplasmic streaming can be observed in a wide variety of plant cells. The mode and the speed of the streaming vary considerably among cells (Kamiya 1959). One of the most important physiological roles of cytoplasmic streaming is the intracellular transport of molecules and organelles in plant cells. Studies on cytoplasmic streaming were started about 200 years ago by Corti (see Kamiya 1959). Systematic analysis of the mechanism of motive-force generation was initiated in Characeae cells and in the plasmodium of the true slime mold, *Physarum*, by Kamiya and his coworkers. At present, the mechanism of motive-force generation for cytoplasmic streaming is attributed to actomyosin in most plant cells. In some plant cells, involvement of microtubules in cytoplasmic streaming has been suggested. The physiological and molecular mechanisms of motive-force generation in Characeae and the *Physarum* plasmodium are the focus of this chapter.

Amoeboid movement is one of the most common motile phenomena in animal cells, from the protozoa to mammals. Such movement is also observed during the amoeboid stage of cellular slime molds and true slime molds. Recent progress in studies of amoeboid movement is reviewed here.

2 Rotational Streaming in Characeae

2.1 Structure of the Internodal Cells of Characeae

The most intensively studied genera of Characeae are *Chara* and *Nitella*. Characeae grow in fresh water, for example, in ponds, lakes, and rice fields. Figure 1 shows a *Chara* plant, which is composed of nodal cells, branchlets, and internodal cells. The plant is anchored by rhizoid cells (not shown) to the substratum of its habitat. Internodal cells are generally used for experiments, and their structure is shown schematically in Fig. 1. The internodal cell is a large cylindrical cell several hundred micrometers in diameter with a length that sometimes exceeds 10 cm. The outermost layer is the cell wall which is mainly composed of cellulose. The cell wall is almost totally permeable to substances of low molecular weight but it is impermeable to macromolecules. The actual barrier to diffusion between the exterior and interior of the cell is the plasma membrane. The central part of the cell is occupied by a vacuole, which is surrounded by a tonoplast (vacuolar membrane). The vacuole occupies more than 90% of the total volume of the full-grown cells and contains the cell sap (vacuolar sap). Cytoplasm is sandwiched between the plasma membrane and the tonoplast. Chloroplasts are attached to the gel ectoplasm. Sol endoplasm is actively streaming at a rate of 50–100 μm/s. The direction of the streaming is reversed on either side of an indifferent line (white line), where chloroplasts are absent. Two indifferent lines run in parallel along the longitudinal axis of the cell. Therefore, the endoplasm rotates in the cell in the longitudinal direction.

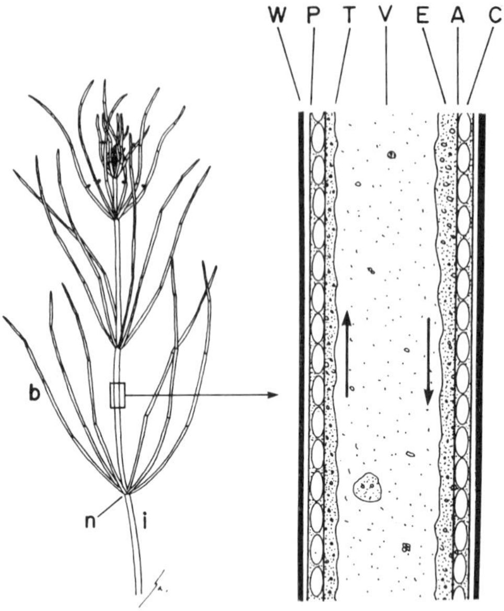

Fig. 1. Morphology of a characean plant and internodal cell. *b* Branchlet; *n* node; *i* internodal cell; *W* cell wall; *P* plasma membrane; *T* tonoplast; *V* vacuole; *E* endoplasm; *A* actin cable; *C* chloroplast. (Tazawa and Shimmen 1987)

2.2 Mechanism of Motive-Force Generation

A clue to the mechanism of motive-force generation was obtained experimentally by Kamiya and Kuroda (1956). They analyzed the distribution of the velocities of streaming in the cell and concluded that the motive force is equivalent to the shearing force exerted at the inner surface of stagnant chloroplasts, i.e., streaming occurs via the active sliding of the sol endoplasm along the inner surface of chloroplasts (sol-gel interface). Stimulated by this pioneering study, attention has been focused on the elucidation of this sliding mechanism. Kamitsubo (1966) discovered cables on the inner surface of chloroplasts (Fig. 2). The cables are indispensable for streaming (Kamitsubo 1972). Nagai and Rebhun (1966) found that the cable is composed of 80–100 microfilaments. When HMM (heavy meromyosin: a proteolytic fragment of skeletal muscle muosin) is applied in the absence of ATP, arrowhead structures are formed around the microfilaments (Palevitz et al. 1974; Williamson 1974; Palevitz and Hepler 1975; Kersey and Wessels 1976; Kersey et al. 1976). Antibodies raised against actin (Owaribe et al. 1979; Williamson et al. 1987) and phallotoxin (Barak et al. 1980; Nothnagel et al. 1981) bind to the microfilaments. Such experiments demonstrate unequivocally that the microfilaments are actin filaments. Cytochalasin B, a specific inhibitor of actin-based motility, reversibly inhibits cytoplasmic streaming, indicating that actin is involved in the the gener-

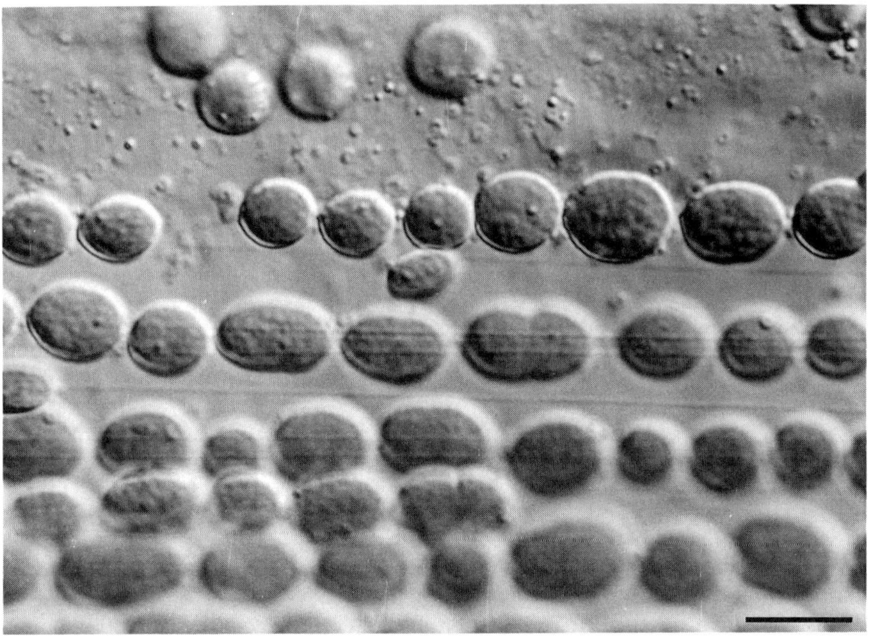

Fig. 2. Actin cables on the inner surface of chloroplasts in an internodal cell of *Chara*. *Bar* = 10 μm. (Tazawa and Shimmen 1987)

ation of the motive force for cytoplasmic streaming in Characeae (Williamson 1972; Bradley 1973; Nagai and Kamiya 1977).

Involvement of actin filaments in motive-force generation necessarily suggests the participation of myosin, which is responsible for the contraction of muscle in cooperation with actin. Kato and Tonomura (1977b) isolated "myosin" from *Nitella flexilis* and its properties were found to be similar to those of skeletal muscle myosin. Since the motive force is generated by sliding of endoplasm along actin filaments, myosin can be assumed to be localized in the flowing endoplasm (Chen and Kamiya 1975, 1981; Kamitsubo 1981). Nagai and Hayama (1979) studied the endoplasmic organelles that bind to actin filaments in the absence of ATP, using an intracellular perfusion technique which will be discussed below. They found extrusions on the surface of unidentified endoplasmic organelles. The globular structures on the extrusions formed cross-bridges with actin filaments. Nagai and Hayama suggested that the globular structures have a myosin-like function. Williamson (1979) reported filamentous structures associated with endoplasmic reticulum. Association of myosin with endoplasmic organelles has also been shown by reconstitution of cytoplasmic streaming (Shimmen and Tazawa 1982), which will also be discussed below. Theoretical analysis supports these experimentally derived conclusions (Nothnagel and Webb 1982).

2.3 Studies Using Demembranated Cell Models

Due to difficulties encountered in attempts to isolate actomyosin from Characeae cells, no biochemical elucidation of the molecular mechanism of motive-force generation was possible initially. These difficulties were overcome by the development of demembranated cell models, in which the chemical composition of the cytoplasm can be easily manipulated. Since the cytoplasm is sandwiched between two diffusion barriers, the plasma membrane and the tonoplast, we can gain access to the cytoplasm by demembranating either the plasma membrane or the tonoplast.

Tonoplast-Free Cells (Fig. 3). Since the internodal cell is a large cylinder, both ends of the cell can be removed by cutting with scissors. When the vacuole is perfused with a medium that contains ethyleneglycol-bis-(β-aminoethylether)N,N,N′,N′-tetraacetic acid (EGTA), the tonoplast disintegrates (Williamson 1975; Tazawa et al. 1976). Although most of endoplasm is effused out after disintegration of the tonoplast, some endoplasmic organelles remain attached to the bundles of actin filaments. The intracellular constituents can then be controlled by intracellular perfusion.

Permeabilized Cell Model. When plasmolysis is induced in ice-cold medium containing EGTA, the plasma membrane is irreversibly permeabilized. In such permeabilized cells, the chemical composition of the cytoplasm can be controlled from the exterior of the cell (Shimmen and Tazawa 1983).

Three other kinds of demembranated cell models have been developed: the longitudinally cut-open cell model (Kuroda 1983; Sheetz and Spudich 1983), the demembranated endoplasmic drop model (Kuroda and Kamiya 1975), and the squeezed-out cytoplasm model (Higashi-Fujime 1980).

When ATP is depleted in tonoplast-free cells, endoplasmic organelles become tightly associated with actin bundles. This association is assumed to be caused by the formation of rigor cross-bridges between actin filaments and the myosin associated with the organelles (Williamson 1975). When ATP is introduced into the ATP-depleted cell, the organelles start movement along the actin bundles. Reactivation does not occur in the absence of Mg^{2+}, indicating that Mg-ATP is the actual substrate. Dependency of the velocity (v) of cytoplasmic streaming on the concentration of ATP ([ATP]) has been studied in tonoplast-free cells. The relationship between [ATP] and v is a saturating curve and the half-maximal [ATP] is 70–90 μM (Shimmen 1978). Cytoplasmic streaming reactivated by ATP is inhibited competitively by ADP, Pi, and PPi (Shimmen 1988a).

Reconstitution of movement on characean actin bundles has been carried out in tonoplast-free cells (Fig. 3). The cytoplasm was squeezed out of the *Chara* cells after removal of the vacuolar sap by intracellular perfusion. Organelles were collected by repeated centrifugation. When *Chara* organelles suspended in Mg-ATP-containing medium were introduced into *Nitella* cells, from which native endoplasm had been removed, the *Chara* organelles moved

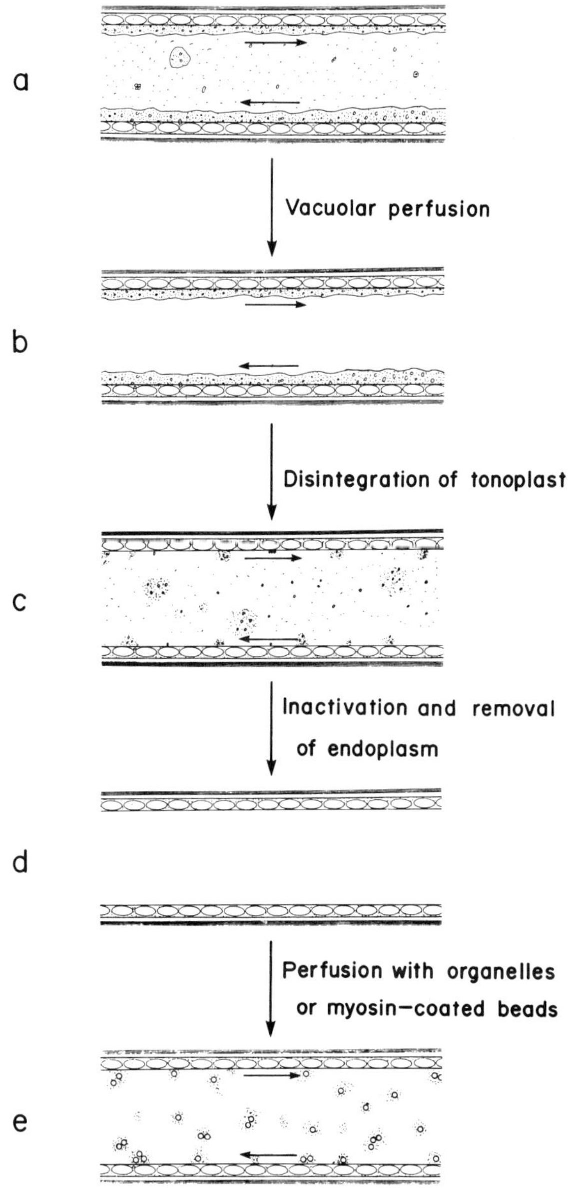

Fig. 3a–e. Schematic representation of the preparation of tonoplast-free cells and reconstitution of the movement of foreign myosin on organelles along characean actin cables. **a** Intact cell; **b** vacuole-perfused cell; **c** tonoplast-free cell; **d** endoplasm-free cell; **e** reconstituted movement. (Shimmen 1988b)

on the *Nitella* actin bundles (Shimmen and Tazawa 1982). Since the organelles moved along actin cables after the soluble fraction had been removed, it was suggested that myosin molecules were bound to the surface of endoplasmic organelles.

When latex heads coated with skeletal muscle myosin were applied to characean actin bundles, they moved in a Mg-ATP-dependent manner (Shimmen and Yano 1984). Reconstitution of myosin-coated beads on characean actin cables can be also carried out using the longitudinally cut-open cell model mentioned above (Sheetz and Spudich 1983). Movement of identified myosin on characean actin cables supports the hypothesis that the molecule in the endoplasm which is responsible for generation of motive force of cytoplasmic streaming is myosin. Movement of skeletal muscle myosin on characean actin bundles occurs in the same direction as native cytoplasmic streaming, i.e., from the pointed to the barbed end. Various kinds of myosin including skeletal muscle myosin can move on characean actin bundles (Shimmen 1988b). The velocity of myosin movement on characean actin bundles is almost consistent with the activity of the actin-activated ATPase of the corresponding myosin, with the exception of myosin I from *Acanthamoeba* (Fig. 4). Since actin filaments of Characeae show no sensitivity to Ca^{2+} (Shimmen and Yano 1986), the sensitivity of myosin movement to Ca^{2+} represents direct sensitivity to Ca^{2+} of the actin-activated ATPase. Thus, a reconstituted system for the movement of foreign myosin on Characean actin cables can provide a novel method to study myosin-based motility (Shimmen 1988b). Organelles isolated from pollen tubes (Kohno and Shimmen 1988) and from *Acanthamoeba* (Adams and Pollard 1986) can also move on characean actin cables (Fig. 4), indicating that myosin is associated with these organelles.

2.4 Regulation of Cytoplasmic Streaming by Action Potentials

An action potential is generated not only by animal cells but also by plant cells, such as those of *Mimosa pudica* and carnivorous plants. Characean cells also generate action potentials in response to electrical or mechanical stimuli. Thus, characean cells are a particularly suitable material for studies of plant action potentials, since each cell is large and single cells can be easily isolated from adjacent cells. When a characean cell is stimulated under the microscope, we noted that cytoplasmic streaming ceases rapidly with the generation of an action potential and recovers with time (excitation-cessation coupling). Depolarization of the membrane potential during generation of the action potential is caused by the activation of Ca^{2+} and Cl^- channels of the plasma membrane (Shiina and Tazawa 1987). With the generation of an action potential, increases in the influx of Ca^{2+}, and efflux of Cl^- and K^+, occur. The total concentration of Ca^{2+} in the cytoplasm is in the millimolar range. However, most of the Ca^{2+} ions are sequestered in some organelles. Therefore, the concentration of free Ca^{2+} in the cytoplasm ($[Ca^{2+}]_c$) is very low in the resting state in both animal and plant cells. Because of the very low concentration in the resting state, a small influx from the cell's exterior or the release of Ca^{2+} from organelles results

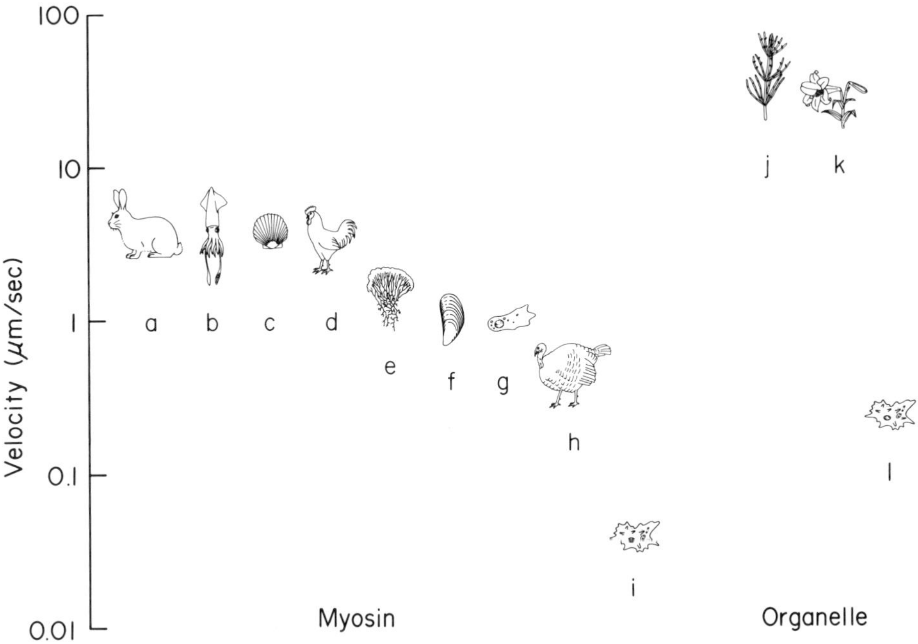

Fig. 4. Competition for movement along characean actin cables by various myosins and organelles. **a** Skeletal muscle myosin from the rabbit, *Oryctolagus*; **b** siphon muscle myosin from the squid, *Loligo*; **c** adductor muscle myosin from the scallop, *Placopecten*; **d** breast muscle myosin from the chicken, *Gallus*; **e** plasmodium myosin of the true slime mold, *Physarum*; **f** anterior byssus retractor muscle myosin from the blue mussel, *Mytilus*; **g** amoeba myosin of the acellular slime mold, *Dictyostelium*; **h** gizzard smooth muscle myosin from the turkey, *Meleagris*; **i** myosin I from *Acanthamoeba*; **j** endoplasmic organelles from Characeae; **k** pollen tube organelles from the lily, *Lilium*; **l** organelles from *Acanthamoeba*. (Shimmen 1988b)

in a drastic increase in $[Ca^{2+}]_c$. Therefore, Ca^{2+} is a suitable intracellular messenger. $[Ca^{2+}]_c$ increases significantly when the membrane of Characeae is excited. $[Ca^{2+}]_c$ at the peak of luminescence was estimated to be 40 μM in *Chara* and 6 μM in *Nitella*. Subsequently, $[Ca^{2+}]_c$ fell rapidly to the resting level (Williamson and Ashley 1982).

Involvement of Ca^{2+} in the cessation of cytoplasmic streaming by membrane excitation was first suggested by experiments designed to modify the extracellular concentration of Ca^{2+}. When extracellular Ca^{2+} is replaced by either Mg^{2+} or Ba^{2+}, internodal cells of *N. axillaris* can generate action potentials. However, cessation of cytoplasmic streaming does not occur (uncoupling of excitation and cessation of streaming). If Ca^{2+} is added to the medium that contains either Mg^{2+} or Ba^{2+}, coupling between excitation and cessation is recovered (Barry 1968). Thus, extracellular Ca^{2+} seems to play an essential role in the cessation of streaming.

Since circumstantial evidence indicates the involvement of Ca^{2+} in excitation-cessation coupling, the effect of Ca^{2+} on cytoplasmic streaming was studied. The demembranated cell model is suitable for this purpose. In the permeabilized cell model, reactivated streaming is reversibly controlled by Ca^{2+}. At $[Ca^{2+}]$ lower than 0.1 μM, v is constant irrespective of $[Ca^{2+}]$. With increases in $[Ca^{2+}]$, v decreases and movement stops completely when $[Ca^{2+}]$ is higher than 1 μM (Tominaga et al. 1983). Thus, it can be concluded that cessation of cytoplasmic streaming upon membrane excitation is caused by an increase in $[Ca^{2+}]_c$.

The question arise as to whether sensitivity to Ca^{2+} is a property of actin or myosin. This question can be answered by a reconstitution experiment. Movement of latex beads coated with skeletal muscle myosin on characean actin bundles is completely independent of $[Ca^{2+}]$. The reconstituted system of movement is not inhibited at all even at 1 mM Ca^{2+} (Shimmen and Yano 1986). In skeletal muscle, regulation by Ca^{2+} is exerted via the troponin-tropomyosin complex that is incorporated into the actin filaments. Therefore, skeletal muscle myosin shows no direct sensitivity to Ca^{2+}. Absence of sensitivity to Ca^{2+} in the reconstituted system of movement indicates that characean actin bundles are insensitive to Ca^{2+}. It is suggested that the sensitivity to Ca^{2+} of native cytoplasmic streaming is associated with myosin. Involvement of the phosphorylation of myosin in the cessation of cytoplasmic streaming has been suggested from experiments with demembranated cell models (Tominaga et al. 1987). When the permeabilized cells were treated with adenosine-5′-O-(3-thiotriphosphate) (ATP-γ-S), in the presence of Ca^{2+}, cytoplasmic streaming was irreversibly inhibited. ATP-γ-S has been used for the thiophosphorylation of proteins by protein kinase. Thiophosphorylated proteins are not dephosphorylated by protein phosphatase (Morgan et al. 1976; Gratecos and Fischer 1974). It was postulated that cytoplasmic streaming of permeabilized cells would be irreversibly inhibited by the irreversible Ca^{2+}-dependent thiophosphorylation of myosin. The results of the following experiments support this hypothesis. Protein phosphatase I restores cytoplasmic streaming even in the presence of Ca^{2+}. By contrast, an inhibitor of protein phosphatase, α-naphthylphosphate, inhibits cytoplasmic streaming in the absence of Ca^{2+} (Tominaga et al. 1987).

It is probable that the dephosphorylation of myosin is involved in the recovery of cytoplasmic streaming after the cessation of streaming induced by an action potential. Since antagonists of calmodulin inhibit the recovery of cytoplasmic streaming after its interruption by Ca^{2+} in permeabilized cells, the involvement of a calmodulin-regulated protein phosphatase in the recovery process has been suggested (Tominaga et al. 1985).

3 Shuttle Streaming in the Plasmodium of *Physarum*, a True Slime Mold

3.1 Cell Cycle of a True Slime Mold

True slime molds change their form drastically during their life cycle. The sessile fruiting body is composed of a stalk and a head. At this stage, cells are covered with cell walls and the appearance is plant-like. After spores are spread from the fruiting body, amoebae hatch from the spores. The amoebae change their form to become flagellates under moist conditions. The reverse change occurs when the water content of the environment is decreased. Two haploid amoebae fuse

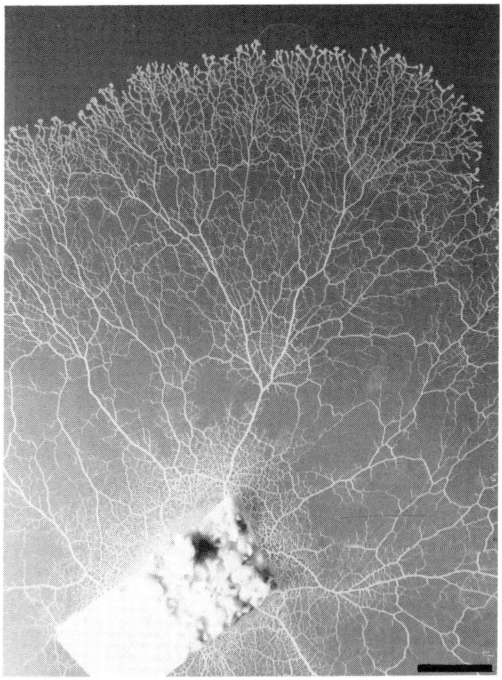

Fig. 5. Plasmodium of *Physarum polycephalum*. *Bar* = 1 cm. The plasmodium is advancing upwards

to form a zygote. As a result of the uncoupling of nuclear division from cell division, a multinucleated plasmodium is formed. The plasmodium migrates as a fan-shaped, slimy mass, sometimes exceeding 10 cm in length. In Fig. 5, the plasmodium is migrating upwards. It's posterior region is composed of a mesh-work of strands. In each strand, the active movement of organelles can be observed and their speeds sometimes reach 1 mm/s. The direction of organelle movement reverses every 2 or 3 min (shuttle streaming). Since larger amounts of organelles are transported in one direction than in the other, the plasmodium migrates in the corresponding direction. Transformation of cytoplasm from a gel to a sol occurs in the posterior region and that from a sol to a gel in the anterior region. *Physarum polycephalum* has been used in many studies of shuttle streaming.

3.2 Mechanism of Motive-Force Generation for Shuttle Streaming

Kamiya (1940) made the most important initial progress in the elucidation of the mechanism of cytoplasmic streaming in the plasmodium. He succeeded in measuring the motive force of the cytoplasmic streaming in the plasmodium using the so-called double-chamber method. Various aspects of cytoplasmic streaming were elucidated by the double-chamber method including the physiological mechanism of motive-force generation.

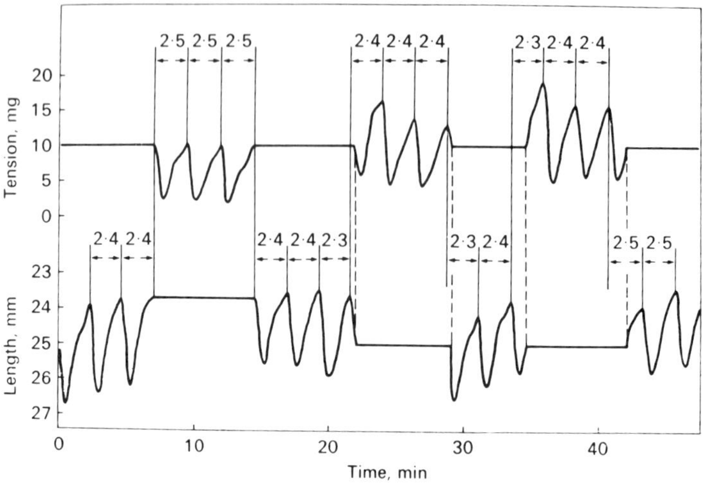

Fig. 6. Isometric and isotonic contraction of a plasmodial strand. *Upper curve*, isometric change in tension; *lower curve*, isotonic change in length. The ordinate for isotonic contractions is shown inverted. The period of each wave is shown (min). (Nagai et al. 1978)

The distribution of velocities in plasmodial strands was analyzed in the double chamber by a cinematographic technique (Kamiya and Kuroda 1958). The form of the velocity distribution was a parabola with an apex that was flattened or almost truncated. The velocity profile obtained by imposing an artificial pressure flow was almost the same as that of natural streaming (Kamiya and Kuroda 1958). Such a velocity profile is obtained when a fluid with structural viscosity is made to move through a capillary tube by an applied difference in pressure. It was concluded that the flow of endoplasm in the plasmodium is caused by local differences in pressure within the plasmodium. The pressure differences in the plasmodium are caused by contraction of the gel ectoplasm.

To characterize the contractile mechanism of the gel ectoplasm, a tension meter was developed with which the contraction of plasmodial strands can be measured (Kamiya et al. 1972). Using the tension meter, the contraction of the plasmodial strands can be measured either in the isometric or isotonic mode. Isotonic or isometric measuring modes can be interchanged instantaneously (Yoshimoto and Kamiya 1978a). In Fig. 6, the length of a strand in the isotonic mode is shown increasing downward and decreasing upward, since the changes in tension and length have opposite signs. The phase of the production of maximal tension in the wave of isometric contraction coincides with the shortest phase in the wave of isotonic contraction and not with the phase at which the rate of contraction is maximal. When the strand is stretched under isometric conditions, the tension increases. By contrast, when the tension is increased under isotonic conditions, the amplitude of the oscillation increases. It seems, thus, that the plasmodium generates extra force against stretching or loading (Kamiya et al. 1972; Yoshimoto and Kamiya 1978a).

In tensiometric studies, such as those mentioned above, a strand isolated from the posterior region of the plasmodium is generally used. It takes 10–20 min for the strand to begin to contract after isolation from the plasmodium. However, if a slender, rectangular piece is isolated from the anterior part of an advancing plasmodium, it begins a cyclic display of contraction and relaxation without any appreciable lag (Yoshimoto and Kamiya 1978c). It is interpreted that the active site of the contraction-relaxation cycle in an intact plasmodium is localized in the anterior region. This possibility is also supported by the morphological studies discussed below.

3.3 Morphological Approaches

Nakajima and Allen (1965) found fibrous structures, with positive birefringence, oriented orthogonally in the advancing front of the plasmodium. The birefringent fibrils showed strong-weak cyclic changes in birefringence in the anterior part of the plasmodium, which were coupled with the cyclic alternations in the direction of streaming (Kamiya 1973). Three types of birefringent fibrils were observed (Ishigami et al. 1981). The first type is localized in the advancing anterior part and is oriented usually in parallel with, or perpendicular to, the front edge of the plasmodium. The second type is localized in the strand region of the small plasmodium and runs parallel or at an angle to the axis of streaming. The third type is localized in the strand region always perpendicular to the axis of streaming. The second and third types of fibril do not show remarkable rhythmic changes, while the first type of birefringent fibril shows remarkable rhythmic changes in brightness that are synchronized with the alternate direction of streaming of the endoplasm, as reported by Kamiya (1973). The birefringence increases when the endoplasm streams away from the anterior region and decreases when it flows back. It appears, then, that the birefringence increases when the front region of the plasmodium contracts. These observations support the previous conclusion that the active site of the contraction-relaxation cycle of the ectoplasm is localized in the anterior region of the plasmodium.

The proposed involvement of the birefringent structure in the contraction of the plasmodium was further supported by the results of the following experiment. Ishigami et al. (1987) prepared small fragments of plasmodium, 100–200 μm in diameter. Each fragment contracted and relaxed in a cyclic manner. They bulged during the contraction phase and flattened during the relaxation phase. The extent of contraction was quantified by measuring the area of the plasmodium fragment. The amounts of birefringent fibril were measured by an image-processing system. The peak of birefringent fibril content corresponded to the maximum contraction (smallest area of the fragment).

The presence of actin filaments in the plasmodium has been demonstrated both biochemically (Hatano and Oosawa 1966) and morphologically (Alléra et al. 1971; Nagai and Kato 1975). The birefringent fibrils are composed of microfilaments (Fig. 7) (Ishigami et al. 1981). The microfilaments in the fibril are decorated with rhodamine-heavy meromyosin and the birefringent fibrils disappear after treatment with KI, indicating that the microfilaments that make up

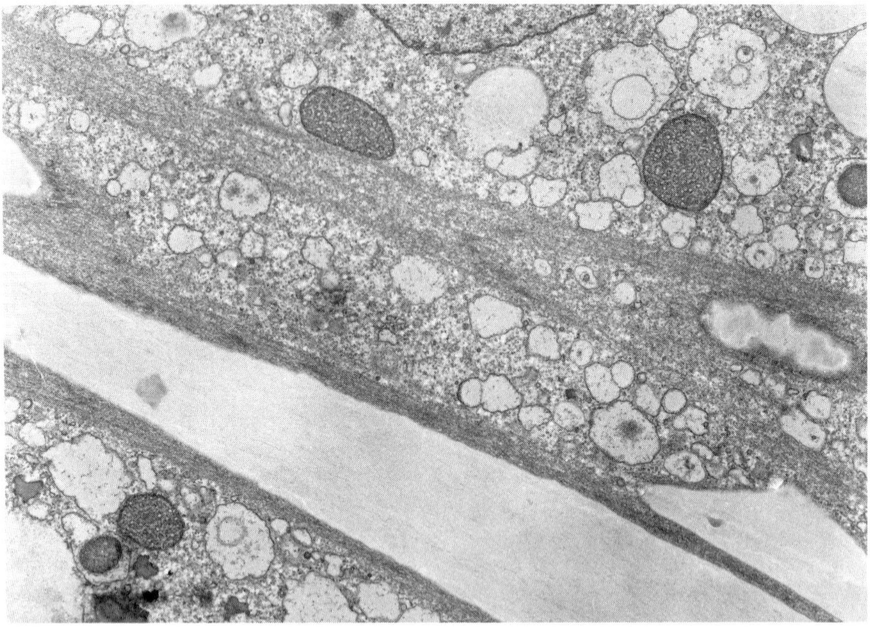

Fig. 7. Microfilaments in a plasmodium. The plasmodial strand was fixed at its maximum length during an isotonic contraction-relaxation cycle; ×12500. (Courtesy of Prof. R. Nagai)

the birefringent fibrils are actin filaments (Ogihara and Kuroda 1979). Nagai et al. (1978) studied the cyclic change in the organization of actin filaments in the plasmodial strands. Plasmodial strands were set on the tension meter and the contraction was measured in the isotonic mode. The strands were fixed at various phases of the contraction-relaxation cycle by application of OsO_4 vapor. In the shortening phase, actin filaments are arranged in parallel and form large, compact bundles. When the strand length reaches a minimum, the actin filaments become kinky and form a network. In the elongation phase, new loose bundles develop from the network. These cyclic changes are shown schematically in Fig. 8. Ishigami (1986) also showed cyclic changes in the organization of actin filaments in the anterior region of the plasmodium, using nitrobenzooxadiazole-phallacidine (NBD-phallacidine). At the early stage of the contraction, slender and flabby fibrils are observed. As contraction progresses, the fibrils become thicker and straighter, and finally they form a dense network. In the relaxation phase, fibrils gradually disintegrate and disappear completely. Although fibrillar structures disappear as relaxation proceeds, Ishigami (1986) argued that the actin filaments are not completely converted to G-actin (monomers), since the plasmodium, when fully relaxed, could still be weakly stained with NBD-phallacidin.

Wohlfarth-Botterman and Fleischer (1976) proposed possible involvement of G-actin/F-actin (actin filament) transformation in fibrillogenesis. When phalloidin was injected into the plasmodium, actin bundles increased in number

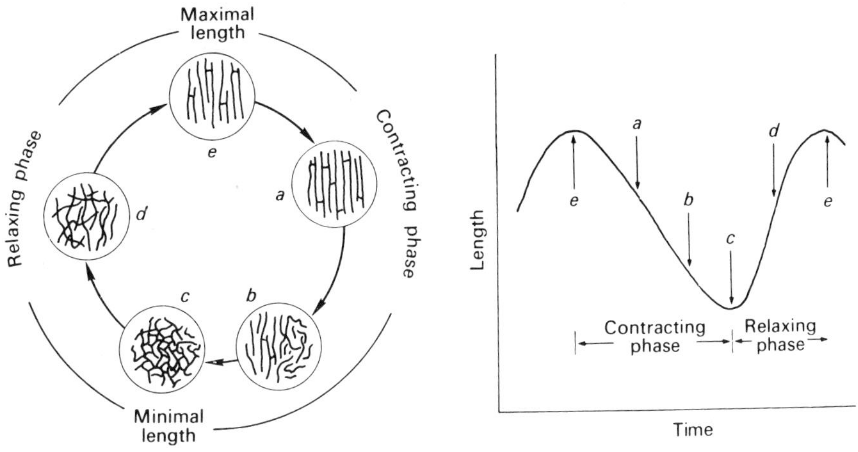

Fig. 8. Schematic drawing of changes in the organization of microfilaments during an isotonic contraction-relaxation cycle. (Nagai et al. 1978)

and condensed. In addition, the contraction-relaxation cycle was irreversibly inhibited (Götz von Olenhusen and Wohlfarth-Bottermann 1979). Phalloidin irreversibly shifted the equilibrium from G-actin to the formation of F-actin (Lengthfield et al. 1974; Wieland 1977). The results of Nagai et al. (1978) and Ishigami (1986) do not support the hypothesis of the G-actin/F-actin transformation. Although some discrepancies remain in the above reports, it appears that the cyclic transformation or organization of actin filaments seems responsible for the cyclic contraction-relaxation cycle of the ectoplasm that results in shuttle streaming.

Morphological studies of the localization of myosin in the plasmodium are scanty. Immunofluorescence microscopy studies using antimyosin antibodies revealed that the distribution of myosin is closely related to that of actin. Immunofluoresence was observed over the entire plasmodium (Osborn and Weber 1983). Thick filaments, assumed to be myosin aggregates, were co-localized with microfilaments (Kessler 1972). Thus, regarding the co-localization of myosin filaments and actin filaments it is possible that contraction of the ectoplasm may be due to actin-myosin sliding.

3.4 Studies with Demembranated Cell Models

To study the mechanism of plasmodial contraction, various models have been developed. The characteristics of reactivated motility vary considerably in the various models. In the present review, results obtained with two models will be discussed. Plasmodial strand was demembranated with saponin and the tension was measured under isometric conditions (Yoshimoto and Kamiya 1984). When ATP was added to the external medium, the tension increased gradually and about 10 min were required for the tension to reach a maximum. The ATP-induced contraction of the saponin model was dependent on Ca^{2+}. The optimal [Ca^{2+}] was 0.1 μM and the tension decreased at [Ca^{2+}], both above or below this value.

When the plasmodium is treated with caffeine, small fragments are formed (caffeine drops). In the caffeine drops, the cytoplasm flows in a manner similar to the cytoplasmic cycle of contraction and relaxation (Hatano 1970; Hatano and Oosawa 1971). When caffeine drops were treated with saponin at a low concentration, to make small pores in the surrounding membrane (Sato et al. 1981), motility was maintained but ATP was not necessary for motility, and the addition of ATP was, in fact, inhibitory. It was suggested that the concentration of ATP in the drops was sufficient to maintain motility and that the addition of extra ATP caused substrate inhibition of actomyosin. The motility of the caffeine drops was also sensitive to Ca^{2+}. At [Ca^{2+}] lower than 0.1 μM, the motility was drastically reduced, but it was activated at [Ca^{2}] higher than 1 μM. Thus, the dependency of motility on Ca^{2+} varies considerably among models.

3.5 Biochemical Approaches

3.5.1 Actin and Related Proteins

Actin from a nonmuscle cells was isolated from the plasmodium for the first time by Hatano and Oosawa (1966). Their work initiated the biochemical study of actin in nonmuscle cells. The actin of the plasmodium has subsequently been studied extensively. Its amino acid sequence has also been determined (Vandekerckhove and Weber 1978). Actin comprises 15–25% of the total protein in the plasmodium (Kessler et al. 1976). F-actin and G-actin are in equilibrium:

$$\text{F-actin} \rightleftharpoons \text{G-actin}. \tag{1}$$

In physiological saline in vitro, almost all actin forms filaments (F-actin). In the plasmodium, however, a considerable fraction of the actin is in an unpolymerized form (Isenberg and Wohlfarth-Bottermann 1976). Some mechanism for shifting the equilibrium toward G-actin must be operative in the cell. Ozaki et al. (1981) isolated profilin from the plasmodium, and its amino acid composition and molecular weight (13 kD) are similar to those of profilin from other organisms. Profilin from *Physarum* forms a 1:1 complex with actin and shifts the equilibrium in Eq. (1) in the right-hand direction by binding to G-actin. The dissociation constant for the binding of *Physarum* profilin to *Physarum* actin is 1.4–3.7 μM. Profilin decreases the rate of elongation of F-actin and also decreases the extent of polymerization. Profilin depolymerize F-actin in a concentration-dependent manner (Ozaki and Hatano 1984).

Fragmin (42 kD) reacts with both G- and F-actin and makes short F-actin. When fragmin is added before polymerization, it accelerates the initial rate of polymerization. Fragmin forms a complex with actin and promotes the formation of nuclei for the polymerization of actin. In the presence of fragmin, the final number of F-actin increases but the average length of the F-actin decreases. When fragmin is added to F-actin, it cuts the filaments into fragments at random positions. Fragmin binds to the barbed end of F-actin (Sugino and Hatano 1982) and is active at [Ca^{2+}] higher than 1 μM (Hasegawa et al. 1980).

Another actin-binding protein whose effect on actin is quite similar to that of fragmin has been isolated from the plasmodium (Furuhashi and Hatano 1989). It binds to the barbed ends of actin filaments and makes short F-actin. However, the molecular weight (60 kD), the tryptic peptide map, and the cross-reactivities with polyclonal fragmin-specific antibodies are different from those of fragmin. This actin-binding protein has been named fragmin 60. Fragmin 60 is also activated by Ca^{2+}.

An actin-binding protein of 36 kD binds 7 mol of actin per mole and makes actin filaments curl and aggregate. It inhibits the actin-activated ATPase activity and the superprecipitation of actomyosin. The activity of this actin-binding protein is not affected by Ca^{2+} (Ogihara and Tonomura 1982). A high molecular weight protein (~230 kD) has been identified that forms cross-links with F-actin and makes a gel-like structure (Sutoh et al. 1984).

3.5.2 Myosin and the Regulation of Its Activity by Ca^{2+}

Myosin comprises 0.77% of the total protein of the plasmodium (Kessler et al. 1976). *Physarum* myosin is composed of six subuits, two heavy chains (230 kD), two phosphorylatable light chains (P-Lc, 18 kD), two Ca^{2+}-binding light chains (Ca-Lc, 16 kD). The heavy chains are composed of a globular head attached to a single fibrous tail, as in skeletal muscle myosin. Two heavy chains span the entire length of the molecule and the light chains are associated with the head region of heavy chains. By treating myosin with trypsin and subsequent gel filtration, heads of myosin molecules (S1: subfragment 1) can be isolated. S1 can hydrolyze ATP, indicating that the globular heads of myosin are the active site. The tail region is responsible for the formation of the bipolar thick filament. At higher ionic strength, myosin disperses as monomers but makes bipolar thick filaments in solutions of physiological ionic strength. The length of the tail region of *Physarum* myosin (170 nm) is greater than that of skeletal muscle myosin (150 nm) (Kohama 1987).

The actin-activated ATPase activity of myosin is regulated by Ca^{2+}. At the beginning of studies on the regulation of actin-activated myosin ATPase by Ca^{2+}, it was reported that the activity was enhanced by Ca^{2+} (Kato and Tonomura 1975; Nachamias and Ash 1976). However, it was found later that the apparent activation by Ca^{2+} of ATPase was caused by contamination of a Ca^{2+}-activated ATP pyrophosphohydrolase (see Kohama and Kendrick-Jones 1986). The superprecipitation of actomyosin purified from the plasmodium is inhibited by Ca^{2+} (Kohama et al. 1980). The inhibition by Ca^{2+} was further supported by results obtained by Ogihara et al. (1983). They isolated two forms of actomyosin. The inactive form has low superprecipitation activity both in the presence and absence of Ca^{2+}. The active form has a high superprecipitation activity in the absence of Ca^{2+} and the activity is inhibited by Ca^{2+}. The active form is converted to the inactive form by treatment with potato phosphatase. The inactive form and the phosphatase-treated active form are converted to the active form by treatment with ATP-γ-S. It has been suggested that protein phosphorylation is involved in controlling the activity of actomyosin. After treatment of dephosphorylated actomyosin with [8-^{35}S]ATP-γ-S, it was found that the myosin heavy chain is phosphorylated. Mg-ATPase activity is activated by F-actin when heavy chains are phosphorylated but not when they are dephosphorylated. The phosphorylation of heavy chains also affects the ability to form thick filaments in medium of physiological ionic strength. Phosphorylated myosin forms thick filaments but dephosphorylated myosin does not (Ogihara et al. 1983; Takahashi et al. 1983). Although the light chain (P-Lc) is also phosphorylated, its functional significance is not known (Kohama 1987). In the absence of actin, the Mg-ATPase activity of *Physarum* myosin is low, irrespective of [Ca^{2+}]. On addition of skeletal muscle actin, the activity is significantly increased. The activity of actin-activated myosin ATPase is highest in the absence of Ca^{2+} and decreases with increases in [Ca^{2+}]. Since skeletal muscle actin per se is insensitive to Ca^{2+}, it appears that the inhibition of actin-activated myosin ATPase by Ca^{2+} is linked to a property of *Physarum* myosin (Kohama and Kohama 1984). Involvement of Ca-Lc has been suggested in the myosin-linked regulation by Ca^{2+} (Kohama et al. 1986).

3.6 Mechanism of the Contraction-Relaxation Cycle

Since actin filaments and myosin filaments are co-localized in the ectoplasm (Kessler 1972), contraction of the gel layer is assumed to reflect the contraction of fibrillar structures composed of actin and myosin filaments. This possibility has been supported by experiments in vitro. Strong, flexible, and uniform threads were prepared from actomyosin purified from the plasmodium (Matsumura et al. 1980). The thread segments generated tension as a result of the addition of ATP. The maximum tension was generated at concentrations of ATP above 10 μM. The concentration of ATP for half-maximal tension was 2–3 μM. Thus, contraction of the fibrillar structure, composed of actomyosin, in the ectoplasm of the plasmodium should generate the internal pressure in the plasmodium.

The ectoplasmic layer of the plasmodium repeats the cycle of contraction and relaxation. Therefore, the actomyosin system must be capable of relaxation and contraction. Using actomyosin threads (Matsumura et al. 1980) and demembranated cell models (Achenbach and Wohlfarth-Bottermann 1986), the cycle of contraction and relaxation can be repeated by the addition and removal of ATP. Oscillations of the concentration of ATP in the cytoplasm have been reported (Yoshimoto et al. 1981b). A plasmodial strand was mounted on a tensiometer and the plasma membrane was rendered leaky by treatment with caffeine and arsenate. Leakage of ATP was measured by addition of luciferin-luciferase in parallel with the application of isometric tension. The period and the phase of oscillation of ATP leakage corresponded closely to those of tension production. The maximum leakage of ATP coincides closely with the maximum of tension. In another experiment, luciferin and luciferase were microinjected into the plasmodial strand and the intracellular luminescence was measured together with tension generation (Yoshimoto et al. 1981b). The period, phase, and maximum luminescence coincide well with those of tension. When the plasmodium was homogenized at room temperature, the concentration of ATP was 4–20 μM, as calculated on the basis of the original volume of the sample. When the homogenate was heated, the measurable concentration of ATP increased to 0.2–0.7 mM. It seems likely that large amounts of ATP are sequestered in organelles and that the concentration of "free ATP" in the plasmodium actually oscillates (Yoshimoto et al. 1981b).

Ca^{2+} is one of the most likely candidates regulating actomyosin-based motility. The contraction of demembranated models is regulated by Ca^{2+} (Sato et al. 1981; Yoshimoto and Kamiya 1984). The activity of actomyosin in vitro is also controlled by Ca^{2+} via myosin-linked regulation (Kohama 1987). Fragmin could also be a Ca^{2+}-linked regulator of actomyosin. Fragmin co-localizes with actin and myosin in the plasmodium (Osborn and Weber 1983). Actomyosin threads prepared by Matsumura et al. (1980) showed no sensitivity to Ca^{2+}. When fragmin was incorporated into actomyosin threads, the contraction was regulated by Ca^{2+}. In the presence of Ca^{2+}, generation of tension by actomyosin threads containing fragmin was inhibited. Fragmin diminished the number of birefringent fibrils in the Triton model in the presence of Ca^{2+}, but more than 30 min were required for their complete disappearance. The birefringent fibrils weakened by a 30-min treatment with fragmin disappeared immediately on addition of ATP (Ishigami and Hatano 1986). Binding of fragmin to actin is practically irreversible in vitro (Hasegawa et al. 1980). If fragmin is responsible for the cyclic change in the organization of actin, the binding of actin to fragmin must be reversible in vivo.

The biochemical results discussed thus indicate that Ca^{2+} inhibits the activity of *Physarum* actomyosin. However, the effect of Ca^{2+} on the contraction of the demembranated models is variable. The concentration of Ca^{2+} in the plasmodium oscillates (Ridgway and Durham 1976). When aequorin was microinjected into plasmodial strands, the emission of light oscillated, but a relationship between the oscillation in levels of Ca^{2+} and the contraction-relaxation cycle was not evident. Yoshimoto et al. (1981a) simultaneously measured the generation of tension and the leakage of Ca^{2+} through a plasma membrane made permeable to Ca^{2+}. The plasmodial strand was mounted on a tensiometer and the plasma membrane was made permeable to Ca^{2+} by applying a Ca^{2+} ionophore. Leakage of Ca^{2+} was measured by monitoring the luminescence of aequorin which was added to the external medium. The period and phase of the changing luminescence was consistent with those of the changes in tension. Leakage of Ca^{2+} was minimum when maximum tension was generated. It was suggested that the amount of leaking Ca^{2+} was a reflection of the intracellular concentration of free Ca^{2+}. This result suggests that Ca^{2+} acts as an inhibitory agent in contraction. Microsomes that transport Ca^{2+} in an ATP-dependent manner have been isolated from the plasmodium (Kato and Tonomura 1977a; Zubrzycka-Gaarn et al. 1979). The membrane system may be involved in the control of the concentration of free Ca^{2+} in the plasmodium.

The contraction-relaxation cycle of the plasmodium may be caused by cyclic changes in actin-myosin sliding and/or the changes in the organization of actomyosin fibrils. The cyclic changes in the activity of the actomyosin system may be caused by oscillations in levels of ATP and/or Ca^{2+}. In addition to changes in levels of ATP and Ca^{2+}, oscillations of cytoplasmic pH (Nakamura and Kamiya 1985) and of levels of NAD(P)H (Mori et al. 1987) have been reported. However, the cause-and-effect relationships between these oscillating parameters are not clear. In general, demembranated models of the plasmodium contract monotonically and do not show cycles of spontaneous contraction and relaxation. In the Triton model, however, spontaneous oscillatory contraction occurs over a limited range of Ca^{2+} concentrations. Ogihara (1982) suggested that oscillatory contraction and relaxation require a factor that is activated at Ca^{2+} concentrations above 10^{-6} M and inactivated at 10^{-8} M.

When two independently oscillating plasmodial strands were connected, their oscillation becomes synchronized (Takeuchi and Yoneda 1977; Yoshimoto and Kamiya 1978b). When the shuttle streaming between the two strands is stopped by the double-chamber method, synchronization does not occur, indicating that some factor(s) in the flowing endoplasm is responsible for the regulation of the contraction-relaxation cycle (Yoshimoto and Kamiya 1978b). Involvement of mitochondrial activity in the timekeeping of the contractional rhythm has been suggested (Satoh et al. 1982).

4 Cytoplasmic Streaming in Other Plant Cells

The mechanism of cytoplasmic streaming has been studied most extensively in characean cells and in the plasmodium of *Physarum* because of the ease of manipulation of these materials. In other plant cells, involvement of an actomyosin system in the generation of motive force has also been suggested from various indirect evidence. The presence of actin filaments has been demonstrated in a variety of plant cells (Staiger and Schliwa 1987). In root-hair cells of *Hydrocharis*, a typical pattern of cytoplasmic streaming is observed, namely, reversed fountain streaming. The endoplasm streams along the cortical gel layer toward the tip and comes back to the base in an organized transvacuolar strand. When the root-hair cells are stained with rhodamine-phalloidin, actin filaments become visible in the cortical layer and the transvacuolar strand. It is likely that these filaments are the tracks along which cytoplasmic streaming occurs (Fig. 9). The involvement of actin filaments in cytoplasmic streaming has also been demonstrated by the inhibition of cytoplasmic streaming by cytochalasin (Staiger and Schliwa 1987). The nucleotide sequence of an actin gene of soybean has been determined (Shah et al. 1982). The actin is composed of 376 amino acids, but the soybean actin is rather different in terms of sequence from actins from other eukaryotic cells. Myosin has been isolated from *Egeria* (Ohsuka and Inoue 1979), tomato fruits (Vahey et al. 1982), pea tendrils (Ma and Yen 1989), and leaf petioles (Sokolov et al. 1986). Few details of the characteristics of such "myosin" are available, presumably because of difficulties in isolating the "myosin" from plant cells. The association of "myosin" with streaming organelles has been demonstrated in pollen tubes. Organelles were isolated by homogenizing the pollen tubes of the lily, *Lilium longiflorum*. When the organelles were introduced into characean cells, from which endoplasm had been removed by intracellular perfusion, the pollen tube organelles moved on the characean actin cables and movement was dependent on Mg-ATP (Kohno

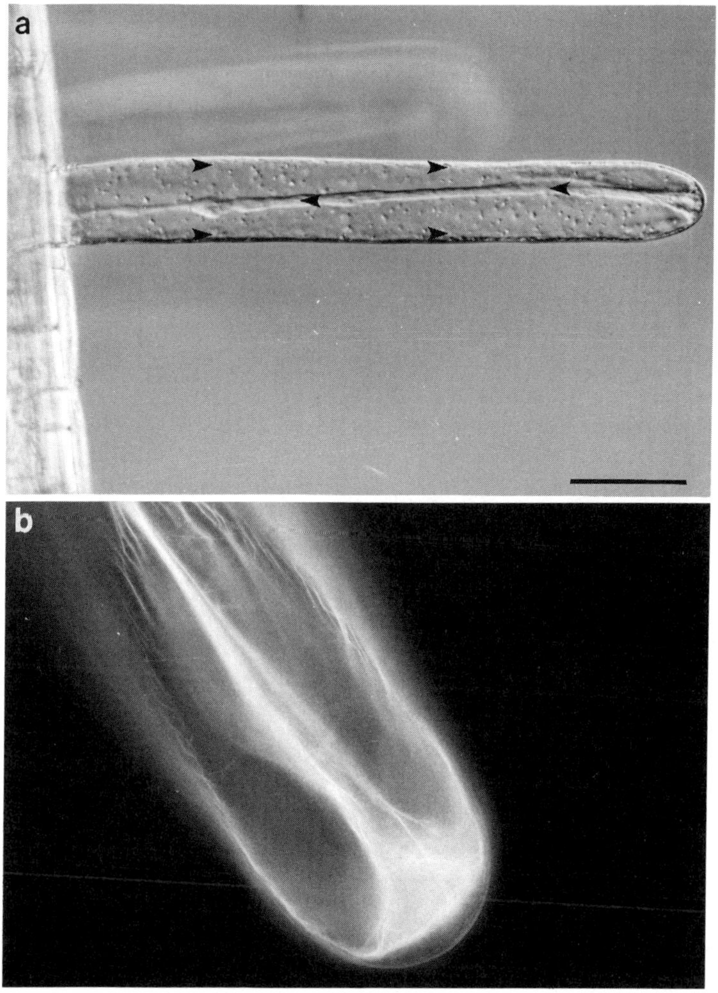

Fig. 9a,b. Organization of actin filament in root-hair cells of *Hydrocharis*. **a** Nomarski micrograph; **b** root-hair cell stained with rhodamine-phalloidin. *Arrowheads* show direction of cytoplasmic streaming. *Bar* = 200 μm in a and 66 μm in **b**

and Shimmen 1988). The ability of the organelles to move along actin cables is impaired by treating the organelles with NEM or heating (50 °C). Cytoplasmic streaming in intact pollen tubes is also inhibited by treatment with NEM or heating. In pollen tubes, actin filaments run in a longitudinal direction. It is likely that cytoplasmic streaming occurs by the sliding of the myosin that is associated with organelles along the actin filaments in the intact pollen tubes. Reconstituted movement of pollen tube organelles in characean cells proceeds from the pointed to the barbed end of the actin filaments (Kohno et al. 1990). The polarity of the myosin movement of pollen tubes is the same as that of skeletal muscle myosin. Antibodies raised against the heavy chains of animal myosin bind to the surface of pollen tube organelles (Heslop-Harrison and Heslop Harrison 1989; and Tang et al. 1989). The association of myosin with organelles has been demonstrated only in characean cells and in pollen tubes. However, it is probable that cytoplasmic streaming in other plant cells that are surrounded by a cell wall occurs by the same mechanism as that in characean cells and pollen tubes.

In the marine alga, *Caulerpa*, involvement of a microtubule-dynein system in the generation of motive force has been reported (Kuroda and Manabe 1983). Both actin and microtubule are likely involved in the generation of motive force of cytoplasmic streaming in *Acetabularia* (Koop and Kiermayer 1980a,b) and in *Bryopsis* (Menzel and Schliwa 1986a,b; Menzel and Elsner-Menzel 1989).

5 Amoeboid Movement

5.1 Types of Amoeboid Movement and Hypotheses for Generation of Motive Force

Amoeboid movement is observed in a variety of organisms; protozoa, mammalian cells, amoeboid cells of true slime molds, and cellular slime molds. In amoeboid movement, the cell moves by the use of pseudopodia or as a result of cytoplasmic streaming. In this sense, the locomotion of *Physarum* plasmodium might be classified as amoeboid movement. The mode of amoeboid movement varies considerably among cell types and has been divided into three classes (Taylor and Condeelis 1979). Class 1; cells show cyclic and extensive changes in cytoplasmic structure and rapid cytoplasmic streaming occurs in the cells. The movements of giant amoebae, *Chaos carolinensis* and *Amoeba proteus*, are included in this class. In class 2, locomotion occurs as a result of relatively slow cytoplasmic streaming and by the extension and retraction of filopodia. The movement of *Acanthamoeba castellanii* and of the amoeboid cells of *Dictyostelium discoidium* are included in this class. In class 3, locomotion is characterized by very slow movement of pharopodia. The movements of mammalian cells in culture and of cells involved in cell sorting during embryogenesis are included in this class (for details, see Taylor and Condeelis 1979).

Many biologists have been interested in the mechanism of locomotion of amoeboid cells. Cytological studies have been mainly carried out using cells of class 1, by taking advantage of the large size of cells. Various hypotheses have been proposed, each being influenced by the prevailing scientific concepts of the time, as summarized in detail by Fukui and Yumura (1986). Two representative hypotheses for motive-force generation are considered here.

1. The posterior contraction hypothesis: the sol endoplasm is passively transported forward by contraction of gel ectoplasm in the posterior region (Mast 1926; Stockem et al. 1982).
2. The frontal contraction hypothesis: The streaming endoplasm is assumed to be a viscoelastic gel and propagated contractions of the endoplasm at the tips of advancing pseudopods transport the endoplasm forward (Allen et al. 1960, 1965; Allen and Cowden 1962).

Although conclusive evidence is lacking, the contraction of cytoplasm and the sol-gel transformation are almost certainly involved in amoeboid movement. Actomyosin and related proteins seem to be involved in these phenomena (Taylor et al. 1979).

Recent studies have been carried out mainly in the cells of class 2, *Acanthamoeba* and the amoeboid cells of *Dictyostelium*. Although these materials are not suitable for classic cytological studies because of the small size of the cell, they are suitable for analysis by biochemical, immunological, and molecular-biological techniques. Only recent progress made with *Acanthamoeba* and *Dictyostelium* will be discussed here.

5.2 Biochemical Studies

Although cytological studies on the mechanism of motive-force generation are difficult in *Acanthamoeba* and *Dictyostelium*, these cells are suitable for large-scale culture and subsequent isolation of contractile proteins. The situation has been summarized by Pollard and Korn (1973a),

> Although our ultimate goal is to understand movement at the cellular level, our immediate approach has been to purify and characterize the presumptive "motility" proteins because the mechanism of movement of an intact cell is clearly too complicated to understand without extensive biochemical and physical knowledge of the molecules involved.

5.2.1 Actin and Actin-Linked Regulation

In physiological solutions in vitro, the G-actin of amoebae polymerizes to form F-actin (Weihing and Korn 1971, 1972; Uyemura et al. 1978). In amoebae, however, more than 50% of the total actin appears to be present in a nonpolymerizable form (Gordon et al. 1977). It has been suggested that profilin influences the equilibrium between G- and F-actin (Reichstein and Korn 1979) as it does in the plasmodium of *Physarum*. In *Amoeba proteus*, polymerization is inhibited by phosphorylation of G-actin (Sonobe et al. 1986). Polymerization of G-actin may also be controlled by phosphorylation.

In addition to profilin, various actin-binding proteins have been isolated. A few of the typical actin-binding proteins isolated from *Dictyostelium* are as follows. "Capping protein" caps actin filaments at the barbed end in a Ca^{2+}-dependent manner (Isenberg et al. 1980; Schleicher et al. 1984). Severin blocks the ends of F-actin and causes the fragmentation and depolymerization of F-actin (Brown et al. 1982; Yamamoto et al. 1982). In addition, various proteins that cross-link F-actin have been isolated, namely, α-actinin (Noegel et al. 1987), the 120-kD cross-linking molecule (Condeelis et al. 1984), and the 30-kD cross-linking molecule (Fechheimer 1987). Some actin filaments are associated with the plasma membrane (Korn and Wright 1973; Pollard and Korn 1973b). Actin filaments may be anchored to the membrane by ponticulin, a membrane-spanning glycoprotein (Wuestehube and Luna 1987).

5.2.2 Myosin and Myosin-Linked Regulation

Two kinds of myosin, myosin I (1A and 1B) and myosin II, have been isolated from both *Acanthamoeba* and *Dictyostelium*. Myosin II is a "conventional", double-headed myosin with a molecular weight of about 400 kD and is composed of two heavy chains and two pairs of light chains (Clarke and Spudich 1974; Maruta and Korn 1977a; Kuczmarski and Spudich 1980). The heavy chain is phosphorylated at or near its tail end (Collins and Korn 1980; Kuczmarski and Spudich 1980; Cote et al. 1981; Peltz et al. 1981; Collins et al. 1982). Actin-activated ATPase activity is inhibited by the phosphorylation of heavy chains (Collins and Korn 1980, 1981; Kuczmarski and Spudich 1980). Formation of thick filaments of myosin II in solutions of low ionic strength is also influenced by phosphorylation of heavy chains. Formation of thick filaments is significant when the heavy chain is in its dephosphorylated form (Kuczmarski and Spudich 1980; Pollard 1982; Kuznicki et al. 1983; Kuczmarski et al. 1987; Sinard and Pollard 1989). In *Dictyostelium*, myosin-heavy chain kinase is inhibited by Ca^{2+}/calmodulin (Maruta et al. 1983). The actin-activated ATPase activity of myosin II is also controlled by phosphorylation of light chains and the activity is higher when the light chain is in the phosphorylated form (Griffith et al. 1987).

Myosin I is a single-headed enzyme with one heavy chain and two light chains. In *Acanthamoeba*, there are two isoforms, IA and IB. Myosin I is incapable of forming oligomers of any size (Pollard and Korn 1973a; Maruta et al. 1979; Albanesi et al. 1983; Cote et al. 1985). The actin-activated ATPase activity is enhanced by the phosphorylation of the heavy chain (Maruta and Korn 1977b; Albanesi et al. 1983; Hammer et al. 1983; Cote et al. 1985). Myosin I has a second actin-binding site in addition to the ATP-dependent binding site. Therefore, myosin I can generate contractile force in collaboration with actin filaments, although they cannot make filament (Albanesi et al. 1985; Fujisaki et al. 1985). Myosin I has the ability to bind directly to membrane lipids and it is associated with the plasma membrane (Gadasi and Korn 1980; Miyata et al. 1989) and organelles (Adams and Pollard 1986).

5.3 Functions of Contractile Proteins in Vivo

Various kinds of contractile proteins and their associated proteins have been demonstrated in amoeboid cells by biochemical methods. However, the physiological functions of many of these proteins remained unknown. Recently, molecular-biological methods have been used in attempts to identify the physiological role of such proteins.

5.3.1 Actin-Associated Proteins

After mutagenesis, a mutant can be selected in two ways. When a specific antibody is available, one can select a mutant that lacks a specific protein. In the second method, mutants can be selected on the basis of a deficiency in cell motility or behavior and then the altered protein can be identified (Gerisch et al. 1989). Mutants of *Dictyostelium* that are defective in actin-associated proteins have been isolated. A mutant lacking α-actinin grows at about the normal rate, and neither cell motility nor chemotaxis are detectably disturbed (Wallraff et al. 1986; Schleicher et al. 1988). The study of a mutant deficient in severin also showed that severin is not essential for either motility or chemotaxis (Andre et al. 1989).

5.3.2 Myosin II

Synthesis of the heavy chains of myosin II has been depressed by transforming *Dictyostelium* with a vector that encoded an RNA complementary to the messenger RNA of the myosin heavy chains (antisense RNA; Knecht and Loomis 1987). The gene for the heavy chain of myosin II has been disrupted by homologous recombination with a plasmid that encoded the amino-terminal half of the myosin heavy chain (Lozanne and Spudich 1987). These cells synthesize a fragment of the myosin II heavy chain. A *Dictyostelium* mutant completely lacking the myosin II heavy chain has also been produced by this method (Manstein et al. 1989). Amoebae lacking the (intact) myosin heavy chains are both viable and motile. The amoebae show cytoplasmic expansion and contraction at roughly half the rate of that of normal cells and move with a rate of translocation that is lower than that of normal cells. In suspension cultures, cytokinesis is defective and the defect results in the frequent appearance of large multinucleated cells. However, the cells are able to divide on Petri dishes.

5.3.3 Myosin I

Organelles isolated from *Acanthamoeba* move along characean actin cables and the reconstituted system is inhibited by antibodies raised against myosin I (Adams and Pollard 1986). This observation suggests that myosin I is responsible for the intracellular transport of organelles. Involvement of myosin I in the formation of pseudopodia has also been suggested (Fukui et al. 1989) and will be discussed below.

5.4 Mechanisms of Locomotion

To understand the mechanism of motility, it is necessary to determine the intracellular localization of contractile proteins. Recent progress in fluorescence microscopy has made it possible to gain an overall view of the localization of contractile proteins in amoeboid cells. Such experiments have been carried out mainly with *Dictyostelium*. Development of the "agar-overlay technique" has allowed clearer images to be obtained by fluorescence microscopy (Yumura et al. 1984). Localization of actin, myosin I, and myosin II in actively moving cells of *Dictyostelium* was studied using antibodies raised against these proteins (Yumura et al. 1984; Fukui et al. 1989; Fig. 10). Actin is localized in both the anterior pseudopodium and in the posterior cortex. Two kinds of pseudopodium are visible, appearing as light and dark regions under a microscope equipped with dark-contrast phase optics. Actin can be observed only in the dark pseudopodium. At the boundary between the light pseudopodium and the ground cytoplasm, a layer of actin is observed (not shown in Fig. 10a). In the ground cytoplasm, diffuse staining by anti-actin antibody is observed. Myosin II is present only in the posterior cortex. By contrast, myosin I is specifically concentrated in the cortex of the leading edge of lamellipodia and pseudopodia. A substantial amount of myosin I is distributed in the ground cytoplasm, particularly in the posterior of the cells. Myosin I is not found in the region behind the leading edge. It is likely that myosin II produces the posterior contractile force in collaboration with actin filaments. Molecular-biological studies have shown

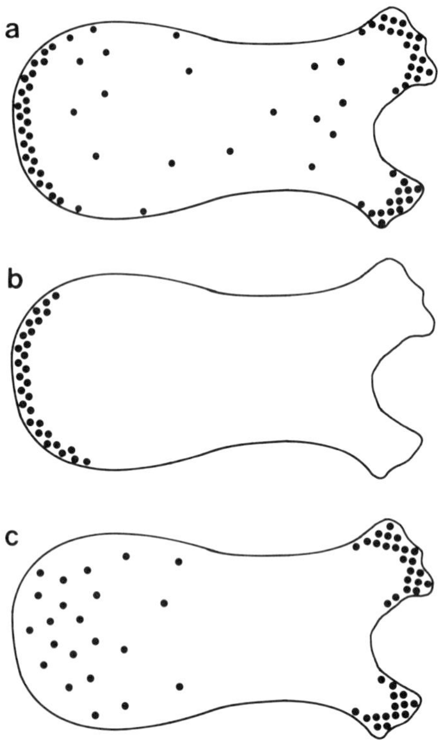

Fig. 10a–c. Localization of actin (**a**), myosin II (**b**), and myosin I (**c**) in amoebae of *Dictyostelium*. Figures were redrawn after photographs of Yumura and Fukui (1985) and Fukui et al. (1989)

that myosin II is required for normal translocation but is not essential for the extension of pseudopodia (Knecht and Loomis 1987; Lozanne and Spudich 1987). The absence of myosin II from the pseudopodia is consistent with the conclusion obtained by molecular-biological methods. Myosin I seems to be involved in the formation of pseudopodia. Myosin I can generate a contractile force in collaboration with actin filaments (Albanesi et al. 1985; Fujisaki et al. 1985).

During amoeboid movement, the cell continuously changes its shape. Therefore, the organization of contractile proteins should also change dynamically. Analysis of the behavior of *Dictyostelium* amoebae in response to chemoattractants allow the characterization of the dynamic changes in the organization of contractile proteins. When an amoeba is stimulated by micromolar levels of cyclic AMP (cAMP), the cell stops moving and rounds by withdrawing pseudopodia within 25 s after stimulation. Then the pseudopodia are extended again within 60 s (Hall et al. 1988). The time course of these events varies among reports (Futrelle et al. 1982).

The extent of actin polymerization in the amoeboid cells of *Dictyostelium* changes dynamically in response to stimulation by cAMP (Condeelis et al. 1988). Between 5 and 20 s after stimulation, rapid polymerization and subsequent depolymerization occur. The peak of polymerization is correlated temporally with withdrawal of the pseudopodia. The peak is followed by a second phase of actin assembly. The second phase is correlated with the extension of the pseudopodia. Actin polymerized in response to cAMP is associated with the cortical cytoskeleton which can be isolated as a Triton-insoluble fraction (McRobbie and Newell 1984). Actin filaments are concentrated in pseudopodia and in surface projections that are extended 120 s after stimulation by cAMP. ABP-120, one of the proteins that cross-link actin, is co-localized with actin filaments. The amount of ABP-120 associated with the Triton-insoluble cytoskeleton begins to increase at about 25 s after stimulation with cAMP. It has been suggested that ABP is responsible for the extension of pseudopodia (Dharmawardhane et al. 1989).

Dynamic translocation of myosin II also occurs after stimulation with cAMP (Yumura and Fukui 1985). Upon the addition of cAMP, translocation of myosin filaments from the endoplasm to the ectoplasm and a subsequent redistribution back to the endoplasm occur (Yumura and Fukui 1985). The accumulation of myosin filaments in the ectoplasm is correlated with a rounding of the cell body (Nachmias et al. 1989). The time course of translocation of myosin filaments parallels the time course of phosphorylation of myosin light and heavy chains (Nachmias et al. 1989). The depolymerization/polymerization of myosin II regulated by the phosphorylation/dephosphorylation of myosin heavy chains (Kuczmarski and Spudich 1980) may be involved in myosin translocation in the cell. The phosphorylation/dephosphorylation of light chains (Griffith et al. 1987) may also be involved in translocation of myosin II (Malchow et al. 1981; Berlot et al. 1985, 1987).

6 Concluding Remarks

The present chapter provides a brief review of the mechanisms of typical non-muscle motilities, cytoplasmic streaming, and amoeboid movement. Although these motilities are phenomenologically different, their basic molecular mechanisms are the same, namely, the sliding of myosin filaments over actin filaments. It is not exaggerating to say that progress in the elucidation of the mechanism of motive-force generation for cytoplasmic streaming in Characeae and *Physarum* was initiated by the analysis of velocity distribution performed by Kamiya and Kuroda (1956, 1958). Movement of organelles in cells that are surrounded by cell walls, including Characeae and higher plants, is the direct manifestation of the sliding of myosin over actin. In the plasmodium of *Physarum*, organelles are passively transported by intracellular differences in pressure caused by the contraction-relaxation of gel ectoplasm that is due to sliding of myosin over actin.

The mechanism of amoeboid movement is much more complicated. It seems that two different macromolecules, myosins I and II, are involved in the extension of pseudopodia and posterior contraction, respectively. From cytological studies, both an anterior contraction hypothesis and a posterior contraction hypothesis have been proposed. Recent studies suggest that active processes occur at both the anterior and posterior regions of amoeboid cells. Cytological studies have been carried out using giant amoebae which belong to class 1 in terms of motility. Biochemical and molecular-biological studies have been carried out using the small amoebae of class 2. These gaps must be filled in future studies.

References

Achenbach F, Wohlfarth-Bottermann K-E (1986) Successive contraction-relaxation cycles experimentally induced in cell-free models of *Physarum polycephalum*. Eur J Cell Biol 42:111–117

Adams RJ, Pollard TD (1986) Propulsion of organelles isolated from *Acanthamoeba* along actin filaments by myosin-I. Nature 322:754–756

Albanesi JP, Hammer JA III, Korn ED (1983) The interaction of F-actin with phosphorylated and unphosphorylated myosins IA and IB from *Acanthamoeba castellanii*. J Biol Chem 258:10176–10181

Albanesi JP, Fujisaki H, Korn ED (1985) A kinetic model for the molecular basis of the contractile activity of *Acanthamoeba* myosins IA and IB. J Biol Chem 260:11174–11179

Allen RD, Cowden RR (1962) Syneresis in ameboid movement: its localization by interference microscopy and its significance. J Cell Biol 12:185–189

Allen RD, Cooledge JW, Hall PJ (1960) Streaming in cytoplasm dissociated from the giant amoeba, *Chaos chaos*. Nature 187:896–899

Allen RD, Francis DW, Nakajima H (1965) Cyclic birefringence changes in pseudopods of *Chaos carolinensis* revealing the localization of the motive force in pseudopod extension. Proc Natl Acad Sci USA 54:1153–1161

Alléra A, Beck R, Wohlfarth-Bottermann KE (1971) Weitreichende fibrillare Protoplasmadifferenzierungen und ihre Bedeutung für die Protoplasmaströmung. VIII. Identifizierung der Plasmafilamente von *Physarum polycephalum* als F-Actin durch Anlagerung von heavy meromyosin in situ. Cytobiologie 4:437–449

Andre E, Brink M, Gerisch G, Isenberg G, Noegel A, Schleicher M, Segall JE, Wallraff E (1989) A *Dictyostelium* mutant deficient in severin, an F-actin fragmenting protein, shows normal motility and chemotaxis. J Cell Biol 108:985–995

Barak LS, Yocum RR, Nothnagel EA, Webb WW (1980) Fluorescence staining of the actin cytoskeleton in living cells with 7-nitrobenz-2-oxa-1,3-diazole-phallacidin. Proc Natl Acad Sci USA 77(2):980–984

Barry WH (1968) Coupling of excitation and cessation of cyclosis in *Nnitella*: role of divalent cations. J Cell Physiol 72(3):153–160

Berlot CH, Spudich JA, Devreotes PN (1985) Chemoattractant-elicited increases in myosin phosphorylation in *Dictyostelium*. Cell 43:307–314

Berlot CH, Devreotes PN, Spudich JA (1987) Chemottractant-elicited increases in *Dictyostelium* myosin phosphorylation are due to changes in myosin localization and increases in kinase activity. J Biol Chem 262:3918–3926

Bradley MO (1973) Microfilaments and cytoplasmic streaming. inhibition of streaming with cytochalasin. J Cell Sci 12:327–343

Brown SS, Yamamoto K, Spudich JA (1982) A 40,000-dalton protein from *Dictyostelium discoideum* affects assembly properties of actin in a Ca^{2+}-dependent manner. J Cell Biol 93:205–210

Chen JCW, Kamiya N (1975) Localization of myosin in the internodal cell of *Nitella* as suggested by differential treatment with N-ethylmaleimide. Cell Struct Funct 1:1–9

Chen JCW, Kamiya N (1981) Differential heat treatment of the *Nitella* internodal cell and its relation to cytoplasmic streaming. Cell Struct Funct 6:201–207

Clarke M, Spudich JA (1974) Biochemical and structural studies of actomyosin-like proteins from non-muscle cells. J Mol Biol 86:209–222

Collins JH, Korn ED (1980) Actin activation of Ca^{2+}-sensitive Mg^{2+}-ATPase activity of *Acanthamoeba* myosin II is enhanced by dephosphorylation of its heavy chains. J Biol Chem 255:8011–8014

Collins JH, Korn ED (1981) Purification and characterization of actin-activatable, Ca^{2+}-sensitive myosin II from *Acanthamoeba*. J Biol Chem 256:2586–2595
Collins JH, Kuznicki J, Bowers B, Korn ED (1982) Comparison of the actin binding and filament formation properties of phosphorylated and dephosphorylated *Acanthamodeba* myosin II. Biochemistry 21:6910–6915
Condeelis J, Vahey M, Carboni JM, DeMey J, Ogihara S (1984) Properties of the 120,000- and 95,000-dalton actin-binding proteins from *Dictyostelium discoideum* and their possible functions in assembling the cytoplasmic matrix. J Cell Biol 99:119s–126s
Condeelis J, Hall A, Bresnick A, Warren V, Hock R, Bennett H, Ogihara S (1988) Actin polymerization and pseudopod extension during amoeboid chemotaxis. Cell Motil Cytoskeleton 10:77–90
Cote GP, Collins JH, Korn ED (1981) Identification of three phosphorylation sites on each heavy chain of *Acanthamoeba* myosin II. J Biol Chem 256:12811–12816
Cote GP, Albanesi JP, Ueno T, Hammer JA III, Korn ED (1985) Purification from *Dictyostelium discoideum* of a low-molecular-weight myosin that resembles myosin I from *Acanthamoeba castellanil*. J Biol Chem 260:4543–4546
Dharmawardhane S, Warren V, Hall AL, Condeelis J (1989) Changes in the association of actin-binding proteins with the actin cytoskeleton during chemotactic stimulation of *Dictyostelium discoideum*. Cell Motil Cytoskeleton 13:57–63
Fechheimer M (1987) The *Dictyostelium discoideum* 30 000-dalton protein is an actin filament-bundling protein that is selectively present in filopodia. J Cell Biol 104:1539–1551
Fujisaki H, Albanesi JP, Korn ED (1985) Experimental evidence for the contractile activities of *Acanthamoeba* myosins IA and IB. J Biol Chem 260:11183–11189
Fukui Y, Yumura S (1986) Actomyosin dynamics in chemotactic amoeboid movement of *Dictyostelium*. Cell Motil Cytoskeleton 6:662–673
Fukui Y, Lynch TJ, Brzeska H, Korn ED (1989) Myosin I is located at the leading edge of locomoting *Dictyostelium* amoebae. Nature 341:328–331
Furuhashi K, Hatano S (1989) A fragmin-like protein from plasmodium of *Physarum polycephalum* that severs F-actin and caps the barbed end of F-actin in a Ca^{2+}-sensitive way. J Biochem 106:311–318
Futrelle RP, Traut J, McKee WG (1982) Cell behavior in *Dictyostelium discoideum*: preaggregation response to localized cyclic AMP pulses. J Cell Biol 92:807–821
Gadasi H, Korn ED (1980) Evidence for differential intracellular localization of the *Acanthamoeba* myosin isoenzymes. Nature 286:452–456
Gerisch G, Segall JE, Wallraff E (1989) Isolation and behavioral analysis of mutants defective in cytoskeletal proteins. Cell Motil Cytoskeleton 14:75–79
Gordon DJ, Boyer JL, Korn ED (1977) Comparative biochemistry of non-muscle actins. J Biol Chem 252:8300–8309
Götz von Olenhusen K, Wohlfarth-Bottermann KE (1979) Evidence for actin transformation during the contraction-relaxation cycle of cytoplasmic actomyosin: cycle blockade by phalloidin injection. Cell Tissue Res 196:455–470
Gratecos D, Fischer EH (1974) Adenosine 5′-0-(3-thiotriphosphate) in the control of phosphorylase activity. Biochem Biophys Res Commun 58:960–967
Griffith LM, Downs SM, Spudich JA (1987) Myosin light chain kinase and myosin light chain phosphatase from *Dictyostelium*: effects of reversible phosphorylation on myosin structure and function. J Cell Biol 104:1309–1323
Hall A, Schlein A, Condeelis J (1988) Relationship of pseudopod extension to chemotactic hormone-induced actin polymerization in amoeboid cells. J Cell Biochem 37:285–299
Hammer JA III, Albanesi JP, Korn ED (1983) Purification and characterization of a myosin I heavy chain kinase from *Acanthamoeba castellanii*. J Biol Chem 258:10168–10175
Hasegawa T, Takahashi S, Hayashi H, Hatano S (1980) Fragmin: a calcium ion sensitive regulatory factor on the formation of actin filaments. Biochemistry 19:2677–2683
Hatano S (1970) Specific effect of Ca^{2+} on movement of plasmodial fragment obtained by caffeine treatment. Exp Cell Res 61:199–203
Hatano S, Oosawa F (1966) Isolation and characterization of plasmodium actin. Biochim Biophys Acta 127:488–498

Hatano S, Oosawa F (1971) Movement of cytoplasm in plasmodial fragment obtained by caffeine treatment. I. Its Ca^{+} sensitivity. J Physiol Soc Jpn 33:589–590

Heslop-Harrison J, Heslop-Harrison Y (1989) Myosin associated with the surfaces of organelles, vegetative nuclei and generative cells in angiosperm pollen grains and tubes. J Cell Sci 94:319–325

Higashi-Fujime S (1980) Active movement in vitro of bundles of microfilaments isolated from *Nitella* cell. J Cell Biol 87:569–578

Isenberg G, Wohlfarth-Bottermann (1976) Transformation of cytoplasmic actin: importance for the organization of the contractile gel reticulum and the contraction-relaxation cycle of cytoplasmic actomyosin. Cell Tissue Res 173:495–528

Isenberg G, Aebi U, Pollard TD (1980) An actin-binding protein from *Acanthamoeba* regulates actin filament polymerization and interactions. Nature 288:455–459

Ishigami M (1986) Dynamic aspects of the contractile system in *Physarum* plasmodium. 1. Changes in spatial organization of the cytoplasmic fibrils according to the contraction-relaxation cycle. Cell Motil Cytoskeleton 6:439–447

Ishigami M, Hatano S (1986) Dynamic aspects of the contractile system in *Physarum* plasmodium. II. Contractility of Triton cell models in accordance with the contraction and relaxation phases of the plasmodia. Cell Motil Cytoskeleton 6:448–457

Ishigami M, Nagai R, Kuroda K (1981) A polarized light and electron microscopic study of the birefringent fibrils in *Physarum* plasmodia. Protoplasma 109:91–102

Ishigami M, Kuroda K, Hatano S (1987) Dynamic aspects of the contractile system in *Physarum* plasmodium. III. Cyclic contraction-relaxation of the plasmodial fragment in accordance with the generation-degeneration of cytoplasmic actomyosin fibrils. J Cell Biol 105:381–386

Kamitsubo E (1966) Motile protoplasmic fibrils in cells of Characeae. II. Linear fibrillar structure and its bearing on protoplasmic streaming. Proc Jpn Acad 42:640–643

Kamitsubo E (1972) A "window technique" for detailed observation of characean cytoplasmic streaming. Expl Cel Res 74:613–616

Kamitsubo E (1981) Effect of supraoptimal temperature on the function of the subcortical fibrils and an endoplasmic factor in *Nitella* internodes. Protoplasma 109:3–12

Kamiya N (1940) The control of protoplasmic streaming. Science 92:462–463

Kamiya N (1959) Protoplasmic streaming. Protoplasmatologia VIII 3a. Springer, Wien

Kamiya N (1973) Contractile characteristics of the myxomycete plasmodium. Proc IV Int Biophysics Congr, Moscow, pp 447–465

Kamiya N, Kuroda K (1956) Velocity distribution of the protoplasmic streaming in *Nitella* cells. Bot Mag., Tokyo 69:544–554

Kamiya N, Kuroda K (1958) Studies on the velocity distribution of the protoplasmic streaming in the myxomycete plasmodium. Protoplasma 49:1–4

Kamiya N, Allen RD, Zeh R (1972) Contractile properties of the slime mold strand. Acta Protozoologica 11:113–124

Kato T, Tonomura Y (1975) *Physarum* tropomyosin-troponin complex. Isolation and properties. J Biochem 78:583–588

Kato T, Tonomura Y (1977a) Uptake of calcium ions into microsomes isolated from *Physarum polycephalum*. J Biochem 81:207–213

Kato T, Tonomura Y (1977b) Identification of myosin in *Nitella flexilis*. J Biochem 82: 777–782

Kersey YM, Wessels NK (1976) Localization of actin filaments in internodal cells of characean algae. J Cell Biol 68:264–275

Kersey YM, Hepler PK, Palevitz BA, Wessels NK (1976) Polarity of actin filaments in characean algae. Proc Natl Acad Sci USA 73:165–167

Kessler D (1972) On the location of myosin in the myxomycete *Physarum polycephalum* and its possible function in cytoplasmic streaming. J Mechanochem Cell Motil 1:125–137

Kessler D, Nachmias VT, Loewy AG (1976) Actomyosin content of *Physarum* plasmodia and detection of immunological cross-reactions with myosins from related species. J Cell Biol 69:393–406

Knecht DA, Loomis WF (1987) Antisense RNA inactivation of myosin heavy chain gene expression in *Dictyostelium discoideum*. Science 236:1081–1086

Kohama K (1987) Ca-inhibitory myosins: their structure and function. Adv Biophys 23:149–182

Kohama K, Kendrick-Jones J (1986) The inhibitory Ca^{2+}-regulation of the actin-activated Mg-ATPase activity of myosin from *Physarum polycephalum* plasmodia. J Biochem 99:1433–1446

Kohama K, Kohama T (1984) Myosin confers inhibitory Ca^{2+}-sensitivity on actin-myosin-ATP interaction of *Physarum polycephalum* plasmodium under physiological conditions. Proc Jpn Acad 60:435–439

Kohama K, Kobayashi K, Mitani S (1980) Effect of Ca ion and ADP on superprecipitation of myosin B from slime mold, *Physarum polycephalum*. Proc Jpn Acad 56:591–596

Kohama K, Takano-Ohmuro H, Tanaka T, Yamaguchi Y, Kohama T (1986) Isolation and characterization of myosin from amoebae of *Physarum polycephalum*. J Biol Chem 261:8022–8027

Kohno T, Shimmen T (1988) Accelerated sliding of pollen tube organelles along *Characeae* actin bundles regulated by Ca^{2+}. J Cell Biol 106:1539–1543

Kohno T, Chaen S, Shimmen T (1990) Characterization of the translocator associated with pollen tube organelles. Protoplasma 154:179–183

Koop HU, Kiermayer O (1980a) Protoplasmic streaming in the giant unicellular green alga *Acetabularia mediterranea*. I. Formation of intracellular transport systems in the course of cell differentiation. Protoplasma 102:147–166

Koop HU, Kiermayer O (1980b) Protoplasmic streaming in the giant unicellular green alga *Acetabularia mediterranea*. II. Differential sensitivity of movement systems to substances acting on microfilaments and microtubuli. Protoplasma 102:295–306

Korn ED, Wright PL (1973) Macromolecular composition of an amoeba plasma membrane. J Biol Chem 248:439–447

Kuczmarski ER, Spudich JA (1980) Regulation of myosin self-assembly: phosphorylation of *Dictyostelium* heavy chain inhibits formation of thick filaments. Proc Natl Acad Sci USA 77:7292–7296

Kuczmarski ER, Tafuri SR, Parysek LM (1987) Effect of heavy chain phosphorylation on the polymerization and structure of *Dictyostelium* myosin filaments. J Cell Biol 105: 2989–2997

Kuroda K (1983) Cytoplasmic streaming in characean cells cut open by microsurgery. Proc Jpn Acad (B) 59:126–130

Kuroda K, Kamiya N (1975) Active movement of *Nitella* chloroplast in vitro. Proc Jpn Acad 51:774–777

Kuroda K, Manabe E (1983) Microtubule-associated cytoplasmic streaming in *Caulerpa*. Proc Jpn Acad 59:131–134

Kuznicki J, Albanesi JP, Cote GP, Korn ED (1983) Supramolecular regulation of the actin-activated ATPase activity of filaments of *Acanthamoeba* myosin II. J Biol Chem 258:6011–6014

Lengthfield A, Low I, Wieland TH, Danka P, Hasselbachi W (1974) Interaction of phalloidin with actin. Proc Natl Acad Sci USA 71:2803–2807

Lozanne AD, Spudich JA (1987) Disruption of the Dictyostelium myosin heavy chain gene by homologous recombination. Science 236:1086–1091

Ma YZ, Yen LF (1989) Actin and myosin in pea tendrils. Plant Physiol 89:586–589

Malchow D, Bohme R, Rahmsdorf HJ (1981) Regulation of phosphorylation of myosin heavy chain during the chemotactic response of *Dictyostelium* cells. Eur J Biochem 117:213–218

Manstein DJ, Titus MA, Lozanne AD, Spudich JA (1989) Gene replacement in *Dictyostelium*: generation of myosin null mutants. EMBO J 8:923–932

Maruta H, Korn ED (1977a) *Acanthamoeba* myosin II. J Biol Chem 252:6501–6509

Maruta H, Korn ED (1977b) *Acanthamoeba* cofactor protein is a heavy chain kinase required for actin activation of the Mg^{2+}-ATPase activity of *Acanthamoeba* myosin I. J Biol Chem 252:8329–8332

Maruta H. Gadashi H. Collins JH, Korn ED (1979) Multiple forms of *Acanthamoeba* myosin I. J Biol Chem 254:3624–3630

Maruta H, Baltes W, Dieter P, Marme D, Gerisch G (1983) Myosin heavy chain kinase inactivated by Ca^{2+}/calmodulin from aggregating cells of *Dictyostelium discoideum*. EMBO J 2:535–542

Mast SO (1926) Structure, movement, locomotion, and stimulation in *Amoeba*. J Morphol Physiol 41:535–542
Matsumura F, Yoshimoto Y, Kamiya N (1980) Tension generation by actomyosin thread from a non-muscle system. Nature 285:169–171
McRobbie SJ, Newell PC (1984) Chemoattractant-mediated changes in cytoskeletal actin of cellular slime moulds. J Cell Sci 68:139–151
Menzel D, Elsner-Menzel CE (1989) Co-localization of particle transport with microtubules in cytoplasmic exudates of the siphonous green alga *Bryopsis*. Bot Acta 102:241–248
Menzel D, Schliwa M (1986a) Motility in the siphonous green alga *Bryopsis*. I. Spatial organization of the cytoskeleton and organelle movements. Eur J Cell Biol 40:275–285
Menzel D, Schliwa M (1986b) Motility in the siphonous green alga *Bryopsis*. II. Chloroplast movement requires organized arrays of both microtubules and actin filaments. Eur J Cell Biol 40:286–295
Miyata H, Bowers B, Korn ED (1989) Plasma membrane association of *Acanthamoeba* myosin I. J Cell Biol 109:1519–1528
Morgan M, Perry SV, Ottaway J (1976) Myosin light chain phosphatase. Biochem J 157: 687–697
Mori Y, Ueda T, Kobatake Y (1987) NAD(P)H oscillation in relation to the rhythmic contraction in the *Physarum* plasmodium. Protoplasma 139:141–144
Nachmias VT, Asch A (1976) A calcium-sensitive preparation from *Physarum polycephalum*. Biochemistry 15:4273–4278
Nachmias VT, Fukui Y, Spudich JA (1989) Chemoattractant-elicited translocation of myosin in motile *Dictyostelium*. Cell Motil Cytoskeleton 13:158–169
Nagai R, Hayama T (1979) Ultrastructure of the endoplasmic factor responsible for cytoplasmic streaming in *Chara* internodal cells. J Cell Sci 36:121–136
Nagai R, Kamiya N (1977) Differential treatment of *Chara* cells with cytochalasin B with special reference to its effect on cytoplasmic streaming. Exp Cell Res 108:231–237
Nagai R, Kato T (1975) Cytoplasmic filaments and their assembly into bundles in *Physarum* plasmodium. Protoplasma 86:141–158
Nagai R, Rebhun L (1966) Cytoplasmic microfilaments in streaming *Nitella* cells. J Ultrastruct Res 14:571–589
Nagai R, Yoshimoto Y, Kamiya N (1978) Cyclic production of tension force in the plasmodial strand of *Physarum polycephalum* and its relation to microfilament morphology. J Cell Sci 33:205–225
Nakajima H, Allen RD (1965) The changing pattern of birefringence in plasmodia of the slime mold, *Physarum polycephalum*. J Cell Biol 25:361–374
Nakamura S, Kamiya N (1985) Oscillation of cytoplasmic pH of *Physarum* plasmodium in relation to motility. Cell Struct Funct 10:133–141
Noegel A, Witke W, Schleicher M (1987) Calcium-sensitive nonmuscle α-actinin contains EF-hand structures and highly conserved regions. FEBS Lett 221:391–396
Nothnagel EA, Webb WW (1982) Hydrodynamic models of viscous coupling between motile myosin and endoplasm in characean algae. J Cell Biol 94:444–454
Nothnagel EA, Barak LS, Sanger JW, Webb WW (1981) Fluorescence studies on modes of cytochalasin and phallotoxin action on cytoplasmic streaming in *Chara*. J Cell Biol 88: 364–372
Ogihara S (1982) Calcium and ATP regulation of the oscillatory torsional movement in a Triton model of *Physarum* plasmodial strands. Exp Cell Res 138:377–384
Ogihara S, Kuroda K (1979) Identification of a birefringent structure which appears and disappears in accordance with the shuttle streaming in *Physarum* plasmodia. Protoplasma 100:167–177
Ogihara S, Tonomura Y (1982) A novel 36,000-dalton actin-binding protein purified from microfilaments in *Physarum* plasmodia which aggregates actin filaments and blocks actin-myosin interaction. J Cell Biol 93:604–614
Ogihara S, Ikebe M, Takahashi K, Tonomura Y (1983) Requirement of phosphorylation of *Physarum* myosin heavy chain for thick filamant formation, actin activation of Mg^{2+}-ATPase activity, and Ca^{2+}-inhibitory superprecipitation. J Biochem 93:205–223

Ohsuka K, Inoue A (1979) Identification of myosin in a flowering plant, *Egeria densa*. J Biochem 85:375–378

Osborn M, Weber K (1983) Immunocytochemistry of the acellular slime mold *Physarum polycephalum*. III. Distribution of myosin and the actin-modulating protein (fragmin) in sandwiched plasmodia. Eur J Cell Biol 29:179–186

Owaribe K, Izutsu K, Hatano S (1979) Cross-reactivity of antibody to *Physarum* actin and actins in eukaryotic cells examined by immunofluorescence. Cell Struct Funct 4:117–126

Ozaki K, Hatano S (1984) Mechanism of regulation of actin polymerization by *Physarum* profilin. J Cell Biol 98:1919–1925

Ozaki K, Sugino H, Hasegawa T, Takahashi S, Hatano S (1981) Isolation and characterization of *Physarum* profilin. J Biochem 93:295–298

Palevitz BA, Hepler PK (1975) Identification of actin in situ at the ectoplasm-endoplasm interface of *Nitella*. J Cell Biol 65:29–38

Palevitz BA, Ash JF, Hepler PK (1974) Actin in the green alga, *Nitella*. Proc Natl Acad Sci USA 71:363–366

Peltz G, Kuczmarski ER, Spudich JA (1981) *Dictyostelium* myosin: characterization of chymotryptic fragments and localization of the heavy-chain phosphorylation site. J Cell Biol 89:104–108

Pollard TD (1982) Structure and polymerization of *Acanthamoeba* myosin-II filaments. J Cell Biol 95:816–825

Pollard TD, Korn ED (1973a) *Acanthamoeba* myosin. I. Isolation from *Acanthamoeba castellanii* of an enzyme similar to muscle myosin. J Biol Chem 248:4682–4690

Pollard TD, Korn ED (1973b) Electron microscopic identification of actin associated with isolated amoeba plasma membranes. J Biol Chem 248:448–450

Reichstein E, Korn ED (1979) *Acanthamoeba* profilin. J Biol Chem 254:6174–6179

Ridgway EB, Durham CH (1976) Oscillations of calcium ion concentration in *Physarum polycephalum*. J Cell Biol 69:223–226

Sato H, Hatano S, Sato Y (1981) Contractility and protoplasmic streaming preserved in artificially induced plasmodial fragments, the "caffeine drops". Protoplasma 109: 187–208

Satoh H, Ueda T, Kobatake Y (1982) Primary oscillator of contractional rhythm in the plasmodium of *Physarum polycephalum*: role of mitochondria. Cell Struct Funct 7: 275–283

Schleicher M, Gerisch G, Isenberg G (1984) New actin-binding proteins from *Dictyostelium discoideum*. EMBO J 3:2095–2100

Schleicher M, Noegel A, Schwarz T, Wallraff E, Brink M, Faix J, Gerisch G, Isenberg G (1988) A *Dictyostelium* mutant with severe defects in α-actinin: its characterization using cDNA probes and monoclonal antibodies. J Cell Sci 90:59–71

Shah DM, Hightower RC, Meagher RB (1982) Complete nucleotide sequence of a soybean actin gene. Proc Natl Acad Sci 79:1022–1026

Sheetz MP, Spudich JA (1983) Movement of myosin-coated fluorescent beads on actin cables in vitro. Nature 303:31–35

Shiina, Tazawa M (1987) Ca^{2+}-activated Cl^- channel in plasmalemma of *Nitellopsis obtusa*. J Memb Biol 99:137–146

Shimmen T (1978) Dependency of cytoplasmic streaming on intracellular ATP and Mg^{2+} concentrations. Cell Struct Funct 3:113–121

Shimmen T (1988a) Cytoplasmic streaming regulated by adenine nucleotides and inorganic phosphates in Characeae. Protoplasma (Suppl) 1:3–9

Shimmen T (1988b) Characean actin bundles as a tool for studying actomyosin-based motilify. Bot Mag Tokyo 101:533–544

Shimmen T, Tazawa M (1982) Reconstitution of cytoplasmic streaming in Characeae. Protoplasma 113:127–131

Shimmen T, Tazawa M (1983) Control of cytoplasmic streaming by ATP, Mg^{2+} and cytochalasin B in permeabilized Characeae cell. Protoplasma 115:18–24

Shimmen T, Yano M (1984) Active sliding movement of latex beads coated with skeletal muscle myosin on *Chara* actin bundles. Protoplasma 121:132–137

Shimmen T, Yano M (1986) Regulation of myosin sliding along *Chara* actin bundles by native skeletal muscle tropomyosin. Protoplasma 132:129–136
Sinard JH, Pollard TD (1989) The effect of heavy chain phosphorylation and solution conditions on the assembly of *Acanthamoeba* myosin-II. J Cell Biol 109:1529–1535
Sokolov OI, Bogatyrev VA, Turkina MV (1986) Myosin from conducting tissues of *Heracleum sosnowskyi*: interaction with muscle actin and formation of filaments. Fiziol Rast 33: 421–431
Sonobe S, Takahashi S, Hatano S, Kuroda K (1986) Phosphorylation of *Amoeba* G-actin and its effect on actin polymerization. J Biol Chem 261:14837–14843
Staiger CJ, Schliwa M (1987) Actin localization and function in higher plants. Protoplasma 141:1–12
Stockem W, Hoffman, H-U, Gawlitta W (1982) Spatial organization and fine structure of the cortical filament layer in normal locomoting *Amoeba proteus*. Cell Tissue Res 221:505–519
Sugino H, Hatano S (1982) Effect of fragmin on actin polymerization: evidence for enhancement of nucleation and capping of the barbed end. Cell Motil 2:457–470
Sutoh K, Iwane M, Matsuzaki F, Kikuchi M, Ikai A (1984) Isolation and characterization of a high molecular weight actin-binding protein from *Physarum polycephalum* plasmodia. J Cell Biol 98:1611–1618
Takahashi K, Ogihara S, Ikebe M (1983) Morphological aspects of thiophosphorylated and dephosphorylated myosin molecules from the plasmodium of *Physarum polycephalum*. J Biochem 93:1175–1183
Takeuchi Y, Yoneda M (1977) Synchrony in the rhythm of the contraction-relaxation cycle in two plasmodial strands of *Physarum polycephalum*. J Cell Sci 26:151–160
Tang X, Hepler PK, Scordilis SP (1989) Immunochemical and immunocytochemical identification of a myosin heavy chain polypeptide in *Nicotiana* pollen tubes. J Cell Sci 92:569–574
Taylor DL, Condeelis JS (1979) Cytoplasmic structure and contractility in amoeboid cells. Int Rev Cytol 56:57–144
Taylor DL, Hellewell SB, Virgin HW, Heiple J (1979) The solation-contraction coupling hypothesis of cell movements. In: Hatano S, Ishikawa H, Sato H (eds) Cell motility: molecules and organization, University of Tokyo Press, Tokyo, pp 363–377
Tazawa M, Shimmen T (1987) Cell motility and ionic relations in characean cells as revealed by internal perfusion and cell models. Int Rev Cytol 109:259–312
Tazawa M, Kikuyama M, Shimmen T (1976) Electric characteristics and cytoplasmic streaming of Characeae cells lacking tonoplast. Cell Struct Funct 1:165–176
Tominaga Y, Shimmen T, Tazawa M (1983) Control of cytoplasmic streaming by extracellular Ca^{2+} in permeabilized *Nitella* cells. Protoplasma 116:75–77
Tominaga Y, Muto S, Shimmen T, Tazawa M (1985) Calmodulin and Ca^{2+}-controlled cytoplasmic streaming in characean cells. Cell Struct Funct 10:315–325
Tominaga Y, Wayne R, Tung HYL, Tazawa M (1987) Phosphorylation-dephosphorylation is involved in Ca^{2+}-controlled cytoplasmic streaming of characean cells. Protoplasma 136:161–169
Uyemura DG, Brown SS, Spudich JA (1978) Biochemical and structural characterization of actin from *Dictyostelium discoideum*. J Biol Chem 253:9088–9096
Vahey M, Titus M, Trautwein R, Scordilis S (1982) Tomato actin and myosin: contractile proteins from a higher plant. Cell Motil 2:131–147
Vandekerckhove J, Weber K (1978) The amino acid sequence of *Physarum* actin. Nature 276:720–721
Wallraff E, Schleicher M, Modersitzki M, Rieger D, Isenberg G, Gerisch G (1986) Selection of *Dictyostelium* mutants defective in cytoskeletal proteins: use of an antibody that binds to the ends of α-actinin rods. EMBO J 5:61–67
Weihing RR, Korn ED (1971) *Acanthamoeba* actin. Isolation and properties. Biochemistry 10:590–600
Weihing RR, Korn ED (1972) *Acanthamoeba* actin. Composition of the peptide that contains 3-methylhistidine and a peptide that contains Nε-methyllysine. Biochemistry 11:1538–1543
Wieland T (1977) Modification of actins by pallotoxins. Naturwissenschaften 64:303–309

Williamson RE (1972) A light-microscope study of the action of cytochalasin B on the cells and isolated cytoplasm of the Characeae. J Cell Sci 10:811–819

Williamson RE (1974) Actin in the alga, *Chara corallina*. Nature 248:801–802

Williamson RE (1975) Cytoplasmic streaming in *Chara*: a cell model activated by ATP and inhibited by cytochalasin B. J Cell Sci 17:655–668

Williamson RE (1979) Filaments associated with the endoplasmic reticulum in the streaming cytoplasm of *Chara corallina*. Eur J Cell Biol 20:177–183

Williamson RE, Ashley CC (1982) Free Ca^{2+} and cytoplasmic streaming in the alga *Chara*. Nature 296:647–651

Williamson RE, McCurdy DW, Hurley UA, Perkin JL (1987) Actin of *Chara* giant internodal cells. Plant Physiol 85:268–272

Wohlfarth-Bottermann KE, Fleischer M (1976) Cycling aggregation patterns of cytoplasmic F-actin coordinated with oscillating tension force generation. Cell Tissue Res 165:327–344

Wuestehube LJ, Luna EJ (1987) F-actin binds to the cytoplasmic surface of ponticulin, a 17-kD integral glycoprotein from *Dictyostelium discoideum* plasma membranes. J Cell Biol 105:1741–1751

Yamamoto K, Pardee JD, Reidler J, Stryer L, Spudich J (1982) Mechanism of interaction of *Dictyostelium* severin with actin filaments. J Cell Biol 95:711–719

Yoshimoto Y, Kamiya N (1978a) Studies on contraction rhythm of the plasmodial strand II. Effect of externally applied forces. Protoplasma 95:101–109

Yoshimoto Y, Kamiya N (1978b) Studies on contraction rhythm of the plasmodial strand. III. Role of endoplasmic streaming in synchronization of local rhythms. Protoplasma 95: 111–121

Yoshimoto Y, Kamiya N (1978c) Studies on contraction rhythnm of plasmodial strand IV. Site of active oscillation in an advancing plasmodium. Protoplasma 95:123–133

Yoshimoto Y, Kamiya N (1984) ATP- and calcium-controlled contraction in a saponin model of *Physarum polycophalum*. Cell Struct Funct 9:135–141

Yoshimoto Y, Matsumura F, Kamiya N (1981a) Simultaneous oscillations of Ca^{2+} efflux and tension generation in the permeabilized plasmodial strand of *Physarum*. Cell Motil 1: 433–443

Yoshimoto Y, Sakai T, Kamiya N (1981b) ATP oscillation in *Physarum* plasmodium. Protoplasma 109:159–168

Yumura S, Fukui Y (1985) Reversible cyclic AMP-dependent change in distribution of myosin thick filaments in *Dictyostelium*. Nature 314:194–196

Yumura S, Mori H, Fukui Y (1984) Localization of actin and myosin for the study of ameboid movement in *Dictyostelium* using improved immunofluorescence. J Cell Biol 99:894–899

Zubrzycka-Gaarn E, Korczak B, Osinska HE (1979) Identification of sarcoplasmic reticulum-like system in *Physarum polycephalum*. FEBS Lett 107:335–339

Chapter 7

Molecular Mechanism of Ciliary and Flagellar Movement

R. Kamiya

Contents

1 Introduction

Cilia and flagella are whip-like motile organelles present in a wide variety of eukaryotes. They are similar to other cellular motility systems in that movement is based on sliding between two types of protein assemblies. They differ from other motile systems, however, in having highly complicated but well-defined structures made up of more than 200 kinds of proteins, and displaying rapid undulating movements with two- or three-dimensional waveforms; their elaborate structures are rather like precision machines. Although such structural complexity has greatly hampered the elucidation of the molecular mechanism underlying their action, several lines of novel approaches are beginning to give us insights into their organization and mechanochemical properties that have been poorly understood before. This chapter describes our current knowledge of the ciliary and flagellar motility mechanism, with emphasis on findings from recent

Department of Molecular Biology, School of Science, Nagoya University, Nagoya, 464-01, Japan

Advances in Comparative and Environmental Physiology
Vol. 12. Edited by H. Sugi

studies using mutants and in vitro motility experiments. For brevity, I will not discuss the regulation of ciliary and flagellar movements, many examples of which in fact have been shown to be regulated by Ca^{2+} and cyclic nucleotides (see, for example, Hyams and Borisy 1978; Tash 1989).

2 The 9+2 Structure

The eukaryotic flagellum (or cilium) is a slender membrane-covered organelle with a diameter of about 0.2 μm and a length of the order of 10 μm. It is evident that the mechanism that produces flagellar or ciliary beating is entirely contained within the internal structure, the axoneme, since flagella or cilia demembranated with detergent, such as Triton X-100, beat almost normally if ATP is supplied under appropriate conditions (Hoffman-Berling 1955; Gibbons and Gibbons 1972).

Axonemes of most species so far studied have a common structural design of two microtubules (the central pair) surrounded by nine microtubules. The nine tubules are called outer doublets because they are composed of two subtubules: an A-tubule with a circular cross-section and a B-tubule with a C-shaped cross-section (Fig. 1). The basal portion of the axoneme is connected to the basal body, which is essentially identical to the centriole, a key cellular organelle that plays a crucial role in chromosome separation in mitosis of most

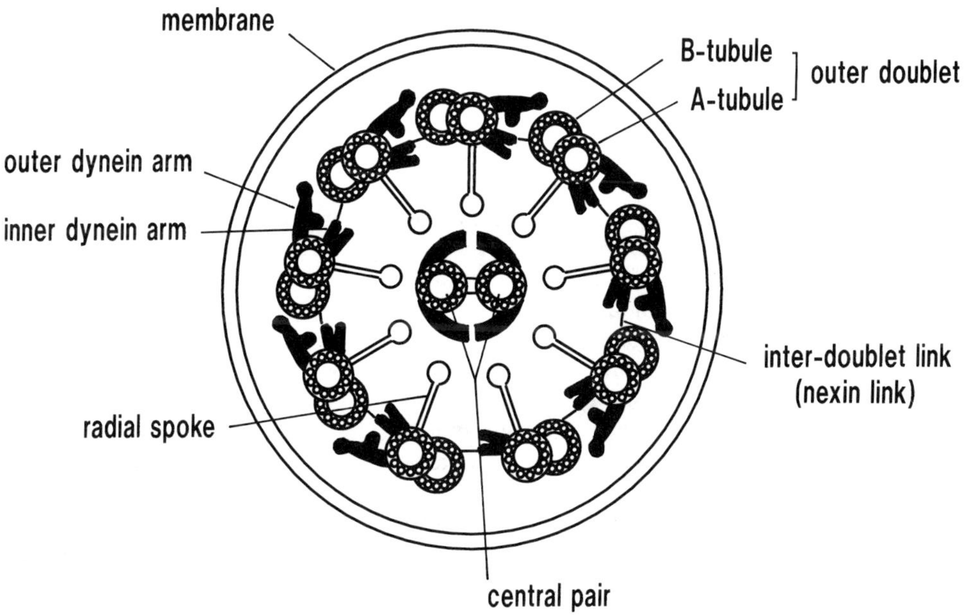

Fig. 1. Schematic cross-section of the 9 + 2 structure, depicting the *Chlamydomonas* axoneme, in which one particular outer doublet lacks the outer dynein arm

animal cells. The basal body has nine short triplet microtubules, which have three, rather than two, fused tubules.

The 9+2 arrangement of microtubules is maintained by several connecting structures. Neighboring outer doublets are connected with each other by inter-doublet links and by inner and outer dynein arms. The central pair is held in the center by radial spokes. These bridges and arms are arranged at precise intervals along the outer-doublet microtubule: one inter-doublet link, two or three radial spokes (the number varies with species), four outer dynein arms, and three inner dynein arms occur within a repeating unit distance of 96 nm. The axoneme, therefore, has a highly regular structure, probably the most ordered of all cellular organelles.

All these bridges and arms make both tight and loose (or dynamic) contacts with microtubules at their opposite ends, thus permitting microtubules to move in certain directions. For example, the radial spokes are firmly attached to outer doublets at one end and loosely interact with the central pair at the other end (spoke head). The central pair, therefore, has the freedom to rotate within the axoneme. The inter-doublet links also appear to make loose bridges (Warner 1983).

Having such a complex structure, the axoneme is naturally made up of many kinds of proteins besides tubulin. Electrophoresis analyses of mutant axonemes of *Chlamydomonas* have demonstrated that the axoneme contains no less than 200 peptides, including 17 for radial spokes, about 20 for inner and outer dynein arms, and about 20 for the central-pair complex (Luck 1984).

3 Microtubule Sliding and Axonemal Beat

3.1 Microtubule Sliding as the Basis for Axonemal Motility

Flagella and cilia of different organisms beat at various frequencies between 1 and 100 Hz, with various waveforms differing in wavelength, symmetry, and curvature. Even the direction of wave propagation can reverse in flagella of some organisms such as *Crithidia*, a protozoa, under certain unusual conditions (Holwill and McGregor 1976). The basic mechanism for axonemal movement, however, should be common to all kinds of cilia and flagella.

In 1963, Gibbons isolated an ATPase enzyme from *Tetrahymena* cilia, and named it dynein after the force unit dyne because it was expected to be the force generator for ciliary motility (see for review, Gibbons 1981). Electron microscopy of axonemes before and after dynein extraction revealed that dynein was localized at two projections on the outer-doublet microtubules, called the inner and outer dynein arms. Later studies have shown that dynein arms contain several similar but different heavy chains with ATPase activities (see below), and that proteins similar to dynein heavy chains are present also in cellular cytoplasm, where they probably work as motors for transport of cellular organelles along the microtubuler cytoskeleton (see, for example, Paschal et al. 1987b). Hence, the axoneme can be regarded as a specially differentiated type of widespread dynein-microtubule motility system.

It is now established that the axonemal beating occurs as a result of regulated sliding between outer doublets driven by dynein arms. This sliding tubule mechanism was first reported by Satir (1968) who observed that outer-doublet microtubules in a cilium had a constant length irrespective of the bend in the axoneme. More direct evidence for sliding as an active process came from studies by Summers and Gibbons (1971), who observed by dark-field microscopy that outer doublets actually slid out of axonemes upon treatment with trypsin and addition of ATP. They suggested that the long-distance sliding occurred because the trypsin treatment destroyed some axonemal structures such as inter-doublet links that would normally restrict the microtubule sliding to a short distance.

Electron microscopy demonstrated further that microtubule sliding occurs in a unidirectional manner (Sale and Satir 1977); dynein arms on an outer doublet always "push" the adjacent one distally, or, in other words, the dynein arms themselves move toward the base of the outer doublet with which they interact. This polarity of movement is determined by an intrinsic property of dynein and the structural polarity of microtubules at the molecular level. The distal end of an outer doublet corresponds to the plus end of the cytoplasmic microtubule, which is defined as the end that grows faster when assembled from tubulin in vitro. All kinds of dynein so far studied, including cytoplasmic types, have been shown to move toward the minus end of microtubules, whereas kinesin, another cytoplasmic microtubule motor, has been shown to move in the opposite direction (Vale et al. 1985).

More recently, the sliding tubule model has been developed and substantiated further by different lines of experiments. First, the force exerted by dynein arms in a disintegrating axoneme was measured directly by use of a fine, flexible glass needle attached to the axoneme (Kamimura and Takahashi 1981; Oiwa and Takahashi 1988). The force measured from the deflection of the needle was about 0.5–1 pN/dynein arm at 4 to 500 μM MgATP. This value is within the range of estimates for the force produced by single myosin head. A second line of development is the direct visualization of microtubule sliding in beating axonemes; Brokaw (1989) found that two fine gold particles attached to a demembranated, but not proteolyzed, axoneme cyclically changed the distance between them, in a manner consistent with the sliding tubule model. Thus, it has been proved that axonemal beating accompanies a shear between outer doublets.

3.2 A Possible Mechanism That Produces Flagellar Bending

Axonemes will not bend if all the outer doublets undergo sliding simultaneously. In other words, the axoneme must have a mechanism that controls where, when, and how much sliding takes place in order to propagate a proper bending wave.

Shingyoji et al. (1977) have demonstrated that iontophoretic application of a small amount of ATP to a midportion of a demembranated sperm flagellum results in the formation of two opposite bends, and interpreted this as being caused by localized sliding (Fig. 2). It is conceivable that real axonemal bending

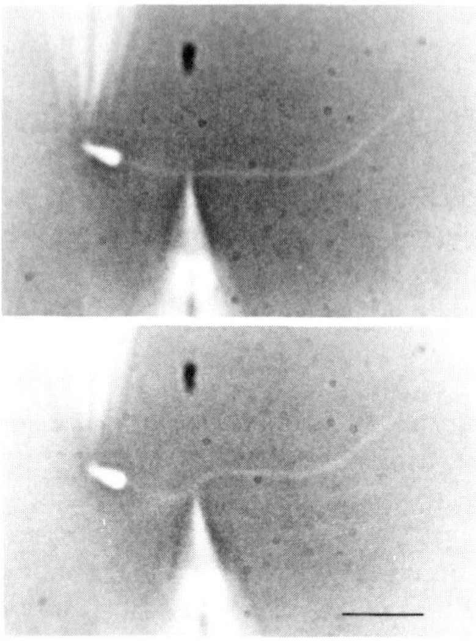

Fig. 2. Formation of bends in a Triton-extracted flagellum by local application of ATP. Before (*upper panel*) and 0.4 s after (*lower panel*) an iontophoretic application of ATP. Scale *bar* = 10 μm. (Shingyoji et al. 1977)

is also caused by such localized activation of sliding. An important difference, however, is that the in vivo mechanism localizing the sliding must operate in the homogeneous presence of ATP.

Because dynein arms induce only unidirectional sliding of microtubules within the cylindrical structure, only those dynein arms contained on one side of the axoneme (with respect to the bending plane running through the axonemal axis) should actively shear outer doublets when the axoneme bends, while those contained on the other side should be at rest. The activation state should reverse when the axoneme bends in the opposite direction; the activated and resting states of each dynein arm should alternate when the axoneme beats.

What causes the oscillation? A most popular idea is that the axonemal mechanochemical system constitutes a feedback loop as a whole, such that the dynein-doublet interaction is regulated by the mechanical state of the axoneme (e.g. curvature), which is in turn determined by the extent of dynein-doublet interaction (for review, see Brokaw 1982). This kind of model seems to work reasonably well in computer simulations, although some aspects of the behavior of beating flagella still remain to be explained (Brokaw 1985)

An oscillatory bending movement which is apparently effected through a feedback mechanism is observed in *Chlamydomonas* axonemes that have been spontaneously frayed in a reactivation solution containing ATP and detergent (Kamiya and Okagaki 1986). The example in Fig. 3 demonstrates that a pair of outer doublets bend and unbend at a frequency of 1–2 Hz in the presence of ATP, indicating that the essential mechanism for the oscillation is contained within outer doublets and dynein arms. Frame-by-frame analyses revealed that the mechanism for this movement can be understood by the assumption that the dynein-tubule interaction shears the two doublets unidirectionally, but that it is turned off at the site where curvature exceeds a certain critical value (for more details, see legend to Fig. 3). The dynein-doublet interaction may be intrinsically

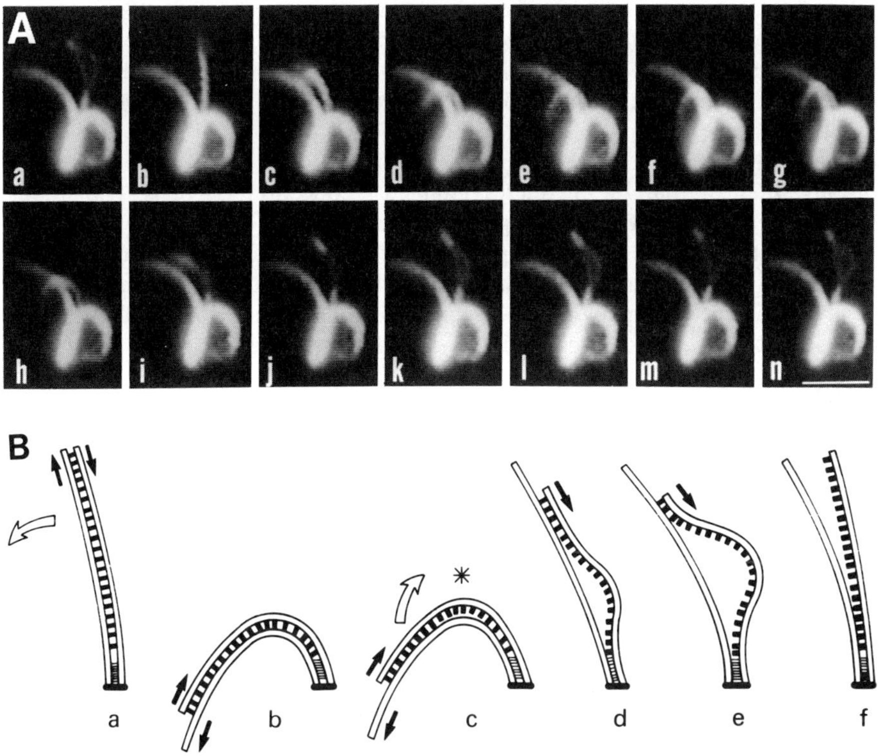

Fig. 3A,B. Bending movements displayed by two outer doublets in a frayed axoneme of (**A**) *Chlamydomonas* and (**B**) a model for this movement. Scale *bar* = 5 μm. In **A**, the axoneme has spontaneously detached from the cell body and split into three bundles of outer doublets in a reactivation solution containing ATP. The finest bundle of the three displayed a cyclic bending movement while the other two remained stuck to the glass slide; dark-field videotaped images taken every 1/15 s (*a–n*). Note that the bending bundle split into two finer fibers, which are most likely single outer doublets. In **B**, this movement can be interpreted by a model that assumes the following: (1) when dynein arms interact with the adjacent outer doublet, the dynein always pull the doublet in a unidirectional manner, (2) the two tubules are firmly fixed at the base; thus a bend is formed by the shear between the two tubules (**a** and **b**), (3) the dynein-tubule interaction is turned off where curvature exceeds a certain limit (marked by an *asterisk* in **c**); when the interaction stops, the two outer doublets dissociate (**d** and **e**), (4) microtubules are elastic so that they tend to assume a straight shape wherever dynein-tubule interaction is absent (**f**). (Kamiya and Okagaki 1986)

sensitive to the mechanical stress on the outer doublet or dynein. Interestingly, the maximal curvature observed here is similar to the one observed in the normal flagellar waveform; this suggests that the same turning-off mechanism is working in normal flagellar beating of a live cell.

The above oscillatory movement displayed by two doublets did not accompany propagation of bending waves. Wave propagation may occur only when three or more outer doublets are organized in a proper structure that provides multiple sets of dynein-doublet interaction sites to produce sliding forces in opposite directions. The minimal number of outer doublets necessary for wave propagation may be three, because an unusual kind of axoneme with only three outer doublets has been found in a protozoa and shown to propagate helical waves at about 1.5 Hz (Prensier et al. 1979).

In the experiment by Shingyoji et al. (1977) shown in Fig. 2, the bends formed upon the iontophoretic application of ATP was observed to propagate distally when the bathing medium contained a concentration of ATP that was too low to cause axonemal beating. This observation suggests that the axoneme has an intrinsic mechanism for propagating preformed bends, and that this mechanism differs from the mechanism for bend formation. Mechanical state (e.g. curvature) in an axonemal segment must influence the dynein-doublet interaction in the adjacent portion in a directional manner.

3.3 Is the Central Pair/Radial Spoke System Involved in Sliding Regulation?

As seen above, the axoneme must have a mechanism that controls the outer-doublet sliding. Is the central pair/radial spoke system somehow involved in this regulation mechanism? *Chlamydomonas* mutants lacking the central pair or radial spokes have been isolated and found to be nonmotile. In vitro experiments examining sliding disintegration, on the other hand, indicated that dynein arms in these mutants are functionally active (Witman et al. 1978). These observations appeared to suggest that the central pair/radial spoke structures are necessary for conversion of microtubule sliding into the bending movement of the axoneme. However, this idea was refuted following the observation that flagella of some organisms, such as eel sperm, do not have the central pair/radial spoke system and yet are fully motile (Gibbons et al. 1985). In fact, the central pair/radial spoke system was eventually shown to be dispensable for flagellar beating in *Chlamydomonas* also; Huang et al. (1982) found an unusual group of mutations called suppressors, which, when combined with central pair-deficient or radial spoke-deficient mutations, can restore motility *without* repairing the structural deficiency. Some of the suppressors were found to be mutations in dynein.

If the central pair/radial spoke system is not important in organizing microtubule sliding into axonemal beating, why is it present in the axoneme at all? Double mutants which lack the central pair/radial spoke system and possess the suppressor mutation show abnormal flagellar beating in that beat frequency is lower and the waveform is more symmetric than those in normal flagella (Brokaw et al. 1982). Therefore, the central pair/radial spoke system may be important for axonemes to beat efficiently with proper waveforms. In this context, it should be noted that the central pairs of some kinds of cilia and flagella have been observed to rotate within the axoneme. Omoto and Kung (1980), examining *Paramecium* cilia by electron microscopy, found that the orientation of the central pair varied continuously up to 360° depending on the phase of the ciliary beating cycle. The obvious conclusion was that the central pair rotates 360° per beat. Rotation of the central pair can be directly observed under a microscope in detergent-extracted "cell models" of *Chlamydomonas*;

the central pair often spontaneously comes out of the axoneme and continuously rotates around the axonemal axis in the presence of ATP (Kamiya 1982). The extruded central pair has a left-handed helical shape, indicating that it must be twisted and strained when contained within the axoneme. The rotation of a helical central pair may facilitate propagation of regular bending waves within the axoneme by propagating helical waves.

Since axonemal beating can occur without the central pair/radial spoke system, the mechanism that produces undulating movement should reside in the outer parts of an axoneme. These would include outer-doublet microtubules, inner-arm dynein, outer-arm dynein, and inter-doublet links. For an understanding of the axonemal motility mechanism, therefore, studies on these parts should be important. In the following sections I will discuss what is known about the properties of inner- and outer-arm dyneins; I will not discuss the properties of the inter-doublet links, of which very little is known so far, except that they also are likely to be dynamic structures that can transiently detach from one of the two adjacent outer doublets (Warner 1983).

4 Dynein: Structure and Function

4.1 General Features

Cross-section electron micrographs of ordinary axonemes show two projections attached to each outer doublet, protruding toward the adjacent outer doublet. The two projections, the inner and outer dynein arms, have been considered to be similar in structure and properties, because they look similar in the axonemal cross-section. Recent studies, however, have indicated that the two types of arms differ significantly in molecular composition and arrangement within the axoneme. In particular, the so-called inner arm was found to comprise three or more different subspecies, each containing two heavy chains, whereas the outer arm comprises only one species containing two or three heavy chains. It remains to be studied what specific functions are performed by different species of dynein arms and their heavy chains.

Besides such an observed significant difference, the inner and outer dynein arms have been found to also have certain common features. First, the shapes of the heavy chains in isolated arms observed by electron microscopy are similar; the outer arm, when isolated, has a bouquet-like structure with two or three globular "heads" connected by slender "stems" to a common "base" (Fig. 4); the inner arm also has a basic stem-head configuration similar to that of the outer arm (Goodenough et al. 1987). Second, the intramolecular structures of all the heavy chains in both arms seem to be related. In 1986, Gibbons and his colleagues found that dynein heavy chains can be cleaved by irradiation with UV light in the presence of ATP and vanadate, a strong inhibitor of dynein ATPase. With the Aα heavy chain of sea urchin outer-arm dynein, the cleavage yielded two peptides with almost identical molecular weights of about 200 kD; the photochemical reaction seems to cleave the peptide at its midportion, where the ATP molecule is presumably bound (Lee-Eiford et al. 1986). Furthermore,

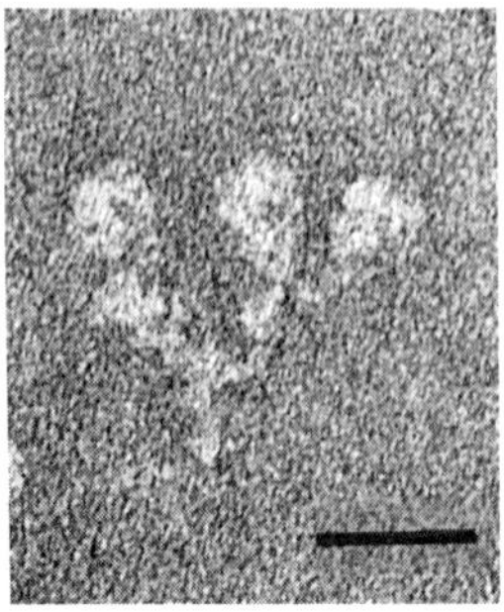

Fig. 4. Three-headed outer-arm dynein isolated from *Tetrahymena* cilia and observed by negative-stain electron microscopy. *Bar* = 25 nm. (Toyoshima 1987)

the UV/vanadate treatment was found to cleave all kinds of heavy chains from the outer arms, inner arms, and cytoplasmic dynein to yield peptides of similar size. Therefore, it is likely that these heavy chains share a common structural organization. More precise information on the primary structure of dynein subunits will soon become available, since cDNA for several heavy chains and intermediate chains have been cloned (Mitchell 1989; Foltz and Asai 1990).

Note added in proof: The primary structure of the β heavy chain of sea urchin dynein was recently determined from its cDNA clones [Gibbons IR, Gibbons BH, Mocz G, Asai DJ (1991) Nature 352:640–643; Ogawa K (1991) Nature 352:643–645].

4.2 Outer-Arm Dynein

The best studied of all kinds of dynein is axonemal outer-arm dynein, in particular, outer arms of sea urchin sperm, *Tetrahymena*, and *Chlamydomonas*. These dynein arms occur as projections located every 24 nm along the outer doublets, i.e., at every third tubulin dimer. This kind of dynein can be extracted with high-salt-concentration solutions and purified by sucrose density-gradient centrifugation. The sedimentation coefficients are 21–23 S.

When isolated and sprayed onto electron microscope grids, the outer arms show up as bouquet-like structures (Johnson and Wall 1983; Goodenough and Heuser 1984; Toyoshima 1987), i.e., two- or three-headed structures bundled by a proximal stem (Fig. 4). There are two heads in sea urchin sperm and three in the other two species, each coinciding with the number of the high-molecular-weight subunits contained. It is not known how the bouquet-like outer arms are folded in situ in the axoneme, although most probably the proximal stems are attached to the A-tubule of a doublet microtubule and all the heads are associated with the B-tubule of the adjacent doublet.

Goodenough and Heuser (1984), using an innovative electron microscopy technique called the quick-freeze deep-etch replica method, have revealed a thin fibrous structure connecting each dynein head with the B-tubule of the adjacent outer doublet. This structure has been observed even under conditions where dynein is thought to be dissociated from the adjacent tubule, i.e., "relaxed" in the presence of ATP and vanadate (for the effect of vanadate, see below). Although the real functional importance of the fibrous structure is unknown at present, it casts a doubt on a popular model for dynein-microtubule motility that postulates an attachment/detachment cycle of dynein arms.

Chlamydomonas outer-arm dynein is composed of three high-molecular-weight peptides (α, β, and γ heavy chains; MW = 415–480 k), two intermediate chains with molecular weights of 69 and 78 kD, and ten light chains of molecular weights ranging from 8 to 22 kD (Piperno and Luck 1979b; Pfister et al. 1982; King and Witman 1989). All the heavy chains appear to have ATPase activities. The composition of the intermediate and light chains differ greatly among outer arms from different organisms.

The identification of these peptides has been greatly facilitated by the isolation of mutants lacking the entire outer dynein arm. The first of such mutants was nonmotile (Huang et al. 1979). However, outer arm-deficient mutants isolated later (Kamiya and Okamoto 1985; Mitchell and Rosenbaum 1985; Kamiya 1988) were found to be motile, although their swimming rates were lower than that of the wild type. This finding, together with the finding that outer arm-extracted sea urchin sperm axoneme can be reactivated to beat at lower frequencies (Gibbons and Gibbons 1973), indicates that the outer arm is not absolutely necessary for axonemal beating. Rather, the outer arm appears to have the function to enhance the basic level of motility that can be caused by inner arms only.

Recently, a novel *Chlamydomonas* mutant has been isolated and shown to lack the α heavy chain but retain the β and γ heavy chains (Sakakibara et al. 1991). The swimming velocity of this mutant is intermediate between those of wild type and a mutant lacking the entire outer arm. Hence, the outer-arm dynein without the α chain appears to retain some functional activity. In other words, the α chain seems to have the function to amplify the activity of the outer arm composed of β and γ chains. The functional amplification may be brought about through a simple addition of activities of the three heavy chains or through regulatory action of the α chain on β and γ functions. Nevertheless, it appears that the α chain is redundant for the outer arm function, just as the entire outer arm is redundant for axonemal beating. Why such redundant components have survived evolution is not understood.

4.3 Inner-Arm Dynein

The structure and biochemical properties of inner-arm dynein are less clear than those of the outer arm. The inner arm has been assumed, explicitly or inexplicitly, to be similar to the outer arm. However, recent studies indicate that the similarity cannot be taken too far. Biochemical analyses of *Chlamydomonas* dynein-deficient mutants have shown that the protein compositions of the two kinds of arms differ strikingly. Although the exact list of subunits that make up the inner arm is yet to be determined, the evidence so far obtained indicates that the inner arm contains at least six high-molecular-weight peptides, an intermediate chain with a molecular weight of 140 kD, and a few light chains. Interestingly, actin is contained as an inner-arm subunit (Piperno and Luck 1979a). There is no common subunit in the two types of arms.

The presence of so many heavy chains is due to the presence of heterologous inner arm species. Heterogeneity among inner arms was first indicated by Goodenough and Heuser (1985), who found that there are two morphologically different types of inner arms: one with a two-headed structure (dyad) and the other with a three-headed structure (triad). Furthermore, they found that the arrangement of the inner arm within the axoneme differs from that of the outer arm; whereas outer arms are located at even intervals of 24 nm, inner arms are located at the proximal end of radial spokes, i.e., at uneven intervals of 24, 32, and 40 nm. A later biochemical study on *Chlamydomonas* inner arm mutants (Piperno et al. 1990) presented a modified view that the three inner arms within the 96 nm repeat distance are actually three different subspecies (called I1, I2, and I3), each containing discrete sets of two different heavy chains. A still new line of evidence, obtained from electrophoresis analysis of inner arms in flagella of different lengths, indicated that the I2 and I3 subspecies may be replaced by types I2′ and I3′ in distal and proximal portions of the axoneme, respectively (Piperno et al. 1990). In this case, I2, I3, I2′, and I3′ are assumed to be homodimers of four different heavy chains. Although this conclusion must be confirmed by more direct methods such as immunolabeling, the indication that the inner arms located at the proximal portion are specialized should be important in considering the flagellar beating mechanism, since the proximal portion must be responsible for the initiation of bend formation, a process that is believed to be somewhat different from bend propagation (Shingyoji et al. 1977).

Despite these new findings, the exact structure and arrangement of the inner-arm dyneins still remain to be established. This is partly because it is technically difficult to distinguish different inner-arm heavy chains that have

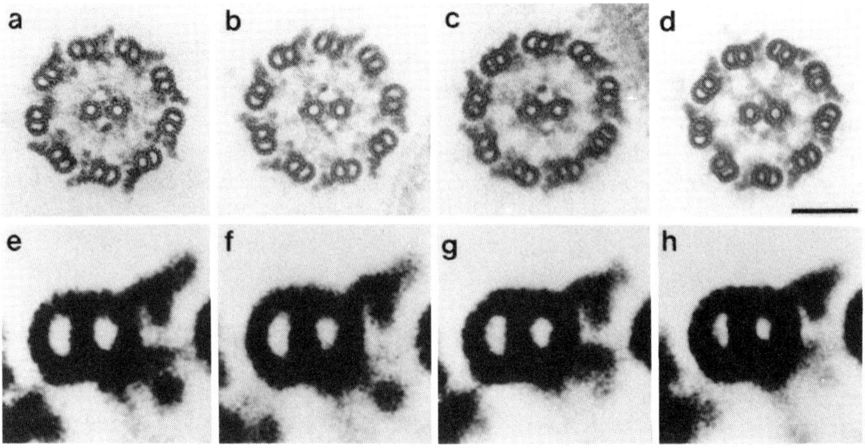

Fig. 5a–h. Axonemes of *Chlamydomonas* mutants lacking subspecies of inner-arm dynein. **a** Wild type; **b** idaA-type mutant lacking subspecies I1; **c** idaB-type mutant lacking subspecies I2; and **d** double mutant between the two types. **e–f** results of averaging of the outer doublet images in the above micrographs with an image processor. Note that the wild-type inner arm has two projections (**e**), whereas the inner arm in idaA (**f**) and idaB (**g**) lacks one of them and the double mutant (**h**) lacks both. *bar* = 0.1 μm

similar molecular weights in a narrow range of 400–450 kD. Future improvements in electrophoresis techniques may lead to the identification of more than six heavy chains. Another reason for the ambiguity is that it is difficult to observe inner arms in situ by electron microscopy, because they tend to be shielded from view by other axonemal structures such as outer doublets and outer arms. To lessen this difficulty, Muto et al. (1991) have recently carried out an extensive electron microscope study using a tilting specimen stage and mutants that lack outer arms and spokes, and reached an unexpected conclusion: the inner arms are arranged in two discrete rows. This indicates that the inner-arm arrangement is more complicated than has been believed before. Their conclusion has been supported by another line of study using new mutants that lack specific inner-arm subspecies (Kamiya et al. 1991). As shown in Fig. 5, the cross-section image of the wild-type *Chlamydomonas* axoneme has two projections (e), whereas mutants lacking the I1 or the I2 subspecies lack one of the projections (f, g). Although this latter study has not identified the location of the third species I3, it is evident that different inner-arm subspecies are aligned not in a straight line, but in two discrete rows or in a staggered row. A further tilt-series analysis on these new mutants will enable one to locate the inner-arm subspecies within the axoneme.

4.4 Inner-Arm vs Outer-Arm Functions

As seen above, recent studies indicate that the inner and outer dynein arms differ greatly in structure and organization. Do they differ in function also?

Gibbons and Gibbons (1973) showed that outer dynein arms can be removed from demembranated sperm by extraction with high-salt-concentration solutions and that the extracted sperm axoneme beats in the presence of ATP at about half the frequency of the normal value. From these observations, they suggested that the two kinds of arms have almost identical functions, the effects being additive. Other work showed that the microtubule sliding rate in disintegrating axonemes is also halved after removal of the outer arm (Yano and Miki-Noumura 1981). These experiments demonstrated not only that the outer dynein arm is actually a mechanochemical motor for axonemal beating, but also that axonemes can function without the outer arm. It is not clear, however, whether these results indicate any functional identity between the two kinds of arms, since it was not possible to remove the inner arm alone.

Chlamydomonas mutants, on the other hand, have provided opportunities to examine the properties of axonemes lacking various inner-arm subspecies, as well as those lacking outer arms. Mutants lacking the entire outer arm have been found to swim at about one-half to one-third the velocity of the wild type, with an almost normal flagellar waveform but at a reduced beat frequency (Brokaw and Kamiya 1987). Thus, the outer dynein arm is not necessary for flagellar beating, as in the case of sea urchin sperm axonemes. In addition, *Chlamydomonas* mutants lacking the inner-arm subspecies I1, I2, or both I1 and I2 have been isolated (Kamiya et al. 1991). Among these, the mutants lacking I1 and those lacking I2 are capable of swimming, although more slowly than the wild type. Thus, the inner-arm subspecies I1 and I2 are not absolutely necessary for axonemal motility. However, the reduced swimming rate of the inner-arm mutants was found to be due to a reduced angle of flagellar bend rather than to a change in the beat frequency (Brokaw and Kamiya 1987). This differs from the situation with the outer-arm mutants. Furthermore, mutants lacking both I1 and I2 were nonmotile; this suggests that flagella with a large defect in the inner arms cannot beat. It remains to be seen whether the third inner-arm subspecies I3 is dispensable for flagellar beating. These findings strongly argue against functional similarity between the two arms. Rather, the inner arms are apparently more important than the outer arms for generation of flagellar beat.

Another line of work has revealed that a significant quantitative difference exists between the two types of arms as mechanochemical transducers. As first demonstrated by Summers and Gibbons (1971), outer-doublet microtubules actively slide out of axonemes when ATP is added to the axonemes that have been treated with protease. In similar experiments, the rate of microtubule sliding was measured by initiating the sliding by perfusing axonemes with a solution containing protease and a known concentration of ATP, and recording the whole process with a sensitive TV camera under a dark-field microscope (Yano and Miki-Noumura 1981; Takahashi et al. 1982). With this method the microtubule sliding rate in sea urchin sperm axonemes was found to depend on the ATP concentration but not on the overlap length between sliding microtubules. In addition, the rate was halved after removal of the outer dynein arm. This result was first taken as an indication that the outer and inner dynein arms are functionally equivalent. However, such an interpretation may be incorrect, since probably not much load is imposed upon the sliding outer doublets, and thus a decrease in number of dynein arms should not result in a reduction in the sliding rate if all the dynein arms are equivalent. On the other hand, the sliding rate in *Chlamydomonas* mutant axonemes lacking the outer dynein arm was found to be unexpectedly low: about one-sixth to one-fifth of the wild-type speed (Okagaki and Kamiya 1986; Kurimoto and Kamiya 1991). Furthermore, the sliding rates in mutants lacking inner-arm subspecies I1, I2, or both were almost identical to that of the wild-type axoneme. These results do not support the view that the two types of arms are equivalent; rather, they indicate that the two types of dynein arms have intrinsically different sliding speeds under low-load conditions, and that the faster outer arms determine the overall sliding rate in disintegrating axonemes. Thus, the axoneme appears to be equipped with two or more motors with different intrinsic speeds. How these motors work efficiently remains a puzzling question.

4.5 ATPase Cycle

The properties of dynein as an ATPase enzyme have been investigated extensively using *Tetrahymena* outer-arm dynein (22S dynein) (Johnson 1985). These investigations have benefited greatly from studies on ATP hydrolysis kinetics of myosin (Lymn and Taylor 1971). As in myosin, the kinetic pathway in dynein consists of rapid binding of ATP, moderately rapid ATP hydrolysis, and a slow release of the products ADP and Pi. In the absence of ATP, dynein binds firmly to microtubules, like actin and myosin, forming the so-called rigor complex. In both systems, the firmly bound complex rapidly dissociates when ATP is added (Porter and Johnson 1983).

In order for dynein to work efficiently as a mechanochemical enzyme, ATP hydrolysis must proceed rapidly only when dynein works to cause microtubule sliding. Hence, the ATP hydrolysis cycle must be slow when dynein alone is present but greatly accelerated when dynein interacts with microtubules. This acceleration should be effected through an increase in the rate of the slowest step, i.e., product release. An acceleration of the ATPase cycle by microtubules has actually been observed by Omoto and Johnson (1986). They showed that the ATPase activity in 22S dynein increased three to ten times when microtubules reconstructed from brain tubulin were added up to a concentration of 50 mg/ml. The activation did not reach a plateau even with this high concentration of microtubules, suggesting that an extremely high concentration of microtubules is necessary for a full activation of dynein ATPase. It is not unreasonable that such a high concentration is needed, since the effective microtubule concentration in the vicinity of dynein arms in situ must be extremely high.

Kinetic properties of dynein other than that from *Tetrahymena* outer arms have been less extensively studied. Recent studies using an in vitro motility assay technique (see below) indicate that there are significant differences in the properties among various dynein species. For example, with sea urchin outer-arm dynein, brain microtubules attach to dynein-coated glass surfaces in the presence of ATP, rather than in its absence. There appears to be no equivalent of a "rigor state" under the conditions where in vitro motility is assayed. On the other hand, it has been shown that sea urchin axonemes become stiff when ATP is depleted; this is an indication that there actually is a rigor state in the axoneme. It appears possible, then, that the kinetic pathway of ATP hydrolysis differs from one species of dynein to another, and some (for example, an inner-arm dynein subspecies) bind strongly in the absence of ATP, while some (for example, outer-arm dynein in sea urchin axoneme) do not. This possibility must be examined in the future using isolated samples.

How ATP hydrolysis is coupled with force production in the dynein-microtubule system remains a matter of speculation. In the popular model for muscular contraction, myosin heads are postulated to undergo a cycle of conformational change accompanying association and dissociation with actin filaments. The cycle is assumed to be coupled with the ATP hydrolysis cycle. This cross-bridge model has been applied to explain the dynein-microtubule motility system and has served as a helpful working hypothesis. Recently, however, the cross-bridge mechanism of muscle contraction has been challenged. For example, there is evidence that actin moves along myosin as much as 100 nm upon hydrolysis of one ATP molecule; a distance much greater than the presumptive cross-bridge can span (Yanagida et al. 1985). Another line of evidence suggests that myosin heads do not change their orientation during muscle contraction (see, for example, Cooke et al. 1982). It seems that we now require a new model for the molecular mechanism of actin-myosin motility, as well as for dynein-microtubule motility.

5 Dynein-Microtubule Motility in Vitro

5.1 Microtubule Translocation

As seen above, ordinary axonemes contain as many as eight or nine species of dynein heavy chains, all of which may be mechanochemical enzymes. For an understanding of the axonemal motility mechanism, it is essential to clarify the specific function performed by each species of dynein. This problem is being approached by analysis of mutants that lack specific dynein components (for review, see Kamiya et al. 1989). However, an obvious drawback of this approach is that it is not always possible to obtain appropriate mutants.

An alternative approach recently developed examines various dynein preparations regarding the activity to cause microtubule sliding on a microscope glass slide. In the first example, a cytoplasmic dynein (MAP1C), coated on the glass slide, was shown to translocate microtubules (Paschal et al. 1987b). The outer-arm dynein from sea urchin sperm flagella (22S dynein) also was shown to support gliding of microtubules at a maximal velocity of about 6 μm/s; that is, at a velocity about one-half of the microtubule sliding in disintegrating axonemes (Paschal et al. 1987a). This was the first demonstration that outer-arm dynein alone can cause microtubule sliding. It was further shown that one (Aβ) of the two heavy chains of the 22S dynein is enough to cause this activity; actually, the sliding caused by the Aβ heavy chains was slightly faster than that caused by the entire outer arm (Sale and Fox 1988). On the other hand, the other heavy chain, Aα, has not been observed to cause sliding.

Another study with *Tetrahymena* 14S dynein, a dynein species whose location within the axoneme is not yet established, revealed a further striking phenomenon: microtubule rotation (Vale and Toyoshima 1988). This kind of dynein was found to cause not only translocation of microtubules at a maximal speed of 2–3 μm/s, but also their rotation around the long axis. The degree of rotation was roughly proportional to that of translocation: about 1.8 rotations/μm on average. The physiological importance of the rotatory movement is not understood.

Inner-arm dynein fractions obtained by chromatography from axonemes of an outer-arm lacking *Chlamydomonas* mutant have also been found to have the ability to translocate microtubules in vitro (Kagami et al. 1990). At least three fractions containing different sets of heavy chains caused translocation at 3–5 μm/s, i.e., at speeds similar to the sliding rate in disintegrating axonemes of this mutant. Interestingly, one fraction was observed to induce microtubule sliding and rotation, similar to the observation with the 14S dynein of *Tetrahymena*. This raises the possibility that some inner-arm dynein can exert torque on the outer doublets. Still another inner-arm dynein subspecies, as well as an outer-arm dynein of this organism, were found to have the ability to trap microtubules on the glass surface in the presence and absence of ATP, but cannot translocate them. These dyneins may be able to work only when cooperating with other species of dynein, or when present within an axoneme that undergoes undulation movements. These observations indicate that the axoneme contains discrete types of dyneins that differ in functional and enzymatic properties. Possibly, the presence of such different types of dyneins is important for the regular function of the axoneme.

Dynein-driven microtubule gliding examined so far all occurred such that microtubules proceed in the + direction, i.e., in the direction corresponding to the base-to-tip direction of the axoneme. This direction is identical to that observed in sliding microtubules in disintegrating axonemes.

5.2 One-Dimensional Diffusion and a Ratchet Model

The in vitro microtubule translocation by dynein is inhibited by vanadate, a dynein inhibitor that forms a stable intermediate dynein/ADP/V, which is analogous to an intermediate of the ATP hydrolysis pathway, dynein/ADP/Pi (Shimizu and Johnson 1983). Vale et al. (1989) have found that in the presence of ATP (or ADP) and vanadate, microtubules bind weakly to the glass surface coated with an outer-arm dynein heavy chain, and display irregular back-and-forth movements. They proved that this motion is a thermally driven random movement, or one-dimensional diffusion, by showing that the mean displacement is zero while the mean square displacement is proportional to time. The diffusion constant was inversely proportional to the microtubule length and viscosity, and was found to be as large as about one-eighth of the value calculated for free-floating microtubules. An interesting observation was that this random fluctuation is virtually confined to the direction of microtubule length; random displacements in the direction normal to the microtubule axis was found to be less than 1/100 of the longitudinal ones. Although the physical basis for such a weak interaction that permits diffusion of microtubules in one dimension while keeping them attached is yet to be elucidated, this phenomenon should be important in the mechanism of axonemal motility.

Furthermore, these observations prompt us to speculate that normal microtubule sliding driven by dynein in the presence of ATP may be based on such thermal movements, the energy of ATP hydrolysis being used to select movements in the right direction only; calculations indicate that such a mechanism can produce a sliding velocity of 22 μm/s, which is similar to the actual velocity. In accordance with such an idea, theoretical models have been presented to explain the generation of directional movements from random thermal movements of a ratchet-like structure which have two components at different energy levels (Vale and Oosawa 1990).

5.3 Nanometer-Scale Vibration in Axonemes

Another novel approach to the dynein-microtubule interaction mechanism has been employed by Kamimura and Kamiya (1989), who used an optical instrument that can detect movements on a subnanometer scale. They monitored the displacement of polystyrene microbeads (diameter: about 1 μm) attached to fragmented axonemes, by comparing the light intensities in two halves of the microscope image of a bead. The fragmented axonemes did not beat even in the

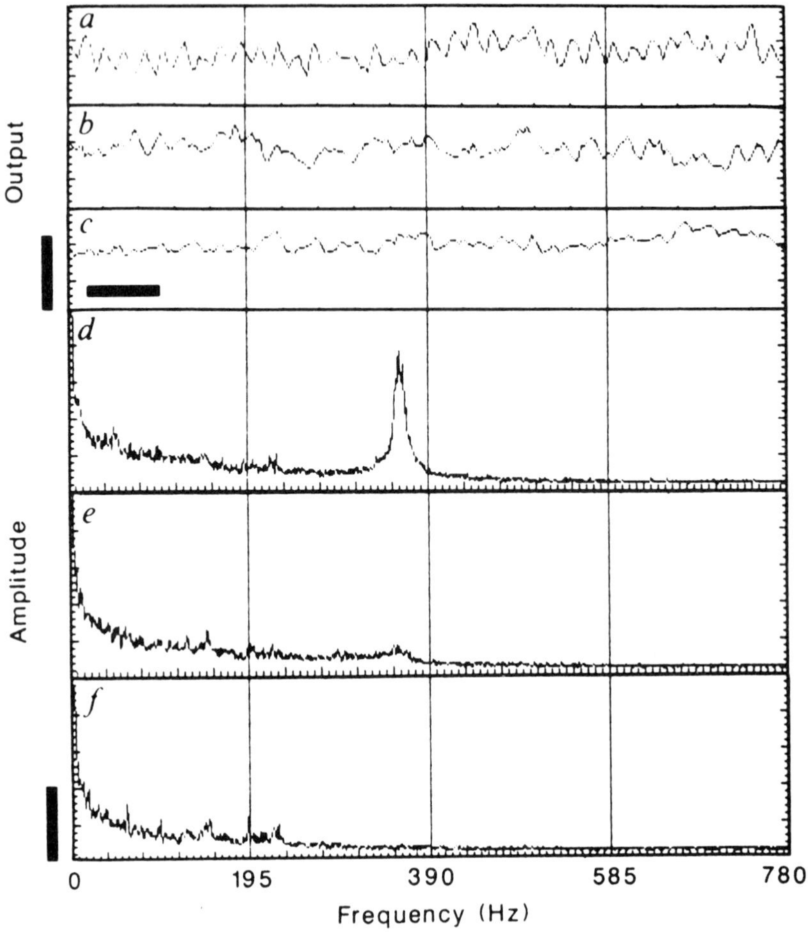

Fig. 6a–f. ATP-dependent minute vibration of a microbead attached to an axoneme fragment. **a** Output from an optical detector showing the vibration in the presence of 1 mM ATP (494 μM MgATP). *Vertical bar* corresponds to 9 nm, and *horizontal bar* 10 ms. **b**, **c** Output for the same bead as in **a** in the presence of 1 mM ATP and 0.5 μM orthovanadate (**b**), or in the absence of ATP (**c**). **d**, **e**, **f** Power spectra for the outputs in **a**, **b**, and **c**, respectively. *Vertical bar* corresponds to 0.18 nm. (Kamimura and Kamiya 1989)

presence of ATP. Instead, they displayed a striking phenomenon: a regular oscillation with a peak-to-peak amplitude of about 4 nm and a frequency as high as about 300 Hz (Fig. 6). The frequency was found to vary with the MgATP concentration in a manner consistent with Michaelis-Menten kinetics. Its apparent Km was about 100 μM, which was similar to the Km obtained from the axonemal beat frequencies in detergent-extracted models reactivated at various MgATP concentrations. A recent measurement using a two-dimensional detector has further indicated that the high-frequency vibration occurs as a back-and-forth shearing between two groups of outer doublets (Kamimura and Kamiya in preparation).

The dynein inhibitor vanadate inhibited the vibration. Unexpectedly, however, vanadate inhibited the vibration by decreasing the amplitude without significantly affecting the frequency. This observation implies that the vibration amplitude is somehow determined by the force produced by the total active dynein arms, since vanadate is known to inactivate dynein by bringing it to a kinetical deadend, analogous to the intermediate dynein/ADP/Pi, as seen above. If this is true, the axoneme fragment under these experimental conditions must be considered to comprise large elastic resistance. The duration of a cycle, 1/300 s, may correspond to twice the duration of the force production by each dynein (it is twice the duration because a cycle contains two active shearing processes occurring in opposite directions). It is not understood why all the dynein arms on a microtubule start and stop producing force every 1/600 s. However, it appears possible that the presumptive large elastic force, induced by active shearing by dynein, somehow works to limit the duration of force production to 1/600 s; dynein may have a property to regulate its force production by the force it receives. If dynein actually has such a property, it may be considered to contain a mechanism for self-oscillation itself (see also Brokaw 1990).

6 Prospects for Future Studies

As seen above, the highly elaborate axoneme of cilia and flagella contains at least eight different kinds of dynein heavy chains, which are suggested by recent experiments to differ rather strikingly in functional properties. Axonemal beating should result from their interplay. Hence, detailed knowledge on the properties of each kind of dynein as a mechanochemical transducer and on their organization within the axoneme must be of vital importance for an understanding of the axonemal motility mechanism.

Information on the force-production properties, such as the load dependence of the sliding velocity or the absolute force produced by individual dynein, will be particularly important. To understand the wave propagation mechanism, it will be interesting to examine how the microtubule-dynein interaction is modulated if the microtubules are artificially bent. Although it may appear extremely difficult to measure such quantities in an object as small as an axoneme or a microtubule, it nevertheless appears possible since force measurements have already been performed with disintegrating axonemes (Kamimura and Takahashi 1981; Oiwa and Takahashi 1988), and with a single actin filament sliding on a myosin-coated glass slide (Kishino and Yanagida 1988). Such microphysiology, in combination with in vitro motility systems with isolated dynein samples, will greatly help us to understand the mysterious mechanism that produces undulating movements.

We will be able to obtain important flagellar mutants relatively easily in the near future. This is because gene clones of dynein heavy chains have already been obtained (Mitchell 1989; Williams et al. 1986; Foltz and Asai 1990), and particularly because an efficient method for transforming *Chlamydomonas* with cloned genes has recently been developed (Kindel 1990). Detailed analyses of a variety of mutants with partial or full defects in dynein will also be one of the logical next approaches to the elucidation of the axonemal motility mechanism.

Acknowledgments. I would like to thank Dr. Hitoshi Sakakibara for providing Fig. 1, Dr. Chikako Shingyoji for Fig. 2, and Dr. Yoko Yano-Toyoshima for Fig. 4. The author's study has been supported ty grants-in-aid from the Ministry of Education, Science and Culture of Japan.

References

Brokaw CJ (1982) Models for oscillation and bend propagation by flagella. Symp Soc Exp Biol 35:313–338

Brokaw CJ (1985) Computer simulation of flagellar movement. VI. Simple curvature-controlled models are incompletely specified. Biophys J 48:633–642

Brokaw CJ (1989) Direct measurements of sliding between outer doublet microtubules in swimming sperm flagella. Science 243:1593–1596

Brokaw CJ (1990) Flagellar oscillation: new vibes from beads. J Cell Sci 95:527–530

Brokaw CJ, Kamiya R (1987) Bending patterns of flagella. IV. Mutants with defects in inner and outer dynein arms indicate differences in dynein arm function. Cell Motil 8:68–75

Brokaw CJ, Luck DJL, Huang B (1982) Analysis of movement of *Chlamydomonas* flagella: the function of the radial spoke system is revealed by comparison of wild type and mutant flagella. J Cell Biol 92:722–732

Cooke R, Crowder MS, Thomas DD (1982) Orientation of spin labels attached to cross-bridges in contracting muscle fibres. Nature (Lond) 300:776–778

Foltz KR, Asai DJ (1990) Molecular cloning and expression of sea urchin embryonic ciliary dynein beta heavy chain. Cell Motil Cytoskel 16:33–46

Gibbons IR (1981) Cilia and flagella of eukaryotes. J Cell Biol 91:107s–124s

Gibbons BH, Gibbons IR (1972) Flagellar movement and adenosine triphosphatase activity in sea urchin sperm extracted with Triton X-100. J Cell Biol 54:75–97

Gibbons BH, Gibbons IR (1973) The effect of partial extraction of dynein arms on the movement of reactivated sea urchin sperm. J Cell Sci 13:337–357

Gibbons BH, Baccetti B, Gibbons IR (1985) Live and reactivated motility in the 9 + 0 flagellum of *Anguilla* sperm. Cell Motil 5:333–350

Goodenough UW, Heuser JE (1984) Structural comparison of purified dynein proteins with in situ dynein arms. J Mol Biol 180:1083–1118

Goodenough UW, Heuser JE (1985) Structure of inner dynein arms, radial spokes, and the central pair/projection complex of cilia and flagella. J Cell Biol 100:2008–2018

Goodenough UW, Gebhart B, Mermall V, Mitchell DR, Heuser JE (1987) High-pressure liquid chromatography fractionation of *Chlamydomonas* dynein extracts and characterization of inner-arm dynein subunits. J Mol Biol 194:481–494

Hoffman-Berling H (1955) Geisselmodelle und Adenosintriphosphat. Biochim Biophys Acta 16:146–154

Holwill MEJ, McGregor JL (1976) Effects of calcium on flagellar movement in the trypanosome *Crithidia oncopelti*. J Exp Biol 65:229–242

Huang B, Piperno G, Luck DJL (1979) Paralyzed flagella mutants of *Chlamydomonas reinhardtii*: defective for axonemal doublet microtubule arms. J Biol Chem 254:3091–3099

Huang B, Ramanis Z, Luck DJL (1982) Suppressor mutations in *Chlamydomonas* reveal a regulartory mechanism for flagellar function. Cell 28:115–124

Hyams JS, Borisy GG (1978) Isolated flagellar apparatus of *Chlamydomonas*: characterization of forward swimming and alteration of waveform and reversal of motion by calcium ions in vitro. J Cell Sci 33:235–253

Johnson KA (1985) Pathway of the microtubule-dynein ATPase and the structure of dynein: a comparison with actomyosin. Annu Rev Biophys Biophys Chem 14:161–188

Johnson KA, Wall JS (1983) Structure and molecular weight of the dynein ATPase. J Cell Biol 96:669–678

Kagami O, Takada S, Kamiya R (1990) Microtubule translocation caused by three subspecies of inner-arm dynein from *Chlamydomonas* flagella. FEBS Lett 264:179–182

Kamimura S, Kamiya R (1989) High-frequency nanometre-scale vibration in "quiescent" flagellar axonemes. Nature (Lond) 340:476–478

Kamimura S, Takahashi K (1981) Direct measurement of the force of microtubule sliding in flagella. Nature (Lond) 293:566–568

Kamiya R (1982) Extrusion and rotation of the central-pair microtubules in detergent-treated *Chlamydomonas* flagella. Cell Motil (Suppl) 1:169–173

Kamiya R (1988) Mutations at twelve independent loci result in absence of outer dynein arms in *Chlamydomonas reinhardtii*. J Cell Biol 107:2253–2258

Kamiya R, Okagaki T (1986) Cyclical bending of two outer-doublet microtubules in frayed axonemes of *Chlamydomonas*. Cell Motil Cytoskel 6:580–585

Kamiya R, Okamoto M (1985) A mutant of *Chlamydomonas reinhardtii* that lacks the flagellar outer dynein arm but can swim. J Cell Sci 74:181–191

Kamiya R, Kurimoto E, Sakakibara H, Okagaki T (1989) A genetic approach to the function of inner- and outer-arm dynein. In: Warner FD, Satir P, Gibbons IR (eds) Cell movement, vol 1. Alan R Liss, New York, pp 209–18

Kamiya R, Kurimoto E, Muto E (1991) Two types of *Chlamydomonas* flagellar mutants missing different components of inner-arm dynein. J Cell Biol 112:441–447

Kindel K (1990) High-frequency nuclear transformation of *Chlamydomonas reinhardtii*. Proc Natl Acad Sci USA 87:1228–1232

King SM, Witman GB (1989) Molecular structure of *Chlamydomonas* outer arm dynein. In: Warner FD, Satir P, Gibbons IR (eds) Cell movement, vol 1. Alan R Liss, New York, pp 61–75

Kishino A, Yanagida T (1988) Force measurements by micromanipulation of a single actin filament by glass needles. Nature (Lond) 334:74–76

Kurimoto E, Kamiya R (1991) Microtubule sliding in flagellar axonemes of *Chlamydomonas* mutants missing inner- or outer-arm dynein: velocity measurements on new types of mutant by an improved method. Cell Motil Cytoskel 19:275–281

Lee-Eiford A, Ow RA, Gibbons IR (1986) Specific cleavage of dynein heavy chains by ultraviolet irradiation in the presence of ATP and vanadate. J Biol Chem 261:2337–2342

Luck DJL (1984) Genetic and biochemical dissection of the eukaryotic flagellum. J Cell Biol 98:789–794

Lymn RW, Taylor EW (1971) Mechanism of adenosine triphosphate hydrolysis by actomyosin. Biochemistry 10:4617–4624

Mitchell DR (1989) Molecular analysis of the alpha and beta dynein genes of *Chlamydomonas reinhardtii*. Cell Motil Cytoskel 14:435–445

Mitchell DR, Rosenbaum JL (1985) A motile *Chlamydomonas* flagellar mutant that lacks outer dynein arms. J Cell Biol 100:1228–1234

Muto E, Kamiya R, Tsukita S (1991) Double-rowed organization of inner dynein arms in *Chlamydomonas* flagella revealed by tilt-series thin-section electron microscopy. J Cell Sci 99:57–66

Oiwa K, Takahashi K (1988) The force velocity relationship for microtubule sliding in demembranated sperm flagella of the sea urchin. Cell Struct Funct 13:193–205

Okagaki T, Kamiya R (1986) Microtubule sliding in mutant *Chlamydomonas* axonemes devoid of outer or inner dynein arms. J Cell Biol 103:1895–1902

Omoto CK, Johnson K (1986) Activation of the dynein adenosine triphosphatase by microtubules. Biochemistry 25:419–427

Omoto CK, Kung C (1980) Rotation and twist of the central-pair microtubules in the cilia of *Paramecium*. J Cell Biol 103:1895–1902

Paschal BM, King SM, Moss AG, Collins CA, Vallee RB, Witman GB (1987a) Isolated flagellar outer arm dynein translocates brain microtubules in vitro. Nature (Lond) 330:672–674

Paschal BM, Shepetner HS, Vallee RB (1987b) MAP1C is a microtubule-activated ATPase which translocates microtubules in vitro and has dynein-like properties. J Cell Biol 105:1273–1282

Pfister KK, Fay RB, Witman GB (1982) Purification and polypeptide composition of dynein ATPase from *Chlamydomonas* flagella. Cell Motil 2:525–547

Piperno G, Luck DJL (1979a) An actin-like protein is a component of axonemes from *Chlamydomonas* flagella. J Biol Chem 254:2187–2190
Piperno G, Luck DJL (1979b) Axonemal adenosine triphosphatases from flagella of *Chlamydomonas reinhardtii*. Purification of two dyneins. J Biol Chem 254:3084–3090
Piperno G, Ramanis Z, Smith EF, Sale WS (1990) Three distinct inner dynein arms in *Chlamydomonas* flagella: molecular composition and location in the axoneme. J Cell Biol 110:379–389
Porter ME, Johnson KA (1983) Transient state kinetic analysis of the ATP-induced dissociation of the dynein-microtubule complex. J Biol Chem 258:6582–6587
Prensier G, Vivier E, Goldstein S, Schrevel J (1979) Motile flagellum with a "3 + 0" ultrastructure. Science (Wash) 207:1493–1494
Sakakibara H, Mitchell DR, Kamiya R (1991) A *Chlamydomonas* outer arm dynein mutant missing the α heavy chain. J Cell Biol 113:615–622
Sale WS, Fox LA (1988) Isolated beta-heavy chain subunit of dynein translocates microtubules in vitro. J Cell Biol 107:1793–1797
Sale WS, Satir P (1977) Direction of active sliding of microtubules in *Tetrahymena* cilia. Proc Natl Acad Sci USA 74:2045–2049
Satir P (1968) Studies on cilia. III. Further studies on the cilium tip and a "sliding filament" model of ciliary motility. J Cell Biol 39:77–94
Shimizu T, Johnson KA (1983) Presteady state kinetics analysis of vanadate-induced inhibition of the dynein ATPase. J Biol Chem 258:13833–13840
Shingyoji C, Murakami A, Takahashi K (1977) Local activation of Triton-extracted flagella by iontophoretic application of ATP. Nature (Lond) 265:269–270
Summers KE, Gibbons IR (1971) Adenosine triphosphate-induced sliding of tubules in trypsin-treated flagella of sea urchin sperm. Proc Natl Acad Sci USA 68:3092–3096
Takahashi K, Shingyoji C, Kamimura S (1982) Microtubule sliding in reactivated flagella. Soc Exp Biol Symp 35:159–177
Tash JS (1989) Protein phosphorylation: the second messenger signal transducer of flagellar motility. Cell Motil Cytoskel 14:332–339
Toyoshima YY (1987) Chymotryptic digestion of *Tetrahymena* ciliary dynein. I. Decomposition of three-headed 22S dynein to one- and two-headed particles. J Cell Biol 105:887–895
Vale RD, Oosawa F (1990) Protein motors and Maxwwell's demons: does mechanochemical transduction involve a thermal ratchet? Adv Biophys 26:97–134
Vale RD, Toyoshima YY (1988) Rotation and translocation of microtubules in vitro induced by dyneins from *Tetrahymena* cilia. Cell 52:459–469
Vale RD, Reese TS, Sheetz MP (1985) Identification of a novel force-generating protein, kinesin, involved in microtubule-based motility. Cell 42:39–50
Vale RD, Soll DR, Gibbons IR (1989) One-dimensional diffusion of microtubules bound to flagellar dynein. Cell 59:915–925
Warner FD (1983) Organization of interdoublet links in *Tetrahymena* cilia. Cell Motil 3: 321–332
Williams B, Mitchell DR, Rosenbaum JL (1986) Molecular cloning and expression of flagellar radial spoke and dynein genes of *Chlamydomonas*. J Cell Biol 103:1–11
Witman GB, Plummer J, Sander G (1978) *Chlamydomonas* flagellar mutants lacking radial spokes and central tubules. Structure, composition, and function of specific axonemal components. J Cell Biol 76:729–747
Yanagida T, Arata T, Oosawa F (1985) Sliding distance of actin filament induced by a myosin crossbridge during one ATP hydrolysis cycle. Nature (Lond) 316:366–369
Yano Y, Miki-Noumura T (1981) Recovery of sliding ability in arm-depleted flagellar axonemes after recombination with extracted dynein. I. J Cell Sci 48:223–239

Chapter 8

Molecular Mechanisms of Mitosis and Cytokinesis

I. Mabuchi[1] and *T.J. Itoh*[2]

Contents

1 Introduction

Eukaryotic cells manifest many types of motility, which are generally based on either microtubules or microfilaments. The cytostructures formed by these filament systems and also by 10-nm filaments are called cytoskeletons. In many cases, the cytoskeleton is a dynamic structure which is cyclically constructed and destroyed in the cell during its lifetime. Typical examples of both of these cytoskeletons are seen during the course of cell division.

Cell division is composed of two sequential and, in some sense (to be described later), independent processes: mitosis or nuclear division, and cytokinesis or cytoplasmic division. The former process assures segregation of the two sets of daughter chromosomes formed through DNA replication. The latter process assures an even distribution of the segregated daughter chromosomes, cell organelles, and cytoplasm into the two daughter cells. It is evident that

[1] Department of Biology, College of Arts and Sciences, University of Tokyo, Komaba, Meguro-ku, Tokyo 153, Japan
[2] Hotani Molecular Dynamic Project, ERATO, JRDC, Morimoto-cho, Shimogamo, Sakyo-ku, Kyoto 606; present address: Department of Bioscience, School of Science and Engineering, Teikyo University, Toyosatodai, Utsunomiya 320, Japan

Advances in Comparative and Environmental Physiology
Vol. 12. Edited by H. Sugi

mitosis is carried out by the microtubular system and cytokinesis is carried out by the microfilament (or actin filament) system. These two processes take place sequentially, although the connection between them is a kind of black box. In this chapter, we will describe recent progress in the studies of these two processes.

2 Mitosis

In mitosis, a construct called a mitotic spindle is temporarily built for the even distribution of daughter chromosomes. The main structural component of the mitotic spindle is the microtubule. Examining mitotic spindles with a polarizing microscope, Inoué and Sato (1967) predicted that the dynamic assembly and disassembly of spindle microtubules are involved in chromosome movement during mitosis. In particular, anaphase chromosome movement toward the spindle poles may be caused by the disassembly of kinetochore microtubules. Since then, many hypotheses have been proposed to explain the mechanisms of the formation of the mitotic spindle and chromosome movement to the pole (for review, see Bajer and Molé-Bajer 1972; Inoué 1981; Pickett-Heaps et al. 1982; McIntosh et al. 1989).

In the past several years, abundant new information about the dynamic properties of the mitotic spindle have been accumulated. Data obtained immunocytochemically indicate that the dynamics of microtubules are essential for chromosome movement to the pole, and the prediction by Inoué and Sato has been supported to some extent. A series of successful reconstructions of a spindle model in vitro provided a conceivable mechanism of chromosome alignment in prometaphase, and also indicated that the disassembly of the microtubule at the kinetochore is necessary for the chromosome translocation toward the spindle poles. The reactivation of the diatom mitotic spindle model helps in understanding the mechanism of pole-to-pole separation at anaphase. Here, we review recent advances in the study of chromosome movement and the dynamics of the mitotic spindle.

2.1 Mitotic Spindle Assembly

Chromosome movement during mitosis can be divided into three processes: (1) the alignment of chromosomes on the equatorial plane of the mitotic spindle to form the so-called metaphase plate, the process that proceeds concomitantly with the formation of the mitotic spindle; (2) the migration of daughter chromosomes from the equatorial plane to each spindle pole (anaphase A); and (3) the mutual separation of two spindle poles (anaphase B).

Process 1 includes the assembly of the mitotic spindle which is initiated concomitantly with the condensation of chromosomes. Polymerization of microtubules is initiated at two spindle poles, or centrosomes. The centrosome cyclically changes its structure and function during the course of the cell cycle (for review, see Mazia 1987; Vandre and Borisy 1989). In interphase, centrosomes organize the stable cytoskeletal network of microtubules. In mitosis, they turn into kinetic centers of the dynamic microtubules of the spindles. The number of microtubules nucleated from them increases dramatically in mitosis. In many animal cells, the centrosome consists of two major components, a pair of centrioles and pericentriolar material. The pericentriolar material was shown to be essential for the organization of the microtubules (Kuriyama 1984). The structural components of the centrosome have been identified in several laboratories (Kuriyama and Borisy 1985; Schatten et al. 1987; Toriyama et al. 1987). Among them, a 51-kD protein purified from sea urchin eggs by Toriyama et al. (1987) was shown to be capable of forming an aster-like structure in vitro. This 51-kD protein was localized in the centrosomes of the mitotic apparatus (Ohta et al. 1988). Moreover, it was recently found to be a GTP-binding protein which is structurally and functionally related to the polypeptide elongation factor 1α form yeast (Ohta et al. 1990).

A microtubule has intrinsic polarity because of the asymmetric structure of the microtubule component, that is, tubulin, so that the behavior of the two ends of the polymer is different (Amos and Klug 1974). These ends of a microtubule can be distinguished in terms of the rate of both assembly and disassembly (Allen and Borisy 1974). One end, the "plus end," grows three times faster than the other end, the "minus end." Microtubules are oriented with their plus end distal to the spindle pole (Bergen et al. 1980; Euteneuer and McIntosh 1981; Haimo and Telzer 1981), and microtubules with opposite polarity are rare in a half-spindle. A structure which binds spindle microtubules, called the kinetochore, is formed on the chromosome concomitantly with the nucleation of microtubules at the centrosomes (for review, see Brinkley et al. 1989). The activity of the kinetochore is independent of the chromosome since it is still functional after the experimental detachment from the chromosome (Brinkley et al. 1988). Although a 80-kD protein in the kinetochore has been identified as the tubulin-binding molecule (Balczon and Brinkley 1987), the molecular composition of the kinetochore is still obscure. The resulting mitotic spindle consists of two populations of microtubules. One is composed of kinetochore microtubules, which interconnect a kinetochore on a chromosome and a centrosome, and the other is composed of non-kinetochore microtubules, the role of which is unclear. The kinetochore microtubules are considered to be important in producing the mechanical force for chromosome movement (Östergren 1951; Hays et al. 1982; Hays and Salmon 1990).

There are two alternative ways to explain the organizing mechanism of the kinetochore microtubules: (1) the kinetochore interacts with preformed microtubules which are nucleated at the centrosomes; (2) the kinetochore itself nucleates the polymerization of kinetochore microtubules. Electron microscopec examinations attempted in many laboratories showed conflicting results. Some results show that kinetochores can capture microtubules nucleated at the spindle pole (Östergren 1951; Molé-Bajer 1975; Rieder 1981, 1982; Sluder and Rieder 1985). However, other observations suggest that the kinetochore nucleates microtubule polymerization (Bajer and Molé-Bajer 1969; Roos 1976; Witt et al. 1981; Steffen et al. 1986).

Although the kinetochore is shown to have the ability to nucleate the polymerization of tubulin in vitro (Mitchison and Kirschner 1985a), the kinetochore-based nucleation does not reflect the real situation in the living cell. Indeed, the polarity of nucleated microtubules was not uniform, being inconsistent with the homopolar kinetochore microtubules observed in vivo (Euteneuer and McIntosh 1981; Haimo and Telzer 1981; Euteneuer et al. 1983). Moreover, the tubulin concentration necessary for the polymerization from the kinetochore was similar to that necessary for the spontaneous polymerization of microtubules in vitro (Mitchison and Kirschner 1984a).

On the other hand, Mitchison and Kirschner (1985b) attempted to reconstruct a model of the spindle from isolated centrosomes, chromosomes, and tubulin. When chromosomes were added after microtubules had been polymerized from centrosomes, the chromosomes captured the microtubules at the kinetochore. These results indicate that kinetochore microtubules are organized through interaction between the chromosomes and the preformed microtubules which are nucleated from centrosomes with the plus end distal. Moreover, the captured microtubules became stable and did not disassemble even under conditions where noncaptured microtubules disassembled completely (Mitchison and Kirschner 1985b). The stability of the captured microtubules is compatible with the observations that the kinetochore microtubules are more resistant against the conditions for microtubule disassembly (Brinkely and Kartwright 1975; Rieder 1981). Capping at the plus end of microtubules by the kinetochore and binding of the centrosome at the minus end may prevent depolymerization from the ends. Further careful study showed that the kinetochore could distinguish the ends of microtubules; the kinetochore captured both the plus and the minus ends of microtubules, but then the number of microtubules attatched at their minus ends decreased (Huitorel and Kirschner 1988). They also suggested that the kinetochore made a lateral attachment near the end of the microtubules in the capture reaction.

In addition to these findings, Mitchison and Kirschner (1985b) also showed that an ATP-dependent alteration of the dynamics of the kinetochore-microtubule complex took place. In the absence of free tubulin molecules, the kinetochore-microtubule complex, which was formed in vitro by the capture reaction, disappeared with the addition of ATP. However, in the presence of both tubulin and ATP, tubulin subunits were inserted into the microtubules proximal to the kinetochore. In this case, the kinetochore appeared to translocate toward the plus direction of the microtubule, that is, away from the centrosome. This ATP-dependent reaction of the kinetochore may represent the chromosome movement at prometaphase, where chromosomes move away from the pole to form the metaphase plate. However, the molecular mechanism for the chromosome alignment is not clear.

Quite recently, Rieder and Alexander (1990) examined the chromosome movement at prometaphase in cultured newt pneumocytes. They found that the chromosome which showed prometaphase poleward movement was associated with one or more microtubules originating from a pole and that these microtubules often extended past the kinetochore. Based on this observation, they suggested that the kinetochore is transported toward the pole along the surface of the microtubules in a manner independent of the microtubule depolymerization at prometaphase. The kinetochore may be an active motor for the chromosome movement.

2.2 Metaphase Spindle

The amount of tubulin assembled into spindle microtubules is relatively constant during metaphase. Spindle microtubules appear in a steady state equilibrium with a cytoplasmic pool of tubulin subunits (Inoué and Sato 1967; Inoué and Ritter 1975). A portion of the spindle microtubules disassembles, while the same amount of free tubulin molecules is incorporated into polymers within the spindle. The dynamic property of the spindle microtubules may account for continuous oscillations of chromosomes in the direction parallel to the spindle axis (Bajer 1982). A model of microtubule dynamics called "treadmilling" was proposed from the in vitro experiment to interpret the dynamic equilibrium (Margolis and Wilson 1978). In this model, tubulin is incorporated into microtubules at one end at a slow rate, and simultaneously dissociates at the other end at the same rate, resulting in a subunit flow from one end to the other. Observations reported from several laboratories of the poleward movement of endogenous and exogenous marks on the living mitotic spindle (Forer 1965; Allen et al. 1969; Hiramoto and Izutsu 1977) are consistent with this model.

The intrinsic rate of tubulin dissociation from the spindle microtubules at metaphase was estimated first by blocking the polymerization with the microinjection of a high concentration of colchicine into dividing sea urchin eggs (Salmon et al. 1984a). The initial rate of tubulin dissociation was in the range of 180–992 molecules/s. The order of this value is quite higher than that predicted by the treadmilling hypothesis (Margolis et al. 1978; Margolis and Wilson 1981). Based on this finding, Salmon et al. (1984a) suggested that the dynamics of microtubules within the metaphase spindle should be somehow different from the previous simple model of end-dependent assembly-disassembly.

Using a different approach, Salmon et al. (1984b) also obtained a high rate constant for the exchange of tubulin which had been incorporated into spindle microtubules. They measured the exchange rate of tubulin by examining the fluorescence redistribution after photobleaching (FRAP). Tubulin labeled with dichlorotriazinylamino fluorescein (DTAF) was injected into fertilized sea urchin eggs. The exchange rate of tubulin was obtained by measuring the half-time for the fluorescence recovery after photobleaching by a laser microbeam. The half-time required for the exchange of tubulin in astral microtubules was 30 to 40 s. This value is also much higher than that expected from the treadmilling model. Wadsworth and Salmon (1986) observed that the bleached regions did not translocate during the fluorescence recovery, and considered that the treadmilling of microtubules did not contribute to the FRAP. In another laboratory, however, a slow poleward migration of the photobleached region during metaphase was reported (Hamaguchi et al. 1987).

A comprehensible explanation for the rapid exchange of subunits in the spindle microtubules came from an in vitro examination by Mitchison and Kirschner (1984a,b), who found that two populations of microtubules coexist at steady state in vitro; the majority of microtubules grow slowly, while the minority shortens rapidly. A transition from the growing to the shortening phase was assumed to occur by the loss of a "cap" of GTP-bound tubulin at the ends of each microtubule (Carlier et al. 1984). This unusual behavior of microtubules was called "dynamic instability." By observing single microtubules directly under a dark-field microscope, this curious property of microtubule dynamics was confirmed to be real (Horio and Hotani 1986; Fig. 1). If astral microtubules of the living mitotic spindle would behave in a manner similar to that in vitro, the microtubules in the bleached region would depolymerize, and the fluorescent tubulin would simultaneously polymerize into microtubules at an adjacent region in the bleached zone. Thus, the rapid rate of exchange of the tubulin subunits within spindle microtubules can be explained by the dynamic instability. According to Salmon et al. (1984b), however, the rate was much higher than that expected from the end-dependent elongation of microtubules. Rather, they thought that tubulin subunits rapidly exchanged along the entire lengths of spindle microtubules, a process which they called "lattice breathing."

When microtubules, which are reconstructed from tubulin labeled with fluorescent dye, are exposed to strong light, the microtubules are easily broken up (Vigers et al. 1988). Similar destruction of microtubules may occur in living spindles irradiated with a laser microbeam. This can accelerate the exchange rate of tubulin subunits in the photobleached region. Therefore, the extremely rapid rate of the subunit exchange obtained from the FRAP experiment can be due to such artificial modification by the laser microbeam irradiation.

In the case of interphase cells, end-dependent elongation of microtubules has been demonstrated in several laboratories by hapten-mediated immunocytochemistry using the microinjection of DTAF- or biotin-labeled tubulin (Soltys and Borisy 1985; Schulze and Kirschner 1986). In both cases, tubulin was incorporated at the distal end of the microtubules in cells, as had been shown in vitro (Bergen and Borisy 1980). Recently, the dynamic instability of microtubules in living cells was directly observed (Cassimeris et al. 1988; Sammak and Borisy 1988; Schulze and Kirschner 1988). There is so far no direct evidence supporting the subunit exchange along the entire wall of the microtubules. Probably the dynamic instability is a general property of microtubules even in the mitotic spindle.

The site of tubulin turnover in the mitotic spindle was not clear in the FRAP experiment because of the limitation of spatial resolution. Mitchison et al. (1986) overcame this problem by electron microscopy and revealed the sites of assembly and disassembly of spindle microtubules. The sites were determined by detecting the localization of biotinylated tubulin which had been microinjected into the cell before metaphase using an anti-biotin antibody. At metaphase, the labeled tubulin was observed to be incorporated into astral microtubules at the end distal to the centrosomes (Fig. 2; Mitchison 1989a). Labeled segments were frequently detected to be continuous to unlabeled segments on a single microtubule with a clear boundary, as had been shown in interphase cells (Soltys and

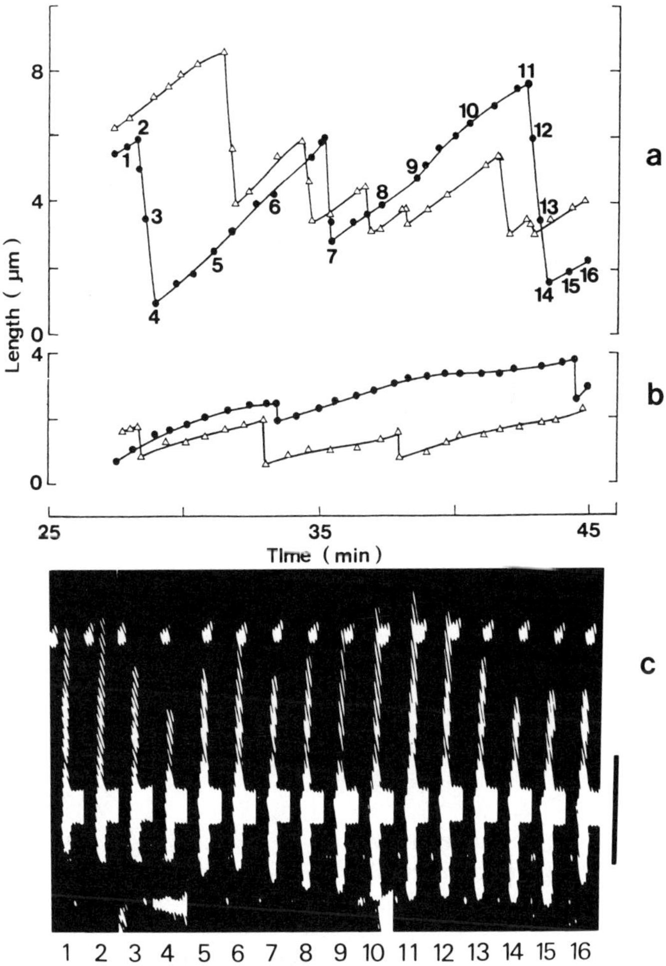

Fig. 1a–c. Interconversion between the growing and shortening phases at the ends single microtubules. The microtubules were observed by dark-field microscopy. **a** Length change at "active" ends. **b** Length change at "inactive" ends. The length change was obtained by measuring the distance between one of the ends and an arbitrarily fixed point on the microtubule. **c** Successive frames obtained from the video image of a microtubule. The *number* below each photograph corresponds to a datum point in **a**. The upper and lower ends of the microtubule in **c** correspond to the active and inactive ends, respectively. *Bar* = 5 μm. (Horio and Hotani 1986)

Borisy 1985; Schulze and Kirschner 1986). The half-time of the tubulin turnover was estimated to be about 30 s. In the case of kinetochore microtubules, the site for the tubulin addition was found at the end, at the kinetochore, and this was compatible with the previous result (Mitchison and Kirschner 1985b). However, the labeled segments of the kinetochore microtubules were considerably shorter than those of astral microtubules. This suggests that the binding of the

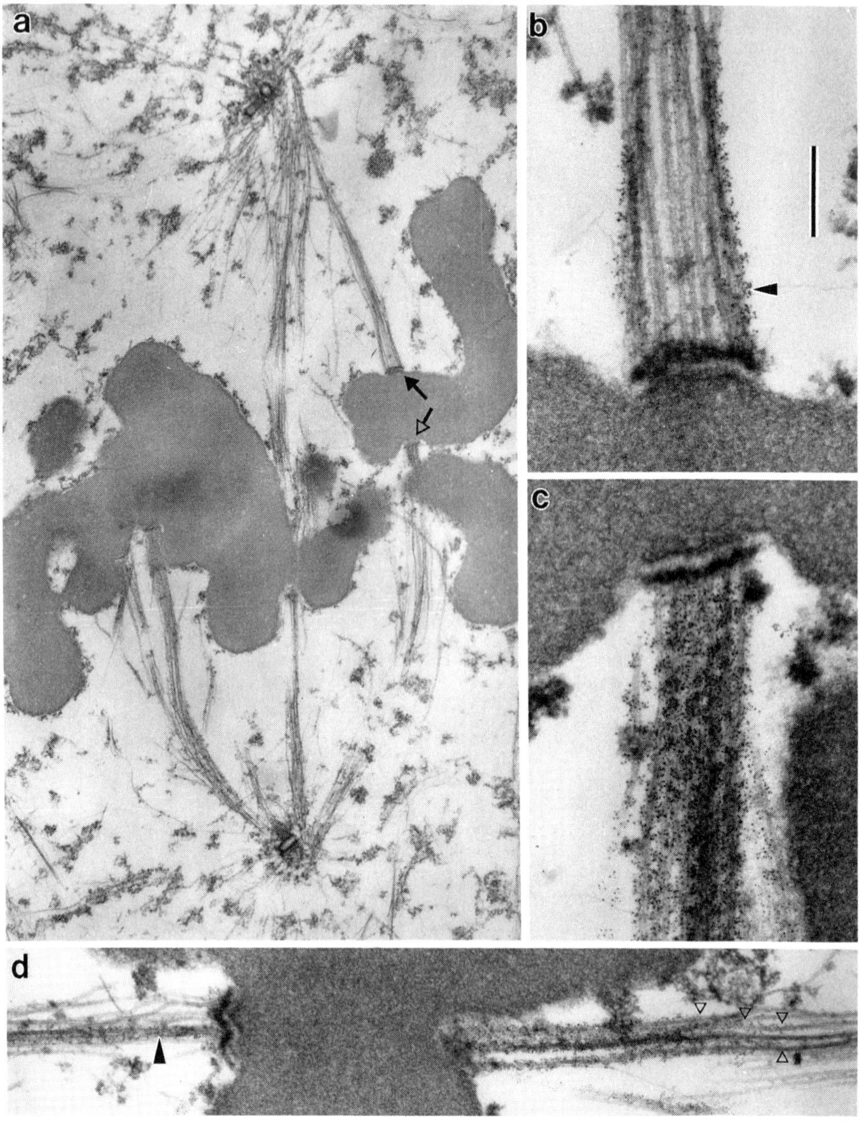

Fig. 2a–d. Visualization of the dynamics of kinetochore microtubules in PtK_2 cells by immunoelectron microscopy. It is clearly shown that the proximal end of kinetochore microtubules is the site of the addition of subunits at metaphase. Cells were injected with biotin-labeled tubulin at metaphase. Microtubules which incorporated biotin-labeled tubulins were visualized by introducing anti-biotin antibodies conjugated with 5 nm-gold particles after permeabilization of the cell. **a** Spindle regions of a cell permeabilized 60 s after injection. **b–d** Kinetochore regions of the same cell at higher magnifications. **b** shows the kinetochore region indicated in **a** by the *filled arrow*. The labeled tubulin was not incorporated into the microtubules except where indicated by *arrowheads*. **c** shows the region of the sister kinetochore (the *open arrow* in **a**) taken from the next section. This region incorporated the labeled tubulin. **d** Kinetochore regions of another chromosome showing a similar event; the labeled tubulin was not incor-

kinetochore to the plus end of the microtubule may suppress the recruitment of tubulin subunits to that end.

Very recently, direct evidence of the poleward flux of a tubulin subunit along the kinetochore microtubules at metaphase was shown (Mitchison 1989b). Utilizing "caged fluorescein," which converts from a nonfluorescent form to a fluorescent form under the exposure to near-UV light, Mitchison succeeded in marking the kinetochore microtubules in living cells and found continuous movement of the fluorescent zone at the velocity of 0.3–0.7 μm/min during metaphase. He considered that the kinetochore microtubules polymerize at kinetochores and depolymerize near the poles throughout metaphase. In this case, the rate of tubulin flux was 8–19 tubulins/s in a microtubule. This rate is quite low compared to that obtained by Salmon et al. (1984a), which was in the range of 180–992 molecules/s. This difference can be ascribed to the contribution of the rapid exchange of nonkinetochore microtubules in the Salmon et al. (1984a) experiment. Indeed, a large fraction of the fluorescence, which was activated by UV flash, decreased within 1 min in Mitchison's (1989b) experiment.

Two populations of microtubules in the metaphase spindle can be distinguished. Nonkinetochore microtubules, the large fraction of whole spindle microtubules, show dynamic instability and rapid turnover. Kinetochore microtubules show an unidirectional poleward flux of subunits. The difference in the dynamic properties of these microtubules can depend on whether or not kinetochores are bound at their ends. The inconsistency between the results of Hamaguchi et al. (1987) and Wadsworth and Salmon (1986) may be due to differences in experimental conditions. Wadsworth and Salmon might have observed the dynamics of the majority of the spindle microtubules, however, Hamaguchi et al. might be able to detect the subunit flux in the kinetochore microtubules.

2.3 Microtubule Dynamics at Anaphase

Kinetochore microtubules are shortened by the loss of subunits during the poleward movement of chromosomes (Nicklas 1971; McIntosh 1985). The translocation of chromosomes has been considered to be due to a pulling force which acts on spindle microtubules. In this case, chromosomes should be passively dragged toward the pole (Bajer 1973; Inoué and Ritter 1975; Margolis and Wilson 1981). This possibility was mainly supported by the report of Forer (1965) in which the areas of reduced birefringence produced by UV irradiation moved toward the pole concomitantly with chromosome movement. Based on the observation of living diatom spindles, however, Pickett-Heaps et al.

porated in the left side but was incorporated in the right side. The *arrowhead* indicates exceptional incorporation. *Triangles* indicate the transition between the labeled proximal and the unlabeled poleward portion of the microtubules. **e** From a cell permeabilized 240 s after injection. *Bar* = 2 μm in **a**; 0.3 μm in **b**, **c**; 0.5 μm in **d**; 1 μm in **e**. The authors would like to thank L. Evans (University of California, San Francisco) for taking the electron micrographs. (Mitchison 1989a)

(1982) suggested the alternative possibility that the kinetochore could actively translocate on the microtubules. Based on the micromanipulation of mitotic spindles of echinoderm eggs, Hiramoto and Shoji (1982) suggested that the force of the chromosome movement was generated near the chromosomes. Their report was the only one which appeared to support the latter possibility at that time.

To understand the mechanism of chromosome movement toward the pole at anaphase A, the experimental discrimination of the above two possibilities is important. If we know the site of microtubule disassembly, we can determine which one is more likely. If chromosomes move passively by pulling force toward the pole, the depolymerization of kinetochore microtubules would occur at the spindle pole or along the microtubule wall concomitantly with the movement. If the kinetochore itself actively moves toward the pole, kinetochore microtubules would depolymerize at the site of the attachment of the kinetochore.

Interesting results contributing an answer to this problem have been obtained in a series of experiments by Mitchison et al. (1986), who observed spindles which were microinjected with biotin-labeled tubulin at metaphase and fixed at anaphase. A significant fraction of kinetochore microtubules did not have a labeled domain at their distal ends. This indicates that the microtubules which were segmentally labeled near the kinetochore during metaphase shortened at that end at anaphase. Based on this observation, Mitchison et al. suggested that the site of disassembly of the kinetochore microtubules is the distal, kinetochore end, or the plus end, and speculated that this microtubule disassembly at the kinetochore is important for anaphase chromosome movement. Switching of the dynamic properties in the kinetochore microtubules was also expected from their data. Before the onset of anaphase, the kinetochore end of the microtubules was the site for the polymerization of tubulin. After the onset, however, the same end turned out to be the site for depolymerization.

Gorbsky et al. (1987) examined microtubule dynamics during anaphase using the photobleaching of fluorescently labeled tubulin. They injected fluorescein-labeled tubulin into cultured cells, which were then incorporated into the spindles at mitosis. At metaphase or anaphase, various sites of the half-spindle were bleached with a laser microbeam. The cells were fixed at various stages of anaphase, and then stained with both anti-tubulin antibody and anti-fluorescein antibody. The bleached region, which was a mark for identifying the microtubule dynamics during chromosome movement, was recognized as the region lacking anti-fluorescein antibody staining. The results clearly showed that the disassembly of the kinetochore microtubules occurred at the kinetochore but not at the centrosome. When the cell was fixed before chromosomes reached the bleached region, the distance between the chromosomes and the bleached region was shortened, while the distance between the bleached region and the spindle pole was constant. In other words, chromosomes approached and invaded the photobleached zone, while the bleached region itself remained stationary with respect to the pole. These findings were compatible with those of Mitchison et al. (1986). By improving the experimental system using X-rhodamine, detection of the bleached zone without fixation became possible (Fig. 3; Gorbsky et al. 1988). Although the fluorescence of the bleached zone

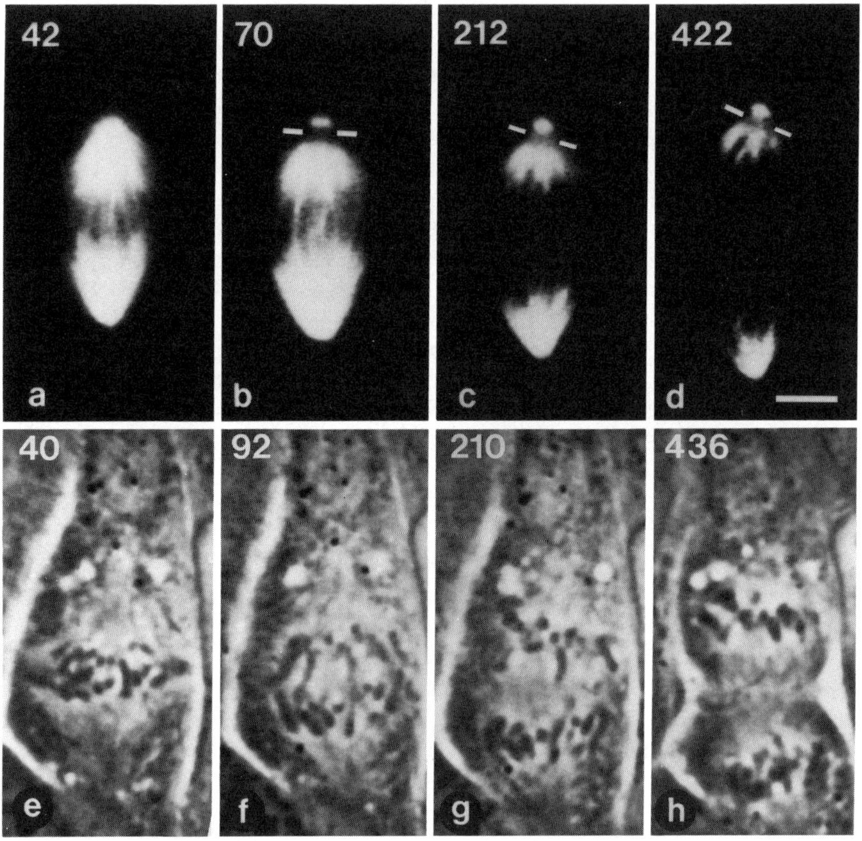

Fig. 3a–h. Photobleaching experiment in a living culture cell at anaphase. **a** Fluorescent and **e** phase-contrast images of as early anaphase cell before photobleaching. The cell was photobleached at 60 s. **b–d** Fluorescent and **f–h** phase-contrast images showing the progression of mitosis after photobleaching. Note that the bleached zone (indicated by *bars*) did not move toward the near pole, while kinetochore microtubules shortened with the poleward movement of chromosomes. Time in seconds after the onset of anaphase is indicated in the *upper left-hand corner*. *Bar* = 5 μm. (Gorbsky et al. 1988)

recovered slightly, the bleached domain did not translocate with respect to the pole during either the poleward chromosome movement or the separation of the two poles. These data disproved the anaphase model, in which chromosomes are passively pulled to the poles by the shortening of microtubules caused by the depolymerization at the poles.

Koshland et al. (1988) developed an in vitro system which is analogous to the anaphase chromosome movement toward the spindle pole. The chromosome was captured at the kinetochore by the plus end of a microtubule polymerized only on the plus end of a chemically cross-linked microtubule seed. The central labile part of the microtubule was divided into two by sequential polymerization of first unlabeled and then biotin-labeled tubulins on the seed. When the solution containing both free tubulin and the construct was diluted in order to induce microtubule depolymerization, the biotin-labeled microtubule protein was shortened at first. In other words, chromosomes were moved to the seeds by the depolymerization at the kinetochore. No nucleotide was required for the shortening of the microtubules. This result suggests that depolymerization of microtubules at the kinetochore produces the force for the poleward movement of chromosomes and that energy supplied by ATP is not necessary for this movement. This experimental system seems to be an excellent model for the chromosome movement. However, it is still uncertain whether this system represents the events which occur in the living cell. A decrease in the tubulin concentration would not occur in vivo. The force produced by the in vitro system has not been measured. The authors did not observe the movement directly. Isolated centrrosomes would be better nuclei than the cross-linked microtubules. At any rate, the mechanism of kinetochore translocation during the depolymerization of microtubules is still unclear.

Very recently, Gorbsky and Borisy (1989) showed that the tubulin subunits of kinetochore microtubules turned over at metaphase but not at anaphase. After the photobleaching of fluorescent spindles, they tried to distinguish the kinetochore microtubules from the nonkinetochore microtubules. Because the kinetochore microtubules are more stable than the nonkinetochore (Brinkely and Kartwright 1975), cells were lysed at a low temperature in a Ca^{2+}-containing buffer to selectively separate the kinetochore microtubules. At metaphase, fluorescence of the bleached zone of the kinetochore microtubules recovered about 70% of the control level. At anaphase, however, no recovery of the fluorescence was detected. These observations support the results of Mitchison et al. (1986) and Mitchison (1989b), which suggest that a continuous addition of tubulin subunits occurs at the kinetochore end of microtubules during metaphase, but that this end becomes the site of depolymerization at the onset of anaphase.

A question arises: what causes the switching of the dynamics of the plus end of the kinetochore microtubules? Investigation of the properties of either end of a microtubule may help in understanding the mode of switching. When single microtubules, which had been reconstructed from purified tubulin, were examined under a dark-field microscope, all the microtubules showed dynamic instability (Horio and Hotani 1986). In a microtubule, both the growing rate and the frequency of phase transition from growing to shortening were higher at the plus end than at the minus end (Horio and Hotani 1986; Walker et al. 1988). Moreover, the shortened length was longer at the plus end than at the minus end (Fig. 1). These observations indicate that the plus end is more labile than the minus end. Probably the length change of a microtubule depends mainly on the stability of the plus end.

The dynamic instability of microtubules is restricted by the presence of high concentrations of microtubule-associated proteins (MAPs) in vitro. While microtubules polymerized from the phosphocellulose column-purified tubulin showed dynamic instability (Horio and Hotani 1986), those polymerized from a crude tubulin preparation which contained MAPs showed treadmilling (Hotani and Horio 1988). Among MAPs, MAP2 is essential to alter the microtubule dynamics (Itoh and Hotani, manuscript in preparation). MAP2-dependent regulation of microtubule dynamics may function in the living mitotic spindle. Phosphorylation of MAP2 or related proteins may be a candidate for the control of microtubule dynamics in the cell. When MAP2 is phosphorylated, it loses the capability of enhancing the polymerization of tubulin in vitro (Jameson et al. 1980; Yamamoto et al. 1983). On the other hand, phosphorylation of a 62-kD protein is reported to induce a depolymerization of microtubules in the mitotic spindle of sea urchin eggs (Dinsmore and Sloboda 1988). Phosphorylation of MAP2-like regulatory protein(s) may be the trigger for the switching of the stability of the plus end of the kinetochore microtubules.

2.4 Mechanism of Pole to Pole Separation

The isolation of a functional spindle which shows anaphase chromosome movement in vitro is an important clue to solving the mechanochemical process of mitosis. Since the first successful isolation of the mitotic apparatus by Mazia and Dan (1952), many attempts to reactivate chromosome separation in isolated mitotic apparatuses have been made. Except for a couple of reports (Sakai et al. 1975, 1976; Sakai 1978a,b; Rebhun and Palazzo 1988), anaphase chromosome movement has not been clearly reactivated. Sakai and his coworkers observed the poleward migration of chromosomes in isolated mitotic apparatuses of echinoderm eggs in the presence of both ATP and free tubulin. Because the movement was inhibited by applying anti-flagellar dynein serum to the mitotic apparatus but not by anti-egg myosin serum, they suggested that dynein ATPase is involved in the movement. However, the velocity of the reactivated chromosome movement was considerably slower than that of the living spindle.

Cande and his collaborators have succeeded in isolating a diatom spindle which retains anaphase spindle elongation (Cande and McDonald 1985; Cande et al. 1989). The diatom spindle is an available model to examine the structural changes of microtubules during anaphase chromosome movement because the microtubules responsible for the spindle elongation are spatially separated from those associated with the chromosome-to-pole movement (McDonald et al. 1977; Pickett-Heaps and Tippit 1978).

Application of ATP to the isolated diatom spindle resulted in the sliding apart of the two half-spindles. The decrease in the extent of birefringence of the overlap zone was associated with spindle elongation, indicating that the number of microtubules in that zone was reduced. Spindle elongation was not related to microtubule polymerization induced in the reactivation medium (Fig. 4; Masuda and Cande 1987; Masuda et al. 1988). From these results, it is conceivable that the force required for the in vitro spindle elongation may be produced by a mechanochemical enzyme that splits ATP and pushes apart antiparallel microtubules. Moreover, phosphorylation of a 205-kD protein may be involved in the reactivation of the diatom spindle elongation (Wordeman and Cande 1987). The protein could be artificially thiophosphorylated and was shown to be located in the microtubule overlap zone. A candidate for the molecular motor which drives the anaphase B movement was discovered quite recently (Shpetner and Vallee

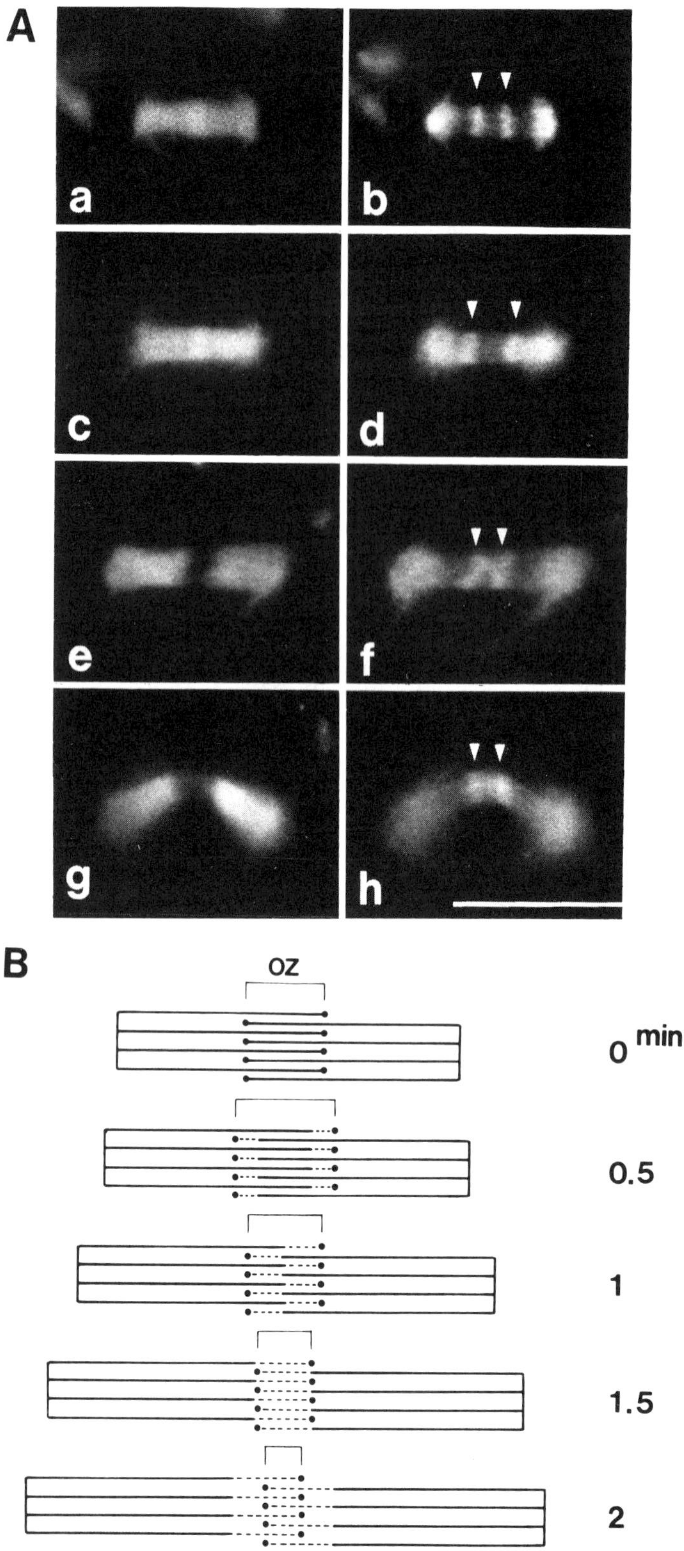

A
a
b
c
d
e
f
g
h
B
OZ
0 min
0.5
1
1.5
2

1989). The mechanochemical enzyme, named dynamin, mediates a sliding between the singlet microtubules in vitro. Quite recently, the sliding of antiparallel microtubules of two half-spindles was also demonstrated to be responsible for the spindle elongation in the fission yeast, *Schizosaccharomyces pombe* (Masuda et al. 1990).

An approach taking advantage of the diatom spindle provided us with the information that mutual sliding of microtubules, the process requiring the hydrolysis of ATP, is involved in anaphase B chromosome separation, as has been suggested in permeabilized culture cells (Cande 1982). Antiparallel sliding of the interdigitated microtubules in the midzone was also suggested to contribute to spindle elongation in lilving PtK_1 cells (Saxton and McIntosh 1987). These results do not agree with the early assumption by Inoué and Sato (1967), who suggested that tubulin incorporation along the entire length of the microtubule wall generates the force necessary for spindle elongation at anaphase B.

3 Cytokinesis

Cytokinesis in the animal cell takes place by constriction of the equatorial cell cortex. This cortical region is called the cleavage furrow, and the contractile machinery in that region is called the contractile ring in the case of equilateral cleavage and the contractile arc in the case of unilateral cleavage. The idea that the cortical gel structure actively constricts to divide the cell was first proposed by Marsland and Landau (1954). The force exerted by the cleavage furrow of the first division of sea urchin or sand dollar eggs has been estimated to be around $0.7–4.5 \times 10^5$ dynes/cm^2 (Rappaport 1967; Hiramoto 1975, 1979). It has been shown that a sea urchin egg microinjected with colchicine to destroy the mitotic apparatus (Swann and Mitchison 1953; Hamaguchi 1975) or an egg from which the mitotic apparatus had been removed by micromanipulation (Hiramoto 1956) can form and advance the cleavage furrow. An egg fragment cut from a dividing sand dollar egg, which contained the cleavage furrow, advanced the furrowing (Rappaport 1969). These experiments show that the cleavage furrow itself exerts a force for its constriction.

The contractile arc or ring has been shown to be mainly composed of actin filaments, and these filaments have been visualized by electron microscopy (Schroeder 1968; Arnold 1969; Tilney and Marsland 1969; and many other papers; reviewed in Mabuchi 1986) and also by fluorescence microscopy using

Fig. 4. A Visualization of the sites of microtubule overlap during the elongation of isolated diatom spindles. Spindles were incubated with biotin-labeled tubulin to show the ends of microtubules. **a**, **c**, **e**, and **g** Spindles stained with anti-plant tubulin; **b**, **d**, **f**, and **h** spindles stained with anti-biotin. **a**, **b** A spindle before the addition of tubulin and ATP; **c**, **d** 30 s after the addition; **e**, **f** 1.5 min after the addition; **g**, **h** 2 min after the addition. The ends of the microtubules in the overlap zone are indicated by arrowheads in **b**, **d**, **f**, and **h**. *Bar* = 10 μm. **B** Diagram of spindle elongation in the presence of 20 μM tubulin, 20 μM taxol, and 1 mM ATP. Microtubules newly polymerized onto the native microtubules in the overlap zone are shown by a *dashed line*. The ends of the microtubules which are labeled by biotin-tubulin are shown by (●●). *OZ* Zone of microtubule overlap. (Masuda et al. 1988) fluorescently labeled phallotoxins (Yonemura and Kinoshita 1986; Schroeder and Otto 1988). Nullifying the function of the actin filaments by introducing cytochalasin B (Schroeder 1970, 1972) or by microinjection of phalloidin (Hamaguchi and Mabuchi 1982) inhibited cleavage.

Myosin is also believed to be involved in the progression of the cleavage furrow since this protein has been shown to accumulate in the cleavage furrow (Fujiwara and Pollard 1976; Aubin et al. 1979), and the microinjection of antibodies against myosin blocked cytokinesis in starfish blastomeres (Mabuchi and Okuno 1977; Kiehart et al. 1982). Thus, the actin-myosin interaction is considered to generate the force for the cleavage.

3.1 Determination of the Cleavage Plane

How the cell determines its cleavage plane is mostly studied with sea urchin or sand dollar eggs. In these eggs, asters are considered to determine the site where the cleavage furrow will be formed (Rappaport 1961). A signal which stimulates the cortex to form the cleavage furrow is hypothesized and is called "cleavage stimulus." The nature of the cleavage stimulus is not known. The rate of the transportation was estimated to be 6.3 μm/min (Rappaport 1973). It may be transported from the inner cytoplasm to the cortex via the astral microtubules as rails. It may stimulate the cortical layer, differentiating to form the furrow only after its local concentration reaches a critical level (Devore et al. 1990). It is also possible that substances which inhibit the cleavage furrow formation are present throughout the cortex, and are sequestered from the anticipated furrow region with the aid of the astral microtubules. A recent electron microscope study by Katoh and Ishikawa (1989) on cellularization followed by cleavage in the *Drosophila melanogaster* embryo showed that microtubules originate from the centrosomes near nuclei, extend to the cortical layer, and appear to attach to the plasma membrane at the early cleavage furrow via an electron-dense material (Fig. 5). Kanbe et al. (1989) also noted the attachment of cytoplasmic microtubules to the plasma membrane at the early septum of dividing fission yeast *Schizosaccharomyces pombe*. These allow the transport of signal materials via the microtubules from the centrosome region or cytoplasm to the cortical layer and vice versa.

Motor proteins which could transport vesicles or organelles along microtubules are found in dividing cells. Dynein, which glides on the surface of a microtubule toward the minus end (Paschal and Vallee 1987), has been found in sea urchin eggs (Weisenberg and Taylor 1968; Mabuchi 1973; Hisanaga and Sakai 1980; Pratt et al. 1980; Scholey et al. 1985a). Kinesin, which glides on the surface of a microtubule toward the plus end (Vale et al. 1985), has been found in sea urchin eggs (Scholey et al. 1985b), and *Dictyostelium discoideum* (McCaffrey and Vale 1989). Therefore, kinesin would support the motility of materials which are associated with this protein from the polar region to the cortex, while dynein would support motility from the cortex to the polar region, since the astral microtubules are arranged so that the distal end is the plus end, although the rates of the gliding of these motor proteins are much faster than

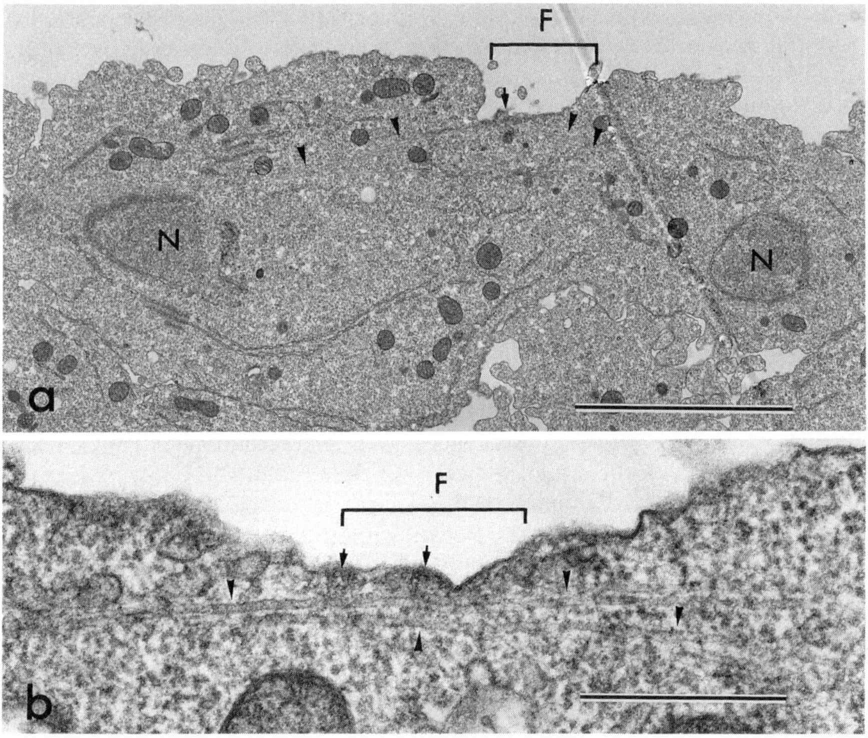

Fig. 5a,b. Attachment of astral microtubules to the early cleavage furrow of a *Drosophila* blastoderm. Microtubules (*arrowheads*) which originate near the nuclei (*N*) extend and attach to the furrow cortex (*F*) through electron-dense materials (*arrows*). **a** *Bar* = 5 μm; **b** *bar* = 0.5 μm. (Katoh and Ishikawa 1989)

that of the cleavage stimulus. Transport of microinjected polystylene beads from the cortex to the poles (1–2 μm/min) is observed in living sand dollar eggs, indicating that the latter movement actually takes place (Hamaguchi et al. 1986). The transport of intracellular granules in the reverse direction is also described. How these motor proteins are related to the transduction of the cleavage stimulus is a very interesting problem.

Dan (1988) considers a mechanism differing from the "chemical signaling" cleavage stimulation. He attached carbon particles to the surface of a sand dollar egg and measured distances between these markers during the course of cleavage. He observed a shortening of the distance when the particles were attached to the equatorial cortex and an expansion when they were attached to other cortical regions (Dan 1943, 1988). He claims that astral microtubules that attach to the cortex cross each other near the equatorial cortex and that they can pull the equatorial cortex when the mitotic spindle elongates at anaphase B. This will cause a shrinking of only the equatorial cortical region, which may organize ordered contractile ring actin filaments. Examination of the images of the anaphase mitotic apparatus of a sea urchin egg visualized by tubulin immunofluorescence through a confocal microscope clearly shows extension of the astral microtubules to the plasma membrane region and crossing over of the astral microtubules near the equatorial cortical layer (Henson et al. 1989). This observation is compatible with Dan's idea. Furthermore, as mentioned above, attachment of microtubules to the furrow membrane was demonstrated in the cellularization in the *Drosophila* embryo (Katoh and Ishikawa 1989; Fig. 5), and in the septum formation in fission yeast (Kanbe et al. 1989). These results may be compatible with both the chemical signaling and the mechanical compression hypotheses.

3.2 Structure of the Contractile Ring or Arc

The contractile ring or arc is mainly composed of actin filaments oriented in a parallel fashion (reviewed by Schroeder 1975; Mabuchi 1986). The maximal diameter depends on the size of the cell, but the width and the thickness of the ring do not differ largely among cell types (reviewed by Mabuchi 1986). The length of these actin filaments is not known. In HeLa cells the interfilament spacing was reported to be 15 nm and is constant during furrow progression (Schroeder 1970). Such spacing of the contractile arc of the newt egg is 10 to 15 nm (Tsukita, Tsukita, Sawai and Mabuchi, unpubl. observ.). Polarity of the actin filaments was investigated by decoration with myosin subfragment 1 for rat kangaroo (PtK2) cells (Sanger and Sanger 1980) and newt eggs (Mabuchi et al. 1988). Actin filaments of opposite polarities are mixed in the contractile ring or arc. This enables the ring or arc to contract via actin-myosin interaction provided that the filaments are firmly attached to the plasma membrane. Moreover, Sanger and Sanger (1980) noted alternating dark and light zones, each about 160 nm in length, in the ring, which resemble those of stress fibers in the PtK2 cells. The nature of these structures is not known. They could represent the periodic presence of actin cross-linking proteins like those demonstrated in the stress fibers of cultured cells (Langanger et al. 1984). The actin cross-linking strands have recently been demonstrated in the contractile arc of the isolated cleavage furrow of newt eggs (Mabuchi et al. 1988). At least two populations of strands of different lengths, one about 15 and the other 30–40 nm, were found in a rapid-frozen and then deeply etched specimen (Fig. 6).

3.3 Proteins Implicated in Furrowing

3.3.1 Myosin

Besides actin filaments, the presence of a number of proteins in the cleavage furrow cortex is confirmed by various methods. Myosin has been demonstrated by immunofluorescence microscopy, and its function has been confirmed as described above. To reconfirm its function, two approaches using gene technology have recently been introduced successfully using *Dictyostelium discoideum* amoeba. One approach is the transformation of the cell with myosin heavy chain antisense RNA vector (Knecht and Loomis 1987). The myosin heavy

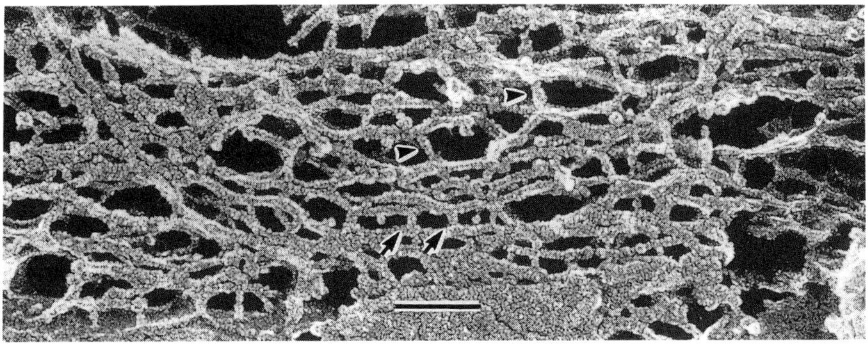

Fig. 6. Ultrastructure of a contractile arc in a cleavage furrow isolated form a newt (*Cynops pyrrhogaster*) egg. The specimen was rapid-frozen, deeply etched, and then replicated. Actin filaments are oriented in parallel and cross-linked with each other by strands 15 nm long (*arrows*) or 30–40 nm long (*arrowheads*). *Bar* = 0.1 μm. (Mabuchi et al. 1988)

chain expression was strongly suppressed in the transformant, and it could not undergo cytokinesis. The other approach is to damage the myosin heavy chain gene by homologous recombination (De Lozanne and Spudich 1987). *Dictyostelium* amoeba has an advantage in the use of this technique since it is a haploid cell and has only one copy for the conventional myosin, which has a tail portion and thus can form filaments in the cell. The transformant, which could only synthesize the globular head portion of myosin heavy chain instead of the entire heavy chain molecule, could not undergo cytokinesis. Even a transformant which could not synthesize myosin heavy chain at all could not undergo cytokinesis (Manstein et al. 1989). It is interesting that other motile activities such as nuclear division, membrane ruffling, chemotaxis, phagocytosis, and cytoplasmic particle movement are not affected in these transformations. These activities may be carried out by other kinds of myosins such as minimyosin (or myosin I in *Acanthamoeba castellanii*), which has no α-helical tail domain and thus cannot form filaments, or in microtubule-based systems. This idea is supported by Fukui et al. (1989), who recently showed by immunofluorescence microscopy that conventional myosin (or myosin II in *Acanthamoeba castellanii*) is concentrated in the cleavage furrow of dividing *Dictyostelium* amoeba but minimyosin (myosin I) is present in the peripheral region of the cell.

Although both the function and presence of myosin in the cleavage furrow have been established as described above, it has not been clear how myosin exists in the furrow. Does it form filaments or small oligomers? Does it exist as monomers? Recent studies using *Dictyostelium discoideum* (Yumura and Fukui 1985) or sea urchin egg cortical layer isolated on a glass surface (Schroeder and Otto 1988; Yonemura 1989; Mabuchi 1990) showed that the distribution of myosin shows a rod-like pattern of punctate pattern, respectively. These observations may indicate that this protein forms filaments in the contractile ring. There are some reports describing myosin filament-like structures in the thin-sectioned contractile region in some cells, e.g., HeLa cells (Maupin and Pollard 1986) or sea urchin egg (Yonemura 1989). However, it is necessary to determine precisely whether these are actually myosin filaments.

It is interesting that myosin is highly concentrated in the furrow while there was no evidence of its presence in the anticipated furrow region before the onset of cleavage or in the bulk cortex throughout the cleavage in sea urchin eggs (Schroeder 1987). On the other hand, Mittal et al. (1987) microinjected fluorescently labeled light chain of smooth muscle myosin into a gerbil fibroma cell and observing its accumulation in the anticipated furrow region before the furrowing. In spite of the discrepancy between these results, myosin seems to come to the furrow region from somewhere in the cytoplasm when the contractile ring forms. This leads to the assumption that myosin may be involved in contractile ring formation as well as in its contraction. The effects of drugs on the cleavage in sea urchin eggs, which are known to inhibit myosin light chain kinase activity, were investigated, and it was found that 1-(5-chloronaphthalene-1-sulfonyl)-1H-hexahydro-1, 4-diazepine hydrochloride (ML-9) prevented contractile ring formation (Mabuchi and Takano-Ohmuro 1990; Fig. 7). Therefore, myosin should be involved in the actin filament concentration in the contractile ring (see Sect. 3.4).

3.3.2 Actin-Modulating Proteins

In nonmuscle cells, actin is a very dynamic protein which forms filaments from monomers and three-dimensional cytoskeleton from the filaments. These dynamic changes of the supramolecular assembly of actin are under the control of various actin-modulating proteins. The dynamic properties of the contractile ring or arc are also considered to be due to actions of the actin-modulating proteins. Some of these proteins are reported to be present in the cleavage furrow region.

Alpha-actinin was revealed by immunofluorescence microscopy to localize in the furrow of chicken fibroblast (Fujiwara et al. 1978). This protein was also seen to localize in the furrow with fluorescently labeled protein into living sea urchin egg (Mabuchi et al. 1985). This protein has been tentatively identified in the isolated cleavage furrows of newt egg (Mabuchi et al. 1988) and sand dollar egg (Yonemura et al. 1991) by two-dimensional gel electrophoresis. One of the strands which cross-links actin filaments in the contractile arc possessed a dimension of the α-actinin molecule (30–40 nm long; see Fig. 6; Mabuchi et al. 1988). However, this protein as well as filamin were not always detected in the furrow region of chick embryo cells by immunofluorescence microscopy (Nunnally et al. 1980). A *Dictyostelium* mutant, which scarcely expresses α-actinin, has been reported to undergo various motile activities including cytokinesis (Wallraffe et al. 1986).

Tropomyosin was demonstrated in the furrow of sea urchin eggs by indirect immunofluorescence microscopy (Ishimoda-Takagi 1979) but it is also present in the mitotic apparatus (Ishimoda-Takagi 1984). This protein has also been demonstrated in the isolated furrow of newt eggs by two-dimensional gel electrophoresis (Mabuchi et al. 1988). A 255–260 kDa protein, which is considered to cross-link actin filaments (Maekawa et al. 1987, Mabuchi and Kane 1987), is also detected in the cleavage furrow of sea urchin eggs by immunofluorescence microscopy (Maekawa et al. 1987). Spectrin is identified in the isolated furrow of sand dollar eggs by two-dimensional gel electrophoresis

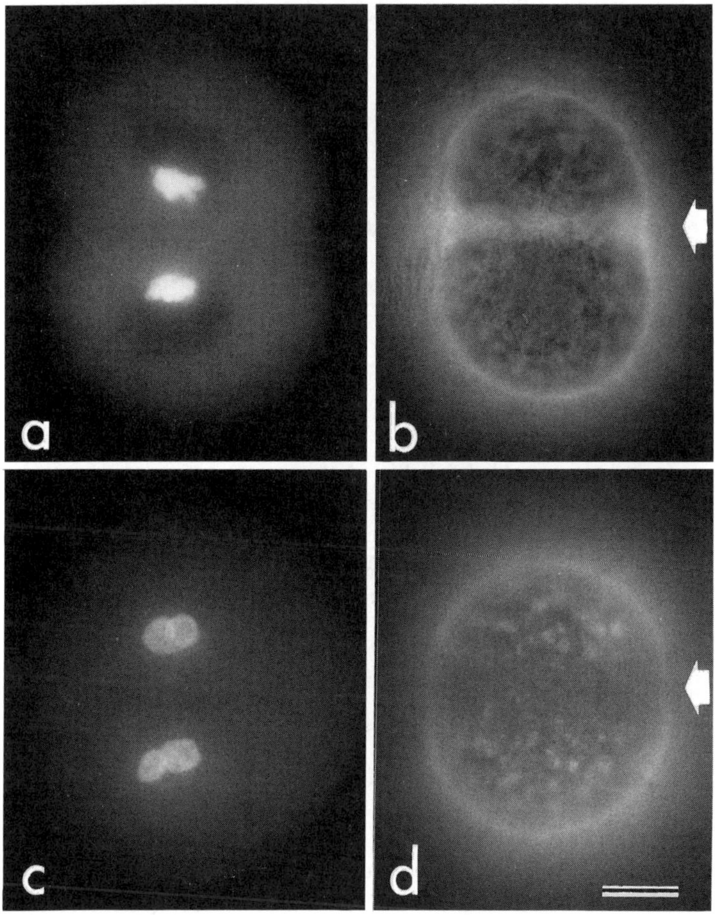

Fig. 7a–d. ML-9 blocks the contractile ring formation. Eggs of the sea urchin (*Hemicentrotus pulcherrimus*), from which fertilization membranes had been removed, were cultured in the absence or presence of 0.2 mM ML-9. The eggs were fixed with 5% formalin at telophase and double-stained with 4′,6-diamidino-2-phenylindole (DAPI) and rhodamine-phalloidin. **a** A control egg stained with DAPI. **b** The same egg as in **a** stained with rhodamine-phalloidin. This egg formed the contractile ring (*arrow*). **c** An egg cultured in the presence of ML-9 stained with DAPI. **d** The same egg as in **c** stained with rhodamine-phalloidin. This egg failed to form the contractile ring and lost the cortical actin filaments in the corresponding region (*arrow*) although it was in the same mitotic stage as the control egg. *Bar* = 20 μm

(Yonemura et al. 1991). Actolinkin, a 20 000 molecular weight, actin barbed end-capping protein, which is present in the isolated cortex (Ishidate and Mabuchi 1988a) and considered to function in the attachment of actin filaments to the plasma membrane, is also concentrated in the furrow region (Ishidate and Mabuchi 1988b). This may reflect the high density of the actin filaments in this region.

Some proteins are present in the isolated furrow but remain undetected or are present in a smaller quantity in the bulk cortex of the newt egg (Mabuchi et al. 1988). Furthermore, some proteins seem to change their isoelectric point in the furrow compared to the cortex. Identification of these proteins is required to clarify the structure and mechanism of formation of the contractile ring and arc.

3.4 Formation of the Cleavage Furrow

The cleavage furrow is formed and advanced by the contraction of the contractile ring in equilaterally dividing cells such as those of sea urchin eggs, *Dictyostelium discoideum*, and various cultured mammalian cells, and of the contractile arc in unilaterally dividing cells such as amphibian eggs. Thus, one must identify the mechanism of the contractile ring or arc formation in order to establish the formation of the cleavage furrow.

The origin of the actin molecules forming the contractile ring actin filaments is not clear. In the sea urchin egg they seemed to be included in the cortical layer since the content of actin in the isolated cortex fraction does not increase after 10-min postfertilization up to the time of the first cell division (Mabuchi et al. 1980). This indicates that the precursor of the contractile ring actin filaments is present in the cortical layer. The microvilli actin filaments do not seem to be the source since the density of the microvilli in the furrow cortex does not differ from that before the furrow formation (Schroeder 1986). Usui and Yoneda (1982) reported that the density of the actin filaments in the equatorial cell cortex increases at late anaphase several minutes before the furrow formation. These actin filaments are reported to be randomly oriented. They are oriented in a parallel fashion 1 min before the furrowing. Yonemura and Kinoshita (1986) noted that a meshwork of actin filaments appears at anaphase which could not be detected at metaphase by a rhodamine-phalloidin staining of cortical layers isolated on a glass surface. Mabuchi (1990) observed an accumulation of actin filaments in the anticipated furrow region at late anaphase about 1 min before the actual furrow formation by the rhodamine-phalloidin staining of whole eggs. Therefore, it is possible that a new polymerization of actin occurs in the equatorial cell cortex prior to contractile ring formation. It is noteworthy that Bonder et al. (1989) demonstrated the presence of nonpolymerized actin in the cortical layer of unfertilized sea urchin egg.

How are the contractile ring or arc actin filaments organized into such a structure? Recent observations suggest that these are organized from the cortical actin filaments by bundling. The contractile arc has an advantage in approaching this problem: it is continuously formed at its tips during cytokinesis. Mabuchi et al. (1988) isolated cleavage furrow cortex from dividing newt eggs and observed the tip region of the furrow by electron microscopy. In the anticipated furrow region, actin filaments were shown to attach to the plasma membrane via a granular mass and to extend into the cytoplasm. The polarity of these filaments was such that the arrowhead structure produced by decoration with myosin subfragment 1 pointed away from the membrane. In the tip region of the advancing contractile arc, these actin filaments were seen to be organized into bundles.

In the sea urchin or sand dollar egg, the actin filaments in the cortical layer can be visualized by staining with fluorescently labeled phalloidin after the cortices are isolated on the protamine-coated glass surface (Yonemura and Kinoshita 1986; Mabuchi 1990). Examination of the contractile ring at various stages of its formation strongly suggests that the ring is formed via bundling of cortical meshwork actin filaments (Fig. 8).

The bundling of actin filaments in the process of contractile arc or ring formation is likely to be mediated by the actin cross-linking proteins since at least two kinds of cross-linking strands were observed in the contractile arc of a newt egg (Mabuchi et al. 1988). In addition, as described in Sect. 3.3.2, it was recently found that the application of inhibitors of myosin light chain kinase to sea urchin or sand dollar eggs prevents contractile ring formation (Mabuchi and Takano-Ohmuro 1990). Therefore, myosin would also be involved in the bundling of the actin filaments into the contractile ring or arc.

3.5 Contraction of the Cleavage Furrow

The contraction of the contractile ring had been considered to start when the contractile ring is constructed. This comes from the observation that no contractile ring actin filament was detected by thin-sectioning electron microscopy when there was no furrow (Schroeder 1972). When the furrow was formed, the contractile ring was there.

However, a somewhat different result has been obtained by careful examination of furrow formation by light microscopy (Mabuchi 1990). When the shape of the sea urchin egg was followed during cytokinesis, it was seen to elongate at late anaphase. It is prominent when the fertilization membrane had been removed. This elongation could be due to the expansion of the mitotic apparatus at anaphase B, since the early elongation is not accompanied by the appearance of the contractile ring as examined by rhodamine-phalloidin staining. However, it could also be due to a very primitive contractile ring which has never been detected by fluorescence or electron microscopy. This elongation causes a reduction in the diameter of the egg at the equator. The reduction advances rapidly until it reaches a maximum and constant velocity (about $50\,\mu$m/min). The noticeable cleavage furrow is formed in the process of acceleration. The reduction in the diameter caused by the elongation of the egg could not be distinguished from that caused by the furrowing. Furthermore, as described above, the formation of the contractile ring as seen by rhodamine-phalloidin staining precedes the onset of furrowing. Thus, it is possible that the contractile ring starts to contract before the furrowing. The complete formation of the contractile ring which contracts at the maximum speed takes some time: about 100 s from the appearance of the ring at early telophase and 60 s from the onset of furrowing. These results suggest that the formation process of the contractile ring, which can contract at the maximum rate, is associated with its contraction.

The cleavage furrow isolated from dividing newt eggs can contract in the presence of ATP (Mabuchi et al. 1988; Fig. 9). This contraction was inhibited by cytochalasins B and D and by phalloidin, indicating that actin filaments are necessary for the contraction. Phalloidin was also inhibitory for the cytokinesis in detergent-lysed cultured cells (Cande 1980). Furthermore, microinjection of

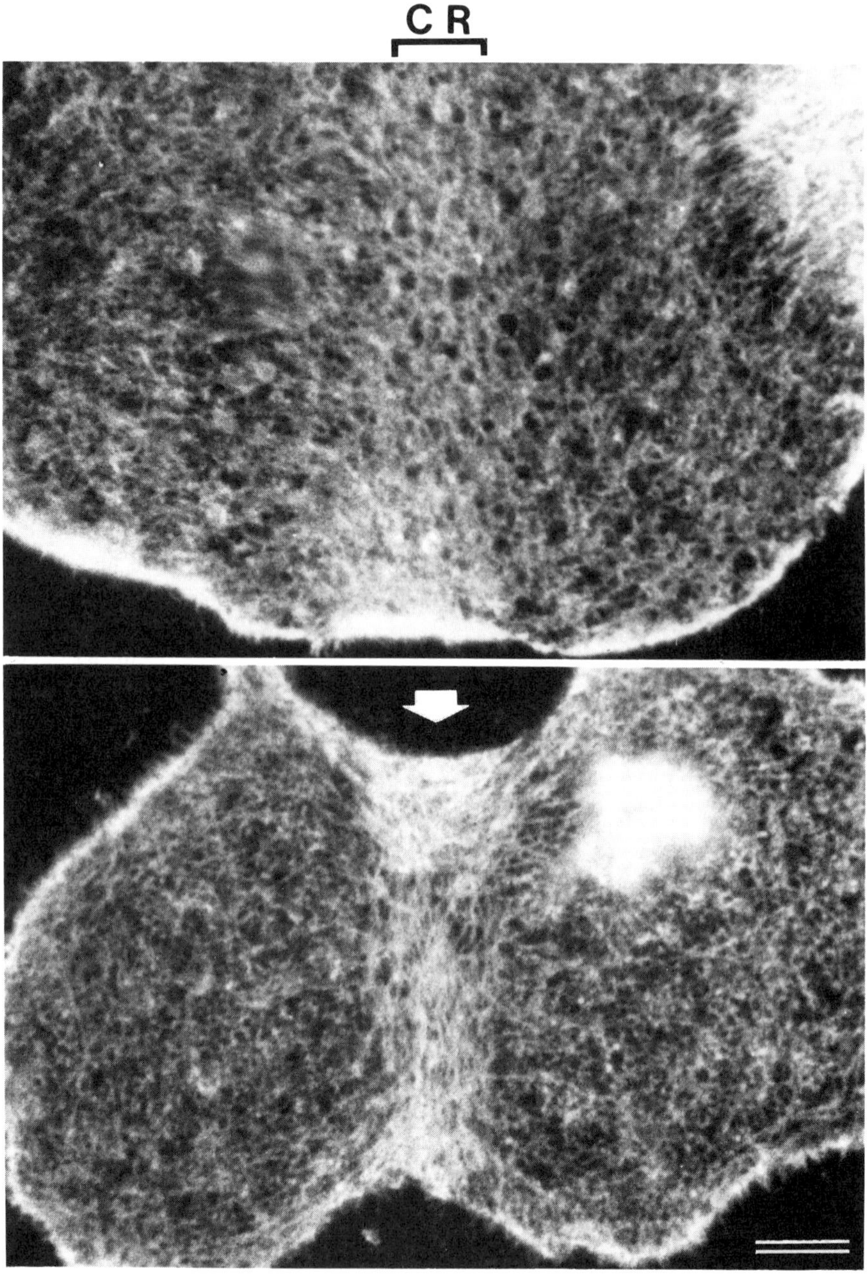

Fig. 8. Formation of contractile ring in sea urchin (*Hemicentrotus pulcherrimus*) eggs. Cortices were isolated on a protamine-coated glass surface and the actin filaments were visualized by rhodamine-phalloidin staining. The contractile ring (*CR*) actin filaments are organized from cortical actin filaments. *Above*, an early stage; *below*, a later stage. *Bar* = 10 μm

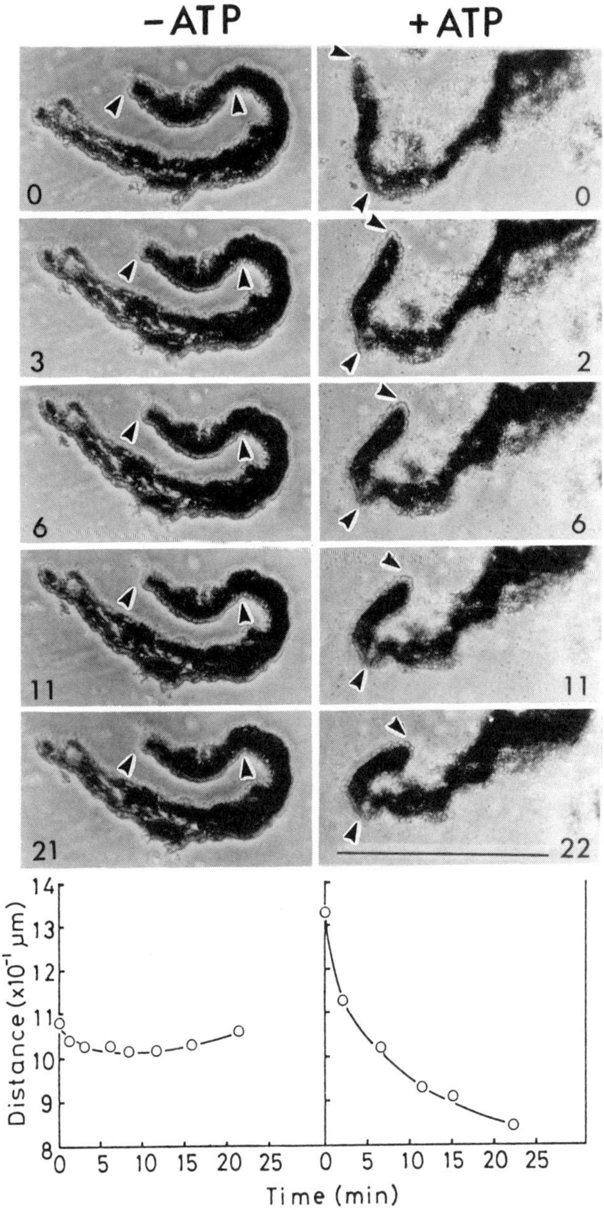

Fig. 9. Contraction of cleavage furrows isolated from newt eggs. Changes over time (*numbers* in micrographs, min) at the distances between two arbitrarily selected points (*arrowheads*) were measured in the absence or presence of 2 mM ATP. *Bar* = 0.2 mm. (Mabuchi et al. 1988)

phalloidin into sea urchin eggs blocked the advance of the furrow (Hamaguchi and Mabuchi 1982). By contrast, labeling of actin filaments by phalloidin in glycerinated muscle fibers did not seem to interfere with the contraction induced by ATP (Yanagida et al. 1985). Gliding of actin filaments on the myosin-attached glass surface was also not hindered by labeling with rhodamine-phalloidin (Kron and Spudich 1986). These conflicting results, using different motile systems, may be explained by the mode of action of phalloidin on actin: it stabilizes actin filaments against depolymerization and reduces the critical concentration of actin for polymerization. The contraction of the contractile ring or arc may be associated with the depolymerization of actin filaments or the interaction of the actin filaments with an unknown component, and the inhibition of either process by phalloidin may lead to the cessation of the contraction of the contractile ring or arc.

The contraction was also inhibited by p-chloromercuribenzoate and N-ethylmaleimide (Mabuchi et al. 1988). Probable inhibition of the ATPase activity of myosin by blocking its SH groups may explain this observation. This is consistent with the previous finding that myosin is implicated in the furrowing.

However, the in vitro contraction of the isolated furrow was slower than that of in vivo contraction, which was more than 30 μm/min (T. Sawai, unpubl. result), comparable to the rate of the contraction of the cleavage furrow in the sea urchin egg. Some unknown factor may be necessary for in vitro contraction. Identification of such a factor, as well as of components unique to the furrow and the mode of association of these components to the contractile ring or arc, would lead to the elucidation of the mechanism of cleavage.

Addendum

Since submitting the draft of this chapter, a number of important papers have appeared in this field. Identificatiojn of γ-tubulin as a component of the spindle pole body (Oakley et al. 1990) provides us with a new understanding of the molecular basis of microtubule nucleation in the mitotic spindle. This new member of the tubulin family has been identified in several animals (Zheng et al. 1991), suggesting that γ-tubulin is a universal component of the microtubule organizing center. Hayden et al. (1990) visualized the capture of kinetochore by astral microtubules during spindle formation in living cells by video-enhanced microscopy. Using *Xenopus* egg extracts, an in vitro system of the reconstruction of the mitotic spindle has been established (Sawin and Mitchison 1991a,b). This system is useful to examine the molecular basis of the regulation of spindle assembly. A factor which rapidly severs microtubules has been found in *Xenopus* egg extracts (Vale 1991). This factor may be involved in the disassembly of the microtubular network prior to the mitotic spindle assembly since it was active only during mitosis. The presence of dynein in the kinetochore region has been observed by immunofluorescence microscopy (Pfarr et al. 1990; Steuer et al. 1990). Molecular genetic approaches have shown that kinesin-like proteins are necessary for chromosome segregation in *Drosophila* (Endow et al. 1990; Zhang et al.

1990) and for spindle pole body separation in *Aspergillus nidulans* (Enos and Morris 1990). The *Drosophila* kinesin-like protein has been shown to be a minus end-directed motor protein (McDonald et al. 1990; Walker et al. 1990). It has recently been shown that chromosomes possess motor element(s) that allow movement on a microtubule in each direction, the switch of which may be regulated by phosphorylation (Hyman and Mitchison 1991). From these results, it is suggested that the anaphase chromosome movement is driven by the well-organized cooperation of some ATP-dependent motors and the depolymerization of microtubules at the kinetochore end. Several centromere components called CENPs have been identified using autoimmune serum from a CREST patient (Pluta et al. 1990).

As to the mechanism of formation of the contractile ring in animal cells, Cao and Wang (1990) microinjected rhodamine-labeled F-actin into cultured cells and observed movement of the actin via the cortical layer to the cleavage furrow, which suggests the presence of a flow of F-actin from the cytoplasm to the furrow region via the cortex. A *Dictyostelium* cell line, from which a myosin heavy chain gene had been deleted and shown to be defective for cytokinesis, has been rescued by introducing an autonomously replicating shuttle vector carrying the myosin gene (Egelhoff et al. 1990), which reconfirmed the role of myosin in cytokinesis. A *Xenopus* mutant embryo which cannot undergo cleavage has been reported (Kubota et al. 1991) and it was shown that two unidentified proteins are missing.

References to Addendum

Cao L-g, Wang Y-l (1990) Mechanism of the formation of contractile ring in dividing cultured animal cells. II. Cortical movement of microinjected actin filaments. J Cell Biol 111: 1905–1911

Egelhoff TT, Manstein DJ, Spudich JA (1990) Complementation of myosin null mutants in *Dictyostelium discoideum* by direct functional selection. Dev Biol 137:359–367.

Endow SA, Henikoff S, Soler-Niedziela L (1990) Mediation of meiotic and early mitotic chromosome segregation in *Drosophila* by a protein related to kinesin. Nature 345:81–83

Enos AP, Morris NR (1990) Mutation of a gene that encodes a kinesin-like protein blocks nuclear division in *A. nidulans*. Cell 60:1019–1027

Hagan IM, Yanagida M (1990) Novel potential mitotic motor protein encoded by the fission yeast $cut7^{+}$ gene. Nature 347:563–566

Hayden JH, Bowser SS, Rieder CL (1990) Kinetochores capture astral microtubules during chromosome attachment to the mitotic spindle: direct visualization in live newt lung cells. J Cell Biol 111:1039–1045

Hyman AA, Mitchison TJ (1991) Two different microtubule-based motor activities with opposite polarities in kinetochore. Nature 351:206–211

Kubota HY, Itoh K, Asada-Kubota M (1991) Cytological and biochemical analyses of the maternal-effect mutant embryos with abnormal cleavage furrow formation in *Xenopus laevis*. Dev Biol 144:145–151

McDonald HB, Stewart RJ, Goldstein LSB (1990) The kinesin-like *ncd* protein of *Drosophila* is a minus end-directed microtubule motor. Cell 63:1159–1165

Oakley BR, Oakley CE, Yoon Y, Jung MK (1990) γ-Tubulin is a component of the spindle pole body that is essential for microtubule function in *Aspergillus nidulans*. Cell 61:1289–1301

Pfarr CM, Coue M, Grissom PM, Hays TS, Porter ME, McIntosh JR (1990) Cytoplasmic dynein is localized to kinetochores during mitosis. Nature 345:263–265
Pluta AF, Cook CA, Earnshaw WC (1990) Structure of the human centromere at metaphase. Trends Biol Sci 15:181–185
Sawin KE, Mitchison TJ (1991a) Mitotic spindle assembly by two different pathways in vitro. J Cell Biol 112:925–940
Sawin KE, Mitchison TJ (1991b) Poleward microtubule flux in mitotic spindles assembled in vitro. J Cell Biol 112:941–954
Steuer ER, Wordeman L, Schroer TA, Sheets MP (1990) Localization of cytoplasmic dynein to mitotic spindles and kinetochores. Nature 345:266–268
Vale RD (1991) Severing of stable microtubules by mitotically activated protein in *Xenopus* egg extracts. Cell 64:827–839
Walker RA, Salmon ED, Endow SA (1990) The *Drosophila claret* segregation protein is a minus-end directed motor molecule. Nature 347:780–782
Zhang P, Knowles BA, Goldstein LSB, Hawleys RS (1990) A kinesin-like protein required for distributive chromosome segregation in *Drosophila*. Cell 62:1053–1062
Zheng Y, Jung MK, Oakley BR (1991) γ-Tubulin is present in *Drosophila melanogaster* and *Homo sapiens* and is associated with the centrosome. Cell 65:817–823

References

Allen C, Borisy GG (1974) Structural polarity and directional growth of microtubules of *Chlamydomonas* flagella. J Mol Biol 90:381–402
Allen RD, Bajer A, LaFountain J (1969) Poleward migration of particles or states in spindle fiber filaments during mitosis in *Haemanthus*. J Cell Biol 43:4a
Amos LA, Klug A (1974) The arrangement of subunits in flagellar microtubules. J Cell Sci 14:523–549
Arnold JM (1969) Cleavage furrow formation in a telolecthal egg (*Loligo pealli*). J Cell Biol 41:894–904
Aubin JE, Osborn M, Weber K (1979) Inhibition of cytokinesis and altered contractile ring morphology induced by cytochalasins in synchronized PtK2 cells. Exp Cell Res 136:63–79
Bajer A (1973) Interaction of microtubules and the mechanism of chromosome movement (zipper hypothesis). Cytobios 8:139–160
Bajer A (1982) Functional autonomy of monopolar spindle and evidence of oscillatory movement in mitosis. J Cell Biol 93:33–48
Bajer A, Molé-Bajer J (1969) Formation of spindle fibers, kinetochore orientation, and behavior of the nuclear envelope during mitosis in endosperm. Chromosoma 27:448–484
Bajer A, Molé-Bajer J (1972) Spindle dynamics and chromosome movements. Int Rev Cytol Suppl 3:1–271
Balczon RD, Brinkley BR (1987) Tubulin interaction with kinetochore protein: analysis by in vitro assembly and chemical cross-linking. J Cell Biol 105:855–862
Bergen LG, Borisy GG (1980) Head-to-tail polymerization of microtubules in vitro: electron microsope analysis of seeded assembly. J Cell Biol 84:141–150
Bergen LG, Kuriyama R, Borisy GG (1980) Polarity of microtubules nucleated by centrosomes and chromosomes of CHO cells in vitro. J Cell Biol 84:151–159
Bonder EM, Fishkind DJ, Cotran NM, Begg DA (1989) The cortical actin-membrane cytoskeleton of unfertilized sea urchin eggs: analysis of the spatial organization and relationship of filamentous actin, nonfilamentous actin, and egg spectrin. Dev Biol 134:327–341
Brinkley BR, Kartwright J (1975) Cold-labile and cold-stable microtubules in the mitotic spindle of mammalian cells. Ann N Y Acad Sci 253:109–123
Brinkley BR, Zinkowski RP, Mollon WL, Davis FM, Pisegna MA, Pershouse M, Rao PN (1988) Movement and segregation of kinetochores experimentally detached from mammalian chromosomes. Nature 336:251–254

Brinkley BR, Valdivia MM, Tousson A, Balczon RD (1989) The kinetochore: structure and molecular organization. In: Hyams JS, Brinkley BR (eds) Mitosis. molecules and mechanisms. Academic Press, New York, pp 77–118

Cande WZ (1980) A permeabilized cell model for studying cytokinesis using mammalian tissue culture cells. J Cell Biol 87:326–335

Cande WZ (1982) Nucleotide requirements for anaphase chromosome movements in permeablized mitotic cells: anaphase B but not anaphase A requires ATP. Cell 28:15–22

Cande WZ, McDonald KL (1985) In vitro reactivation of anaphase spindle elongation using isolated diatom spindles. Nature 316:168–170

Cande WZ, Baskin T, Hogan C, McDonald KL, Masuda H, Wordeman L (1989) In vitro analysis of anaphase spindle elongation. In: Warner FD, McIntosh JR (eds) Cell movement, vol 2. Kinesin, dynein and microtubule dynamics. Alan R Liss, New York, pp 441–452

Carlier M-F, Hill TL, Chen Y-D (1984) Interference of GTP hydrolysis in the mechanism of microtubule assembly: an experimental study. Proc Natl Acad Sci USA 81:771–775

Cassimeris L, Pryer NK, Salmon ED (1988) Real-time observations of microtubule dynamic instability in living cells. J Cell Biol 107:2223–2231

Dan K (1943) Behavior of the cell surface during cleavage. VI. On the mechanism of cell division. J Fac Sci Tokyo Imp Univ Sec 4, 6:323–368

Dan K (1988) Mechanism of equal cleavage of sea urchin egg: transposition from astral mechanism to constricting mechanism. Zool Sci 5:507–517

De Lozanne A, Spudich JA (1987) Disruption of the *Dictyostelium* myosin heavy chain gene by homologous recombination. Science 236:1086–1091

Devore JJ, Conrad GW, Rappaport R (1990) A model for astral stimulation of cytokinesis in animal cells. J Cell Biol 109:2225–2232

Dinsmore JH, Sloboda RD (1988) Calcium and calmodulin-dependent phosphorylation of a 62 kD protein induces microtubule depolymerization in sea urchin mitotic apparatus. Cell 53:769–780

Euteneuer U, McIntosh JR (1981) Structural polarity of kinetochore microtubules in PtK1 cells. J Cell Biol 89:338–345

Euteneuer U, Ris H, Borisy GG (1983) Polarity of kinetochore microtubules in Chinese hamster ovary cells after recovery from a colcemid block. J Cell Biol 97:202–208

Forer A (1965) Local reduction of spindle fiber birefringence in living *Nephrotoma suturalis* spermatocytes induced by ultraviolet microbeam irradiation. J Cell Biol 25:95–117

Fujiwara K, Pollard TD (1976) Fluorescent antibody localization of myosin in the cytoplasm, cleavage furrow, and mitotic spindle of human cells. J Cell Biol 71:848–875

Fujiwara K, Porter ME, Pollard TD (1978) Alpha-actinin localization in the cleavage furrow during cytokinesis. J Cell Biol 79:268–275

Fukui Y, Lynch TJ, Brzeska H, Korn ED (1989) Myosin I is located at the leading edges of locomoting *Dictyostelium* amoebae. Nature 341:328–331

Gorbsky GJ, Borisy GG (1989) Microtubules of the kinetochore fiber turn over in metaphase but not in anaphase. J Cell Biol 109:653–662

Gorbsky GJ, Sammak PJ, Borisy GG (1987) Chromosomes move poleward in anaphase along stationary microtubules that coordinately disassemble from their kinetochore ends. J Cell Biol 104:9–18

Gorbsky GJ, Sammak PJ, Borisy GG (1988) Microtubule dynamics and chromosome motion visualized in living anaphase cells. J Cell Biol 106:1185–1192

Haimo LT, Telzer BR (1981) Dynein-microtubule interactions. ATP-sensitive dynein binding and the structural polarity of mitotic microtubules. Cold Spring Harbor Symp Quant Biol 46:207–218

Hamaguchi MS, Hamaguchi Y, Hiramoto T (1986) Microinjected polystylene beads move along astral rays in sand dollar eggs. Dev Growth Differ 28:461–470

Hamaguchi Y (1975) Microinjection of colchicine into sea urchin eggs. Dev Growth Differ 17:111–117

Hamaguchi Y, Mabuchi I (1982) Effects of phalloidin microinjection and localization of fluorescein-labeled phalloidin in living sand dollar eggs. Cell Motil 2:103–113

Hamaguchi Y, Toriyama M, Sakai H, Hiramoto Y (1987) Redistribution of fluorescently labeled tubulin in the mitotic apparatus of sand dollar eggs and the effects of taxol. Cell Struct Funct 12:43–52

Hays TS, Salmon ED (1990) Poleward force at the kinetochore in metaphase depends on the number of kinetochore microtubules. J Cell Biol 110:391–404

Hays TS, Wise D, Salmon ED (1982) Traction force on a kinetochore at metaphase acts as a linear function of kinetochore fiber length. J Cell Biol 93:374–382

Henson JH, Begg DA, Beaulieu SM, Fishkind DJ, Bonder EM, Terasaki M, Lebeche DL, Kaminer B (1989) A calsequestrin-like protein in the endoplasmic reticulum of the sea urchin: localization and dynamics in the egg and first cell cycle embryo. J Cell Biol 109:149–161

Hiramoto Y (1956) Cell division without mitotic apparatus in sea urchin eggs. Exp Cell Res 11:630–636

Hiramoto Y (1975) Force exerted by the cleavage furrow of sea urchin eggs. Dev Growth Differ 17:27–38

Hiramoto Y (1979) Mechanical properties of the dividing sea urchin egg. In: Hatano S, Ishikawa H, Sato H (eds) Cell motility: molecules and organization. University of Tokyo Press, Tokyo, pp 653–663

Hiramoto Y, Izutsu K (1977) Poleward movement of "markers" existing in mitotic spindles of grasshopper spermatocytes. Cell Struct Funct 2:257–259

Hiramoto Y, Shoji Y (1982) Location of the motive force for chromosome movement in sand dollar eggs. In: Sakai H, Mohri H, Borisy GG (eds) Biological functions of microtubules and related structures. Academic Press, New York, pp 247–259

Hisanaga S, Sakai H (1980) Cytoplasmic dynein of the sea urchin egg. I. Dev Growth Differ 22:373–384

Horio T, Hotani H (1986) Visualization of the dynamic instability of individual microtubules by dark-field microscopy. Nature 321:605–607

Hotani H, Horio T (1988) Dynamics of microtubules visualized by dark field microscopy: treadmilling and dynamic instability. Cell Motil Cytoskelet 10:229–236

Huitorel P, Kirschner MW (1988) The polarity and stability of microtubule capture by the kinetochore. J Cell Biol 106:151–159

Inoué S (1981) Cell division an the mitotic spindle. J Cell Biol 91:131s–147s

Inoué S, Ritter H (1975) Dynamics of mitotic spindle organization and function. In: Inoué S, Stephens RE (eds) Molecules and cell movement. Raven, New York, pp 3–29

Inoué S, Sato H (1967) Cell motility by labile association of molecules. The nature of mitotic spindle fibers and their role in chromosome movement. J Gen Physiol 50:259–292

Ishidate S, Mabuchi I (1988a) A novel actin filament-capping protein from sea urchin eggs: a 20000-molecular weight protein-actin complex. J Biochem 104:72–80

Ishidate S, Mabuchi I (1988b) Localization and possible function of 20 kDa actin-modulating protein (actolinkin) in the sea urchin egg. Eur J Biochem 46:275–281

Ishimoda-Takagi (1979) Localization of tropomyosin in sea urchin eggs. Exp Cell Res 119:423–428

Ishimoda-Takagi (1984) Evidence for the involvement of muscle tropomyosin in the contractile elements of the coelomesophagus complex in sea urchin embryos. Dev Biol 105: 365–376

Jameson L, Frey T, Zeeberg B, Dalldorf F, Caplow M (1980) Inhibition of microtubule assembly by phosphorylation of microtubule-associated proteins. Biochemistry 19:2472–2479

Katoh K, Ishikawa H (1989) The cytoskeletal involvement in cellularization of the *Drosophila melanogaster* embryo. Protoplasma 150:83–95

Kiehart DP, Mabuchi I, Inoué S (1982) Evidence that myosin does not contribute to force production in chromosome movement. J Cell Biol 94:165–178

Knabe T, Kobayashi I, Tanaka K (1989) Dynamics of cytoplasmic organelles in the cell cycle of the fission yeast *Schizosaccharomyces pombe*: three dimensional reconstruction from serial sections. J Cell Sci 94:647–656

Knecht DA, Loomis WF (1987) Antisense RNA inactivation of myosin heavy chain gene expression in *Dictyostelium discoideum*. Science 236:1081–1086

Koshland DE, Mitchison TJ, Kirschner MW (1988) Polewards chromosome movement driven by microtubule depolymerization in vitro. Nature 331:499–504
Kron SJ, Spudich JA (1986) Fluorescent actin filaments move on myosin fixed to a glass surface. Proc Natl Acad Sci USA 83:6272–6276
Kuriyama R (1984) Activity and stability of centrosomes in Chinese hamster ovary cells in nucleation of microtubules in vitro. J Cell Sci 66:277–295
Kuriyama R, Borisy GG (1985) Identification of molecular components of the centrosphere in the mitotic spindle of sea urchin eggs. J Cell Biol 101:524–530
Langanger G, De Mey J, Moeremans M, Daneels G, De Brabander M, Small JV (1984) Ultrastructural localization of α-actinin and filamin in cultured cells with the immunogold staining (IGS) method. J Cell Biol 99:1324–1334
Mabuchi I (1973) ATPase in the cortical layer of sea urchin egg: its properties and interaction with cortex protein. Biochim. Biophys Acta 297:317–332
Mabuchi I (1986) Biochemical aspects of cytokinesis. Int Rev Cytol 101:175–213
Mabuchi I (1990) Cleavage furrow formation and actin-modulating proteins. Ann N Y Acad Sci 582:131–146
Mabuchi I, Kane RE (1987) A 250K-molecular weight actin-binding protein from actin-based gels formed in sea urchin egg cytoplasmic extract. J Biochem 102:947–956
Mabuchi I, Okuno M (1977) The effect of myosin antibody on the cell division of starfish blastomeres. J Cell Biol 74:251–263
Mabuchi I, Takano-Ohmuro H (1990) Effects of inhibitors of myosin light chain kinase and other protein kinases on the first cell division of sea urchin eggs. Dev Growth Differ 32:549–556
Mabuchi I, Hosoya H, Sakai H (1980) Actin in the cortical layer of the sea urchin egg. Changes in its content during and after fertilization. Biomed Res 1:417–426
Mabuchi I, Hamaguchi Y, Kobayashi T, Hosoya H, Tsukita S, Tsukita S (1985) Alpha-actinin from sea urchin eggs: biochemical properties, interaction with actin, and distribution in the cell during fertilization and cleavage. J Cell Biol 100:375–383
Mabuchi I, Tsukita S, Tsukita S, Sawai T (1988) Cleavage furrow isolated from newt eggs: contraction, organization of actin filaments, and protein components of the furrow. Proc Nat Acad Sci USA 85:5966–5970
Maekawa S, Endo S, Sakai H (1987) A high molecular weight actin binding protein: its localization in the cortex of the sea urchin egg. Exp Cell Res 172:340–353
Manstein DJ, Titus MA, De Lozanne A, Spudich JA (1989) Gene replacement in *Dictyostelium*: generation of myosin null mutants. EMBO J 8:923–932
Margolis RL, Wilson L (1978) Opposite end assembly and disassembly of microtubules at steady-state in vitro. Cell 13:1–8
Margolis RL, Wilson L (1981) Microtubule treadmills – possible molecular machinery. Nature 298:705–711
Margolis RL, Wilson L, Kiefer BI (1978) Mitotic mechanism based on intrinsic microtubule behavior. Nature 272:450–452
Marsland D, Landau JV (1954) The mechanism of cytokinesis: temperature-pressure studies on the cortical gel system in various marine eggs. J Exp Zool 125:507–539
Masuda H, Cande WZ (1987) The role of tubulin polymerization during spindle elongation in vitro. Cell 49:193–202
Masuda H, McDonald KL, Cande WZ (1988) The mechanism of anaphase spindle elongation: uncoupling of tubulin incorporation and microtubule sliding during in vitro spindle reactivation. J Cell Biol 107:623–633
Masuda H, Hirano T, Yanagida M, Cande WZ (1990) In vitro reactivation of spindle elongation in fission yeast *nuc2* mutant cells. J Cell Biol 110:417–425
Maupin P, Pollard TD (1986) Arrangement of actin filaments and myosin-like filaments in the contractile ring and of actin-like filaments in the mitotic spindle of dividing HeLa cells. J Ultrastruct Mol Struct Res 94:92–103
Mazia D (1987) The chromosome cycle and the centrosome cycle in the mitotic cycle. Int Rev Cytol 100:49–92
Mazia D, Dan K (1952) The isolation and biochemical characterization of the mitotic apparatus of dividing cells. Proc Natl Acad Sci USA 38:826–838

McCaffrey G, Vale RD (1989) Identification of a kinesin-like microtubule-based motor portein in *Dictyostelium discoideum*. EMBO J 8:3229–3234
McDonald KL, Pickett-Heaps JD, McIntosh JR, Tippit DH (1977) On the mechanism of anaphase spindle elongation in *Diatom vulgare*. J Cell Biol 74:377–388
McIntosh JR (1985) Spindle structure and the mechanism of chromosome movement. In "Aneuploidy: Etiology and Mechanism" Dellarco V, Voytek PE and Hollander A, eds. Plenum, New York, pp 197–229
McIntosh JR, Vigers GPA, Hays TS (1989) Dynamic behavior of mitotic microtubules. In: Warner FD, McIntosh JR (eds) Cell movement, vol 2. Kinesin, dynein and microtubule dynamics. Alan R Liss, New York, pp 371–382
Mitchison TJ (1989a) Chromosome alignment at mitotic metaphase: balanced forces or smart kinetochores? In: Warner FD, McIntosh JR (eds) Cell movement, vol 2. Kinesin, dynein and microtubule dynamics. Alan R Liss, New York, pp 421–430
Mitchison TJ (1989b) Polewards microtubule flux in the mitotic spindle: evidence from photoactivation of fluorescence. J Cell Biol 109:637–652
Mitchison TJ, Kirschner MW (1984a) Microtubule assembly nucleated by isolated centrosomes. Nature 312:232–237
Mitchison TJ, Kirschner MW (1984b) Dynamic instability of microtubule growth. Nature 312:237–242
Mitchison TJ, Kirschner MW (1985a) Properties of the kinetochore in vitro. I. Microtubule nucleation and tubulin binding. J Cell Biol 101:755–765
Mitchison TJ, Kirschner MW (1985b) Properties of the kinetochore in vitro. II. Microtubule capture and ATP dependent translocation. J Cell Biol 101:766–777
Mitchison TJ, Evans L, Schulze E, Kirschner MW (1986) Sites of microtubule assembly and disassembly in the mitotic spindle. Cell 45:515–527
Mittal B, Sanger JM, Sanger JW (1987) Visualization of myosin in living cells. J Cell Biol 105:1753–1760
Molé-Bajer J (1975) The role of centrioles in the development of the astral spindle (newt). Cytobios 13:117–140
Nicklas RB (1971) Mitosis. In: Prescott DM, Goldstein L, McConkey E (eds) Advances in cell biology, vol 2. Appleton-Century-Crofts, New York, pp 225–297
Nunnally MH, D'Angelo, JM, Craig SW (1980) Filamin concentration in cleavage furrow and midbody region: frequency of occurrence compared with that of alpha-actinin and myosin. J Cell Biol 87:219–226
Ohta K, Toriyama M, Endo S, Sakai H (1988) Localization of mitotic-apparatus-associated 51-kD protein in unfertilized and fertilized sea urchin eggs. Cell Motil Cytoskelet 10: 496–505
Ohta K, Toriyama M, Miyazaki M, Murofushi H, Hosoda S, Endo S, Sakai H (1990) The mitotic apparatus-associated 51-kDa protein from sea urchin eggs is a GTP-binding protein and is immunologically related to yeast polypeptide elongation factor 1α. J Biol Chem 265:3240–3247
Östergren G (1951) The mechanism of co-orientation in bivalents and multivalents. The theory of orientation by pulling. Hereditas 37:85–156
Paschal BM, Vallee RB (1987) Retrograde transport by the microtubule-associated protein MAP 1C. Nature 330:181–183
Pickett-Heaps JD, Tippit DH (1978) The diatom spindle in perspective. Cell 14:455–467
Pickett-Heaps JD, Tippit DH, Porter KR (1982) Rethinking mitosis. Cell 29:729–744
Pratt MM, Otter T, Salmon ED (1980) Dynein-like Mg^{2+}-ATPase in mitotic spindles isolated from sea urchin embryos (*Strongylocentrotus droebachiensis*). J Cell Biol 86:738–745
Rappaport R (1961) Experiments concerning the cleavage stimulus in sand dollar eggs. J Exp Zool 148:81–89
Rappaport R (1967) Cell division: direct measurement of maximum tension exerted by furrow of echinoderm eggs. Science 156:1241–1243
Rappaport R (1969) Division of isolated furrows and furrow fragments in invertebrate eggs. Exp Cell Res 56:87–91
Rappaport R (1971) Cytokinesis in animal cells. Int Rev Cytol 31:169–213

Rappaport R (1973) On the rate of the cleavage stimulus in sand dollar eggs. J Exp Zool 183:115–120
Rebhun LI, Palazzo RE (1988) In vitro reactivation of anaphase B in isolated spindles of the sea urchin egg. Cell Motil Cytoskelet 10:197–209
Rieder CL (1981) The structure of cold-stable kinetochore fiber in metaphase PtK1 cells. Chromosoma 84:145–158
Rieder CL (1982) The formation, structure and composition of the mammalian kinetochore and kinetochore fiber. Int Rev Cytol 79:1–57
Rieder CL, Alexander SP (1990) Kinetochores are transported poleward along a single astral microtubule during chromosome attachment to the spindle in newt lung cells. J Cell Biol 110:81–95
Roos U-P (1976) Light and electron microscopy of rat kangaroo cells in mitosis. III. Patterns of chromosome behavior during prometaphase. Chromosoma 54:363–385
Sakai H (1978a) Isolated mitotic apparatus and chromosome motion. Int Rev Cytol 55:23–48
Sakai H (1978b) Induction of chromosome motion in the isolated mitotic apparatus as a function of microtubules. In: Dirkson ER, Prescott DM, Fox CF (eds) Cell reproduction: in honor of Daniel Mazia. Academic Press, New York, pp 425–432
Sakai H, Hiramoto Y, Kuriyama R (1975) The glycerol-isolated mitotic apparatus: a response to porcine brain tubulin and induction of chromosome motion. Dev Growth Differ 17:265–274
Sakai H, Mabuchi I, Shimoda S, Kuriyama R, Ogawa K, Mohri H (1976) Induction of chromosome motion in the glycerol-isolated mitotic apparatus: nucleotide specificity and effects of antidynein and myosin sera on the motion. Dev Growth Differ 18:211–219
Salmon ED, McKeel M, Hays T (1984a) The rapid rate of tubulin dissociation from microtubules in the mitotic spindle in vivo measured by blocking polymerization with colchicine. J Cell Biol 99:1066–1075
Salmon ED, Leslie RJ, Saxton WM, Karow ML, McIntosh JR (1984b) Spindle microtubule dynamics in sea urchin embryos: analysis using a fluorescein-labeled tubulin and measurements of fluorescence redistribution after laser photobleaching. J Cell Biol 99:2165–2174
Sammak PJ, Borisy GG (1988) Direct observation of microtubule dynamics in living cells. Nature 332:724–726
Sanger JM, Sanger JW (1980) Banding and polarity of actin filaments in interphase and cleaving cells. J Cell Biol 86:568–575
Saxton WM, McIntosh JR (1987) Interzone microtubule behavior in late anaphase and telophase spindles. J Cell Biol 105:875–886
Schatten H, Walter M, Mazia D, Biessmann H, Paweletz N, Coffe G, Schatten G (1987) Centrosome detection in sea urchin eggs with a monoclonal antibody against *Drosophila* intermediate filament proteins: characterization of stages of the division cycle of centrosomes. Proc Natl Acad Sci USA 84:8488–8492
Scholey JM, Neighbors B, McIntosh JR, Salmon ED (1985a) Isolation of microtubules and a dynein-like MgATPase from unfertilized sea urchin eggs. J Biol Chem 259:6516–6525
Scholey JM, Porter ME, Grissom PM, McIntosh JR (1985b) Identification of kinesin in sea urchin eggs, and evidence for its localization in the mitotic spindle. Nature 318:483–486
Schroeder TE (1968) Cytokinesis: filaments in the cleavage furrow. Exp Cell Res 53:272–276
Schroeder TE (1970) The contractile ring. I. Fine structure of dividing mammalian (HeLa) cells and the effects of cytochalasin B. Z Zellforsch Mikrosk Anat 109:431–449
Schroeder TE (1972) The contractile ring. II. Determining its brief existence, volumetric changes, and vital role in cleaving *Arbacia* eggs. J Cell Biol 53:419–434
Schroeder TE (1975) Dynamics of the contractile ring. In: Inoué S, Stephens RE (eds) Molecules and cell movement. Raven, New York, pp 305–334
Schroeder TE (1986) The egg cortex in early development of sea urchins and starfish. In: Browder LW (ed) Developmental biology, vol 2. Plenum, New York, pp 59–100
Schroeder TE (1987) Fourth cleavage of sea urchin blastomeres: microtubule patterns and myosin localization in equal and unequal cell divisions. Dev Biol 124:9–22
Schroeder TE, Otto JJ (1988) Immunofluorescent analysis of actin and myosin in isolated contractile rings of sea urchin eggs. Zool Sci 5:713–725

Schulze E, Kirschner MW (1986) Microtubule dynamics in interphase cells. J Cell Biol 102:1020–1031
Schulze E, Kirschner MW (1988) New features of microtubule behavior observed in vivo. Nature 334:356–359
Shpetner HS, Vallee RB (1989) Identification of dynamin, a novel mechanochemical enzyme that mediates interactions between microtubules. Cell 59:421–432
Sluder G, Rieder CL (1985) Experimental separation of pronuclei in fertilized sea urchin eggs: chromosomes do not organize a spindle in the absence of centrosomes. J Cell Biol 100:897–903
Soltys BJ, Borisy GG (1985) Polymerization of tubulin in vivo: direct evidence for assembly onto microtubule ends and from centrosomes. J Cell Biol 100:1682–1689
Steffen W, Fuge H, Dietz R, Bastmeyer M, Muller G (1986) Aster-free spindle poles in insect spermatocytes: evidence for chromosome-induced spindle formation? J Cell Biol 102:1679–1687
Swann MM, Mitchison JM (1953) Cleavage of sea-urchin eggs in colchicine. J Exp Biol 35:506–513
Tilney LG, Marsland D (1969) A fine structural analysis of cleavage induction and furrowing in the eggs of *Arbacia punctulata*. J Cell Biol 44:170–184
Toriyama M, Ohta K, Endo S, Sakai H (1987) 51k Protein, a component of microtubule-organizing granules (MTOGs) in the mitotic apparatus involving in aster formation in vitro. Cell Motil Cytoskelet 9:117–128
Usui N, Yoneda M (1982) Ultrastructural basis of the tension increase in sea urchin eggs prior to cytokinesis. Dev Growth Differ 24:453–465
Vale RD, Reese TS, Sheetz MP (1985) Identification of a novel force-generating protein, kinesin, involved in microtubule-based motility. Cell 42:39–50
Vandre DD, Borisy GG (1989) The centrosome cycle in animal cells. In: Hyams JS, Brinkley BR (eds) Mitosis. Molecules and mechanisms. Academic Press, New York, pp 39–75
Vigers, GPA, Coue M, McIntosh JR (1988) Fluorescent microtubules break up under illumination. J Cell Biol 107:1011–1024
Wadsworth P, Salmon ED (1986) Analysis of the treadmilling model during metaphase of mitosis using fluorescence redistribution after photobleaching. J Cell Biol 102:1032–1038
Walker RA, O'Brien ET, Pryer NK, Soboeiro MF, Voter WA, Erickson HP, Salmon ED (1988) Dynamic instability of individual microtubules analyzed by video light microscopy: rate constants and transition frequencies. J Cell Biol 107:1437–1448
Wallraffe E, Schleicher M, Modersitzki M, Rieger D, Isenberg G, Gerisch G (1986) Selection of *Dyctiostelium* mutants defective in cytoskeletal proteins: use of an antibody that binds to the ends of α-actinin rods. EMBO J 5:61–67
Weisenberg R, Taylor EW (1968) Studies on ATPase activity of sea urchin eggs and the isolated mitotic apparatus. Exp Cell Res 53:372–384
Witt PL, Ris H, Borisy GG (1981) Structure of kinetochore fibers: microtubule continuity and intermicrotubule bridges. Chromosoma 83:523–540
Wordeman L, Cande WZ (1987) Reactivation of spindle elongation in vitro is correlated with the phosphorylation of a 205 kD spindle-associated protein. Cell 50:535–543
Yamamoto H, Fukunaga K, Tanaka E, Miyamoto E (1983) Ca^{2+}- and calmodulin-dependent phosphorylation of microtubule-associated protein 2 and tau factor, and inhibition of microtubule assembly. J Neurochem 41:1119–1125
Yanagida T, Arata T, Oosawa F (1985) Sliding distance of actin filament induced by a myosin crossbridge during one ATP hydrolysis cycle. Nature 316:366–369
Yoneda M, Dan K (1972) Tension at the surface of the dividing sea-urchin egg. J Exp Biol 57:575–587
Yonemura S (1989) Morphological and biochemical analyses of the cortical cytoskeleton and of the contractile ring in the sea urchin egg. Thesis, University of Tokyo
Yonemura S, Kinoshita S (1986) Actin filament organization in the sand dollar egg cortex. Dev Biol 115:171–183
Yonemura S, Mabuchi I, Tsukita S (1991) Mass isolation of cleavage furrows from dividing sea urchin eggs. J Cell Sci 100:73–84
Yumura S, Fukui Y (1985) Reversible cyclic AMP-dependent change in distribution of myosin thick filaments in *Dictyostelium*. Nature 314:194–196

Subject Index